The IMA Volumes in Mathematics and Its Applications

Volume 5

Series Editors
Geroge R. Sell   Hans Weinberger

Institute for Mathematics and Its Applications
IMA

The **Institute for Mathematics and Its Applications** was established by a grant from the National Science Foundation to the University of Minnesota in 1982. The IMA seeks to encourage the developent and study of fresh mathematical concepts and questions of concern to the other sciences by bringing together mathematicians and scientists from diverse fields in an atmosphere that will stimulate discussion and collaboration.

The IMA Volues are intended to involve the broader scientific community in this process.

Hans Weinberger, Director
George R. Sell, Associate Director

IMA Programs

1982–1983 Statistical and Continuum Approaches to Phase Transition

1983–1984 Mathematical Models for the Economics of Decentralized Resource Allocation

1984–1985 Continuum Physics and Partial Differential Equations

1985–1986 Stochastic Differential Equations and Their Applications

1986–1987 Scientific Computation

1987–1988 Applied Combinatories

1988–1989 Nonlinear Waves

Springer Lecture Notes from the IMA

The Mathematics and Physics of Disordered Media
Editors: Barry Hughes and Barry Ninham
(Lecture Notes in Mathematics, Volume 1035, 1983)

Orienting Polymers
Editor: J. L. Ericksen
(Lecture Notes in Mathematics, Volume 1063, 1984)

New Perspectives in Thermodynamics
Editor: James Serrin
(Springer-Verlag, 1986)

Models of Economic Dynamics
Editor: Hugo Sonnenschein
(Lecture Notes in Economics, Volume 264, 1986)

J.L. Ericksen D. Kinderlehrer
Editors

Theory and Applications of Liquid Crystals

With 60 Illustrations

Springer-Verlag
New York Berlin Heidelberg
London Paris Tokyo

J.L. Ericksen
Institute for Mathematics and Its Applications
Minneapolis, Minnesota 55455 USA

David Kinderlehrer
Institute for Mathematics and Its Applications
Minneapolis, Minnesota 55455 USA

AMS Classification: 82 XX

Library of Congress Cataloging in Publication Data
Theory and applications of liquid crystals.
(The IMA volumes in mathematics and its applications ; v. 5)
Includes bibliographies.
1. Liquid crystals. I. Ericksen, J. L.
(Jerald L.), 1924– II. Kinderlehrer, David.
III. Series.
QD923.T46 1987 548′.9 87-9858

Printed and bound by Edwards Brothers, Ann Arbor, Michigan.
Printed in the United States of America.

9 8 7 6 5 4 3 2 1

ISBN 0-387-96546-7 Springer-Verlag New York Berlin Heidelberg
ISBN 3-540-96546-7 Springer-Verlag Berlin Heidelberg New York

The IMA Volumes in Mathematics and Its Applications

Current Volumes:

Volume 1: Homogenization and Effective Moduli of Materials and Media
Editors: Jerry Ericksen, David Kinderlehrer, Robert Kohn, and J.-L. Lions

Volume 2: Oscillation Theory, Computation, and Methods of Compensated Compactness
Editors: Constantine Dafermos, Jerry Ericksen, David Kinderlehrer, and Marshall Slemrod

Volume 3: Metastability and Incompletely Posed Problems
Editors: Stuart Antman, Jerry Ericksen, David Kinderlehrer, and Ingo Müller

Volume 4: Dynamical Problems in Continuum Physics
Editors: Jerry Bona, Constantine Dafermos, Jerry Ericksen, and David Kinderlehrer

Volume 5: Theory and Applications of Liquid Crystals
Editors: Jerry Ericksen and David Kinderlehrer

Volume 6: Amorphous Polymers and Non-Newtonian Fluids
Editors: Constantine Dafermos, Jerry Ericksen, and David Kinderlehrer

Volume 7: Random Media
Editor: George Papanicolaou

Volume 8: Percolation Theory and Ergodic Theory of Infinite Particle Systems
Editor: Harry Kesten

Forthcoming Volumes:

1985–1986: Stochastic Differential Equations and Their Applications

Hydrodynamic Behavior and Interacting Particle Systems
Stochastic Differential Systems, Stochastic Control Theory and Applications

1986–1987: Scientific Computation

Computational Fluid Dynamics and Reacting Gas Flows
Numerical Algorithms for Modern Parallel Computer Architectures
Numerical Simulation in Oil Recovery
Atomic and Molecular Structure and Dynamics

D
A
B
C
E

Contents

Foreword

This IMA Volume in Mathematics and its Applications

AMORPHOUS POLYMERS AND NON-NEWTONIAN FLUIDS

is in part the proceedings of a workshop which was an integral part of the 1984-85 IMA program on CONTINUUM PHYSICS AND PARTIAL DIFFERENTIAL EQUATIONS We are grateful to the Scientific Committee:

Haim Brezis
Constantine Dafermos
Jerry Ericksen
David Kinderlehrer

for planning and implementing an exciting and stimulating year-long program. We especially thank the Program Organizers, Jerry Ericksen, David Kinderlehrer, Stephen Prager and Matthew Tirrell for organizing a workshop which brought together scientists and mathematicians in a variety of areas for a fruitful exchange of ideas.

George R. Sell
Hans Weinberger

Preface

The diversity of experimental phenomena and the range of applications of liquid crystals present timely and challenging questions for experimentalists, mechanists, and mathematicians. The scope of this workshop was to bring together research workers and practitioners in these areas from laboratories, industry, and universities to explore common issues. The contents of this volume vary from descriptions of experimental phenomena, of which our understanding is insufficient, to questions of a mathematical nature and of efficient computation.

Interest in this area is stimulated by problems relating to the many familiar devices as well as by questions which arise in the processing of high strength polymer fibers such as Kevlar. From the standpoint of pure science, our concern is with mesomorphic phases of matter. These had received little or no serious mathematical treatment although the equations governing macroscopic behavior of small molecule liquid crystals are well established.

Among the workshops of the program, this was the most adventurous. In addition to describing recent activity in liquid crystal theory and experiment, our objective was to stimulate mathematical research connected to the discipline. Our thesis was that better mathematical understanding would lead to improved theory and more effective computational methods. Unlike most of the workshop topics, almost no mathematicians were engaged in liquid crystal research in January 1985. The contents of this volume are witness to the fruit of this effort. For example, the papers of Brezis, Cohen *et.al.*, Hardt *et.al.*, and Maddocks all report on investigations undertaken after the workshop took place. The paper of Maddocks attempts to place configurations with line singularities within a framework acceptable from the viewpoint of energetics. Those of Brezis, Cohen *et.al.*, and Hardt *et.al.* establish, among other things, the notion of a stable point defect. This surprising phenomenon was discovered by a combination of analysis and computation and then precisely classified for harmonic mappings into spheres.

A brief introduction to the theory of small molecule liquid crystals is provided in the articles of Leslie. Various aspects of the study of liquid crystal polymers are presented by Berry, Doi, and Ryskin. The reader is introduced to blue phases in the papers of Cladis and Sethna. Phase transitions, especially connected to smectic states, are explored by Huang. The contributions of

Capriz *et. al.*, Choi, Di Benedetto, Miranda, and Spruck discuss mechanical or mathematical issues closely related to those encountered in the study of liquid crystals.

The conference committee greatly appreciates the concerted efforts of the speakers and discussants to make their presentations intelligible to a mixed audience. Especial thanks are due to Robert Hardt and C.-C. Huang for their assistance in the organization of the workshop. Additional thanks are due to James Fergason and Alfred Saupe for their stimulating lectures, which we knew in advance would not be available for publication.

The conference committee would like to take this opportunity to extend its gratitude to the staff of the I.M.A., Professors Weinberger and Sell, Mrs. Pat Kurth, and Mr. Robert Copeland for their assistance in arranging the workshop. Special thanks are due to Mrs. Debbie Bradley, Mrs. Patricia Brick, and Mrs. Kaye Smith for their preparation of the manuscripts. We gratefully acknowledge the support of the National Science Foundation.

J. L. Ericksen

D. Kinderlehrer

Conference Committee

RHEOLOGICAL AND RHEO-OPTICAL STUDIES WITH NEMATOGENIC SOLUTIONS OF A RODLIKE POLYMER: A REVIEW OF DATA ON POLY (PHENYLENE BENZOBISTHIAZOLE)

G.C. Berry

Department of Chemistry
Carnegie-Mellon University
Pittsburgh, PA 15213

Abstract

Rheological and rheo-optical studies on isotropic and nematic solutions of poly (1,4-phenylene-2, 6-benzobisthiazole), PBT, are reviewed. The linear viscoelastic behavior is compared with theoretical models for the isotropic solutions. A BKZ-type constitutive equation coupled with an experimentally determined distribution of relaxation times is found to represent the nonlinear viscoelastic data reported for the isotropic data; comparisons with theoretical models are made. With the nematic solutions, nonlinear viscoelastic behavior is found even for very slow shearing deformations. This behavior may be related to orientation effects at the bounding surfaces. For more rapid shearing deformations, the bulk of the fluid is strongly oriented, similar to the orientation obtained in the isotropic fluid at high rates of shearing deformation. Comparisons are made among a number of rheological properties for isotropic and nematic solutions for the latter flows.

1. Introduction

In previous work [1-6], rheological properties were reported in the linear and nonlinear response range for rodlike macroions in mesogenic solution in protic sulfonic acids. The otherwise uncharged polymers are modified to become macroions in solution through protonation by the sulfonic acid. Here, we will review some rheological properties for solutions of the rodlike polymer poly(1,4-phenylene-2,6-benzobisthiazole), PBT:

The steady-state rheological properties included the dependence on shear rate κ of the steady-state viscosity η_κ, recoverable compliance $R_\kappa^{(S)}$ and flow birefringence function $M_\kappa = \Delta n_\kappa^{(13)}/(\eta_\kappa \kappa)^2$, where $\Delta n_\kappa^{(13)}$ is the flow birefringence in the 1-3 plane. Transient properties obtained after initiation of flow include the nonlinear creep compliance $J_\sigma(t)$ and the viscosity growth function $\eta_\kappa(t)$ defined, respectively, as [7,10]

$$J_\sigma(t) = \gamma(t)/\sigma \tag{1.1}$$

$$\eta_\kappa(t) = \sigma(t)/\kappa \tag{1.2}$$

Here $\gamma(t)$ is the strain at time t after imposition of a shear stress σ and $\sigma(t)$ is the shear stress at time t after imposition of a steady rate of shear κ. Of course, for large t, both $\eta_\kappa(t)$ and $t/J_\sigma(t)$ are equal to η_κ for isotropic fluids.

Transient properties obtained after cessation of steady-state flow include the recoverable compliance function $R_\sigma(t)$, the viscosity relaxation function $\hat{\eta}_\kappa(t)$ and the flow birefringence relaxation function $\hat{M}_\kappa(t)$ defined, respectively, as

$$R_\sigma(t) = \gamma_R(t)/\sigma \tag{1.3}$$

$$\hat{\eta}_\kappa(t) = \sigma(t)/\kappa \tag{1.4}$$

$$M_\kappa(t) = \Delta n_\kappa^{(13)}(t)/(\eta_\kappa \kappa)^2 \tag{1.5}$$

Here $\gamma_R(t)$ is the recovered strain at time t after cessation of steady flow at shear rate κ (shear stress $\sigma = \eta_\kappa \kappa$), $\sigma(t)$ is the shear stress at time t after cessation of steady flow at shear rate κ, and $\Delta n_\kappa^{(13)}(t)$ is the birefringence (in the 1,3 plane) at time t after cessation of steady flow at shear rate κ. For large t, $R_\sigma(t) = R_\kappa^{(s)}$, whereas both $\hat{\eta}_\kappa(t)$ and $\hat{M}_\kappa(t)$ are zero for isotropic fluids. Prior to cessation of steady-flow, the latter are η_κ and M_κ, respectively.

For small κ or σ, linear viscoelastic behavior is expected, at least for isotropic fluids, for which one has the well-known relations [10,11]

$$\lim_{\sigma=0} J_\sigma(t) = J_0(t) = R_0(t) + t/\eta_0 \tag{1.6}$$

$$\lim_{\sigma=0} R_\sigma(t) = R_0(t) \tag{1.7}$$

$$\lim_{\kappa=0} \eta_\kappa(t) = \eta_0(t) = \int_0^t G_0(u)du \tag{1.8}$$

$$\lim_{\kappa=0} \hat{\eta}_\kappa(t) = \hat{\eta}_0(t) = \eta_0 - \eta_0(t) \tag{1.9}$$

The creep compliance $J_0(t)$ and the linear modulus $G_0(t)$ are related through a convolution integral:

$$\int_0^t G_0(u)\, J_0(t - u)du = t \tag{1.10a}$$

$$\int_0^t G_0(u)R_0(t - u)du = t\,\frac{\hat{\eta}_0(t)}{\eta_0} + \tau_c\,\frac{N_0^{(1)}(t)}{R_0^{(S)}} \tag{1.10b}$$

where $\tau_c = \eta_0 R_0^{(S)}$ with $R_0^{(S)}$ the limiting value of $R_\kappa^{(S)}$ for small κ, and

$$N_0^{(1)}(t) = \eta_0^{-2} \int_0^t uG_0(u)du \tag{1.11}$$

Equations 1.10 provide a means to compute $J_0(t)$ from $G_0(t)$ or vice versa [8,11,12], see below. The limiting values $R_0(t)$ of $R_\sigma(t)$ at small σ, and $\eta_0(t)$ of $\eta_\kappa(t)$ at small κ can be expressed in terms of retardation and relaxation times, respectively, as:

$$R_0(t) = R_0^{(s)} - \sum_1^{n-1} R_i \exp(- t/\lambda_i) \tag{1.12}$$

$$\eta_0(t) = \eta_0 - \sum_1^{n} \eta_i \exp(- t/\tau_i) \tag{1.13}$$

where, of course, the set of η_i and τ_i may be determined from the set of R_i and λ_i together with $R_0^{(S)}$ and τ_c, and vice versa.

According to the "stress-optic" law [13] M_κ is related to the first normal stress function $N_\kappa^{(1)}$ (with neglect of $\Delta n^{(23)}$ in comparison with $\Delta n^{(13)}$):

$$M_\kappa \approx 2C'\, N_\kappa^{(1)} \tag{1.14}$$

Here, $N_\kappa^{(1)} = \nu^{(1)}/2(\kappa\eta_\kappa)^2$, where $\nu^{(1)}$ is the first normal stress difference for steady flow at shear rate κ, and the coefficient C' is about equal to the ratio C of the principal components of the refractive index and stress ellipsoids: [13,14]

$$C' = \left(\frac{\cot 2\chi}{N_\kappa^{(1)} \quad \eta_\kappa \kappa} \right) C \approx C \tag{1.15}$$

with χ the extinction angle locating the cross of isocline. As indicated in Eqn. 1.15, the term in parenthesis is expected to be nearly unity. With Eqn. 1.14, the limiting value M_0 of M_κ for small κ is given by $M_0 R_0^{-1} \approx 2C'$. With the stress-optic law, C' is expected to be independent of polymer concentration or molecular weight for rodlike chains. For small κ, data on the limiting value of the flow berefringence $\Delta n^{(12)}$ in the 1-2 flow plane provide an alternative measure of C', with $\Delta n^{(12)}/\eta_0 \kappa \approx 2C'$.

With many isotropic materials, the single-integral relation for the shear stress $\sigma(t) = \sigma_1(t)$ or the first normal stress difference $\nu^{(1)}(t) = \sigma_2(t)$ can be expressed in the form [8,10,15,16]

$$\sigma_S(t) = \sum \eta_i \tau_i^{-2} \int_0^\infty [\Delta\gamma(t,u)]^S F[\Delta\gamma(t,u)] \exp(-u/\tau_i) du \tag{1.16}$$

permitting estimation of η_κ, $R_\kappa^{(S)}$ and $N_\kappa^{(1)}$, as discussed in references 5 and 8. Here $\Delta\gamma(t,u) = \gamma(t) - \gamma(t-u)$ and $F(\gamma)$ is given by the expression [5,8]

$$F(\gamma) = \exp[-m(|\gamma| - \gamma')/\gamma''] \tag{1.17}$$

where m is zero for $|\gamma| < \gamma'$ and unity otherwise, with γ' and γ'' material parameters. Of course, if $F(\gamma)$ is unity (i.e., $|\gamma| < \gamma'$), and Eqn. (1.16) reduces to the result of the linear Boltzmann constitutive equation. [10,11] With Eqn. (1.16-1.17) [5],

$$\eta_\kappa = \sum \eta_i (1 - q_{\kappa,i}) \tag{1.18}$$

$$\eta_0 \eta_\kappa R_\kappa^{(S)} = \sum \eta_i \tau_i (1 - q_{\kappa,i} (\exp - \eta_\kappa R_\kappa^{(S)}/\tau_i)) \tag{1.19}$$

$$\eta_\kappa^2 N_\kappa^{(1)} = \sum \eta_i \tau_i (1 - q_{\kappa,i} p_{\kappa,i}) \tag{1.20}$$

$$\hat{\eta}_\kappa(t) = \sum \eta_i (1 - q_{\kappa,i}) \exp(-t/\tau_i) \tag{1.21}$$

$$\eta_\kappa^2 \hat{N}_\kappa^{(1)}(t) = \sum \eta_i \tau_i (1 - q_{\kappa,i} p_{\kappa,i}) \exp(-t/\tau_i) \tag{1.22}$$

where $g_i = \gamma'/\tau_i\kappa$, $\alpha = \gamma'/\gamma''$, $f_i^{-1} = 1 + \alpha/g_i$, and

$$q_{\kappa,i} = (1 + \alpha f_i - f_i^2)\exp(-g_i) \qquad (1.23)$$

$$p_{\kappa,i} = 1 + (1 - f_i)[(f_i + g_i)^2 + f_i^2]/2(1 + \alpha f_i - f_i^2) \qquad (1.24)$$

Based on studies [5] of $J_\sigma(t)$, γ' is found to be inversely proportional to c. Mechanistic models with results similar to Eqns. 1.16-1.17 are discussed below.

2. Experimental

The polymers and concentration ranges studied are identified in Table 1 -- details are given in references 4-6. Methane sulfonic acid was distilled prior to use and stored away from contamination by atmospheric moisture. Polymers were dried in vacuo prior to use. Extreme caution was taken to prevent contamination of the solutions by moisture. Even modest amounts of water can lead to intermolecular association in acidic solution of heterocyclic polymers, with substantial effects on rheological properties [1]. Methods used are discussed in reference 4.

A wire-suspension cone-and-plate rheometer [17] and a flow birefringence apparatus [4] constructed in our laboratory were used for most of the work reported here. A few data were also obtained with a Rheometrics model RMS 7200 rheometer, equipped with a Birnboim Correlator model DAS-IV, principally to permit estimation of the linear steady-state recoverable compliance. The apparatus was modified to retard the rate of contamination by moisture by use of the protective ring assembly described (e.g., see Fig. 1 of ref. 4).

The temperature for the onset of the ordered nematic phase was determined by observation of the transmitted light with the sample between crossed polaroids in a microscope (approximately 100 × magnification). The sample was held in a special cell fabricated from rectangular glass tubing. (Vitro Dynamics, Inc., Rockaway, NJ). The sample thickness 0.4mm; use of cells ca. 0.02 mm thick produced a marked increase of the transition temperature, but no such effect was found for cells of the thickness used. The sample was sealed in the cell to prevent contamination by moisture. After a rapid temperature scan (ca. 0.01K s^{-1}) to provide an approximate estimate for the transition temperature, the temperature was adjusted to give a nematic sample about 5-10 degrees below the transition temperature. After equilibration, the temperature was slowly increased by increments (ca. 1K), allowing the necessary time for equilibration between increments; equilibration times varied from ca. 10 min to several hours, depending on the viscosity. Stability of the anisotropic texture was taken as the criterion of equilibration. The transition temperature was taken as the temperature for the disappearance of the last nematic domains. After conversion to the isotropic state, the process was reversed, with slow cooling to confirm the transition temperature. The process is more difficult, and it is easy to obtain supercooling, by as much as 5K for viscous samples.

3. Discussion

3.1 Isotropic Solutions

A plot of $\eta_0/\eta_s M[\eta]\alpha^{*3}$ versus $cL/M_L\alpha^*$ is given in Fig. 1a [4,5]. Here, $c = w\rho$, with the solution density ρ approximately equal to the solvent density ρ_s for the weight fraction w of polymer interest here, $M_L = M/L$ is the mass per unit contour length and α^* is an empirical parameter chosen to fit the experimental data by the relation [18]

$$\eta_0/\eta_s = KN_A^2 M[\eta]\alpha^{*3}\; X^3(1 - BX)^{-2} \qquad (3.1)$$

where $X = cL/M_L\alpha^*$. The plot in Fig. 1a was constructed using the data on $[\eta]$, L_η and $M_\eta = L_\eta M_L$ given in Table 1. The data are well fitted by Eqn. 3.1, with the values of $w^*/B = \alpha^* M_L/L_\eta \rho B$ given in Table 2 and $KB^{-3} = 1.5 \times 10^{-4}$. These values are conveniently deduced by comparison of bilogarithmic plots of $\eta_0/\eta_s M_\eta[\eta]$ versus cL_η/M_L with a similar plot of $(BX)^3(1 - BX)^{-2}$ versus BX (eqn. 3.1). If these have the same shape, the vertical and horizontal 'shifts' required to superpose the curves are equal to $\log KN_A^2(\alpha^*/B)^3$ and $\log \alpha^*/B$, respectively. As may be seen in Table 2, values of w_c and w^*B^{-1} are equal within experimental error. The correspondence of w_c and w^*B^{-1} and the temperature dependence of w_c provides a basis for the observed negligible temperature dependence of $\eta_0/\eta_s w_c^3$ see below. The observed value of KB^{-3} is smaller than the original estimate [18], but in accord with subsequent calculations [19].

Plots of $\ln \eta_0/\eta_s$ and $\ln w_c$ versus T^{-1} are given in Fig. 2 [5]. Here, w_c is the concentration for conversion from an isotropic state $(w < w_c)$ at T to a nematic state $(w > w_c)$ at T and η_s is the solvent viscosity. For methane sulfonic acid η_s is 7.67 mPa's at 313K and $\partial \ln \eta_s / \partial T^{-1} = 2800K$. As may be seen in Fig. 2, $\partial \ln(\eta_0/\eta_s)/\partial T^{-1} = 1265K$ for the isotropic samples studied. The data on w_c may also be fitted by an Arrheneus relation with $\partial \ln w_c / \partial T^{-1} = -433K$, and w_c/gkg^{-1} equal to 32.2, 32.3 and 34.9 at 313K for PBT 72, 53 and 43, respectively. With these results, $\partial \ln(\eta_0/\eta_s w_c^3)/\partial T^{-1} = 34K$ is nearly negligible. The product $[\eta]w_c\rho$ is equal to 81, 67 and 47 for PBT 72, 53 and 43, respectively (T = 313K).

Results [5] for $J_\sigma(t)$ and $R_\sigma(t)$ for isotropic solutions reveal linear behavior $J_0(t)$ for small enough σ, but also show that $J_\sigma(t) = J_0(t)$ for $t < t^*$, and $J_\sigma(t) > J_0(t)$ for $t > t^*$, where t^* decreases with increasing σ, e.g., Fig. 3. The strain $\gamma(t^*) = \sigma J_\sigma(t^*)$ is found [5] to be nearly independent of σ, and universely proportional to c, see Fig. 4. This behavior is consistent with Eqn. (1.16), with $F(\gamma) = 1$ for $\gamma(t) < \gamma^*$. Similar behavior has been reported with flexible chain polymers [7,8], and analyzed with Eqn. (1.16-1.17). Values of λ_i/τ_c and $R_i/R_0^{(S)}$ computed by use of "Procedure X" [20] and Eqn. 1.12 with data on $R_0(t)$ (e.g., see Fig. 3) are given in Table 3, along with τ_i/τ_c and η_i/η_0 computed from $R_0^{(S)}$, τ_c and the distribution of retardation times (i.e., λ_i/τ_c and $R_i/R_0^{(S)}$) using a method described elsewhere [8,12]. It would be difficult to extract the contribution with the longest relaxation time τ_1 by direct analysis of $\hat{\eta}_0(t)$ by, for example, use of Procedure X owing to experimental limitations at large t. In general, the reduced functions λ_i/τ_c, etc. are independent of temperature.

The linear viscoelastic data discussed above indicate that the distribution of retardation (or relaxation) times is broad [5]. For example, for the average relaxation times $\tau^{(k)}$ defined by [21]

$$\tau^{(k)} = \sum' \eta_i \tau_i^{k-1} / \sum' \eta_i \tau_i^{k-2} \qquad (3.2)$$

it is found that $\tau^{(2)}/\tau^{(1)} \sim 10$ for typical data [5]. Here the primes indicate that the summation is limited to the "terminal" relaxation time regime (e.g., the n terms of interest here). In the terms used above $\tau^{(2)} \sim \tau_c$ and $\tau^{(1)} \sim \eta_0(\sum \eta_i \tau_i^{-1})^{-1} = \eta_0(R_0^{(S)} - \sum R_i)$.

A constitutive equation obtained by Doi and Edwards [22] for isotropic solutions of rodlike chains based on a mechanistic model similar to that leading to Eqn. 3.1 leads to a particularly simple result for $G_0(t)$: one relaxation time (e.g., $\eta_1 = \eta_0$ and $\tau_1 = \tau_c$). The considerable disparity between this prediction and the behavior observed here may reflect the effects of molecular weight distribution. Alternatively, relaxation modes not included in the theoretical treatment may also contribute to $G_0(t)$. Possibilities for the latter may include the motions contributing to smaller K in Eqn. 3.1 than the theoretical expectation, or fluctuation in the local density of interchain interactions related to fluctuation in the local concentration.

The mechanistic treatment of Doi and Edwards leads to a single-integral constitutive equation of the type employed here [5,8], but with an exponential $G_0(t)$ and with $F(|\gamma|)$ that is fitted to within 10% by $(1 + \gamma^2/5)^{-1}$. For the calculations of interest here, Eqn. 1.17 provides a satisfactory fit to the theoretical estimate [6], provided $\gamma' = 0.6$ and $\gamma'' = 2.13$ (e.g., $\beta^{-1} = 2.38$). Consequently, Eqns. 1.16-1.17 provide close representation of the expressions that would be obtained with the theoretical single-integral constitutive equation if the latter is generalized to use the empirical η_i, τ_i set and the experimental values of γ' and γ''. For this constitutive equation, with deformation at constant κ, $\sigma(t)$ exhibits a maximum for time t^+ such that $\gamma^+ = \kappa t^+$ is equal to γ'', see below. As shown in Figs. 5 and 6, the functions η_κ/η_0, $R_\kappa^{(S)}/R_0^{(S)}$ and $M_\kappa/M_0 \sim N_\kappa^{(1)}/R_0^{(S)}$ computed in this way are in good accord with experiment; functions computed with the theoretical single relaxation time and $\beta^{-1} = 2.38$ do not correspond to experiment. The theoretical constitutive equation discussed above employs the so-called "independent alignment approximation". For rodlike molecules, the theoretical results obtained with and without the use of this approximation are numerically quite similar for the functions of interest here [23]. Behavior similar to that embodied in Eqns. 1.16-1.17 is attributed to molecular weight heterogeneity in a mechanistic calculation [24].

As with moderately concentrated solutions of flexible-chain polymers [8], the observed $\gamma' \sim \gamma^{*}$ exceed 0.6, and are proportional to c^{-1}, whereas the observed $\gamma'' \sim \gamma^{+}$ are not too far from 2.13, and do not depend on c, see Fig. 4. With flexible chain polymers it was suggested that the discrepancy between γ' and 0.6 might be attributed to a looseness in the pseudo-entanglement network caused by the finite chain length of the polymers studied (e.g., $cM/\rho M_C \sim 10$, where M_C is the critical chain length for which $\partial \ln \eta_0 / \partial \ln cM$ changes from 1 to 3.4). A similar effect may be operative here such that γ' would approach 0.6 only if the rodlike chains were very long. In the latter case, however, with rodlike chains, an ordered state would develop, negating the comparisons being made here. Thus, with rodlike chains, the "universal" behavior may be unattainable.

Whereas molecular weight distribution may be a principal source of the distribution of τ_i noted above for the samples studied here, contributions to the distribution could also arise from the postulated looseness of the network constraints to which values of γ' larger than 0.6 are attributed. Thus, D_R, from which the relaxation time of the mechanistic model arises, is closely linked with the translational motion of a rodlike chain along its axis. Fluctuations in the distance over which this translation must occur might contribute to a distribution of τ_i, and such fluctuations could be enhanced if the chains are short, similar to the effect postulated for γ'.

Date on $\hat{\eta}_\kappa(t)/\eta_\kappa$ and $\hat{M}_\kappa(t)/M_\kappa$ for two solutions at several temperatures are given in Fig. 7 [5]. The data are represented in reduced form versus t/β_κ, where the reduction factors β_κ are shown as a function of $\tau_\kappa = \eta_\kappa R_\kappa^{(S)}$ in the inserts. For each function, the reduced curves superpose over the range of κ studied. The rate of relaxation of $\hat{M}_\kappa(t)/M_\kappa$ is considerably less than that of

$\hat{n}_\kappa(t)/n_\kappa$; similar behavior has been reported [25] for flexible chain polymers.

As should be expected, the relaxations $\hat{n}_\kappa(t)$ and $\hat{N}^{(1)}_\kappa(t)$ are both much faster with the theoretical one-relaxation time model than is observed experimentally. With the theoretical model, $N^{(1)}_\kappa(t/\tau_c)/N^{(1)}_\kappa$ is independent of κ, whereas experimentally $N^{(1)}_\kappa(t/\beta_\kappa)/N^{(1)}_\kappa$ is essentially independent of κ, with $\beta_\kappa \sim \tau_\kappa < \tau_c$. In effect, τ_κ is a measure of the relaxation time for the deformed sample. With τ_κ given by Eqn. 1.19 and use of the approximation $1 - q_{\kappa,i} \approx \exp - 2/5\ \beta\kappa\tau_i$, one obtains

$$\tau_\kappa/\tau_c \approx 1 - \hat{N}^{(1)}_0(\tau_\kappa + 2(5\ \beta\kappa)^{-1})/R^{(S)}_0 \qquad (3.3)$$

where $\hat{N}^{(1)}_0(t)$ is given by Eqn. 1.11. As κ is increased, τ_κ decreases owing to successive suppression of contributions from the largest τ_i.

3.2 Nematic Solutions

The threshold volume fraction ϕ_c or concentration c_c for incipient separation of the ordered phase from the isotropic solution is expected to depend on chain length L and diameter d according to a relation of the form [26-29]

$$\phi_c = (6A/L_w)k_D f(L/d) \qquad (3.4a)$$

$$c_c[\eta]\frac{\ell n L_n/d}{L_n/d} = A(M_z/M_w)^{1/2}k_D f(L/d) \qquad (3.4b)$$

where k_D is a polydispersity factor, equal to unity for a monodispersed polymer, $f(L/d) \sim 1$ for large L/d, A is a constant and $L_n \sim (L_z L_w)^{1/2}$. In expressing Eqn. 3.4b, use is made of the relation [30]

$$M_L[\eta] = \pi N_A L_n^2/24\,\ell n(L_n/d) \qquad (3.5)$$

and c_c is calculated from the threshold volume fraction ϕ_c as $c_c = \bar{v}_2\phi_c$ where the partial specific volume $\bar{v}_2$ is equal to $\pi N_A d^2/4M_L$; data in reference 31 give $d \sim 0.5$ nm for PBT. For monodispersed polymers, theories of Onsager [26] and Flory [27] give A equal to 5/9 and 4/3, respectively. Calculations have been given for chains with a most probable [28] and a gaussian distribution [29] of L, with results, respectively, that can be represented by the expressions (for large L/d), $A(M_z/M_w)^{1/2}k_D \sim 6^{-1/2}$ and $Ak_D \sim (4/3)(M_w/M_n)^{1/3}$. The experimental estimates of c_c may be higher than values given by Eqn. 3.4 owing to the limitations inherent in observing the onset of a birefringent phase. Experiental values of $c_c[\eta](d/L_n)\ell n(L_n/d)$ obtained here (for $T = 313K$), calculated with $d = 0.5$ nm, range from 0.9 to 1.1. Thus, the experimental values of c_c seem to be in a range predicted by theory given the molecular weight distribution that obtains [32] with PBT and related polymers.

As may be seen in Fig. 8, in some cases η_κ did not go to a limiting value at small κ with the nematic solutions -- this feature is discussed further below. In such cases, $R_\kappa^{(S)}$ did tend to a constant value $R_0^{(S)}$ at small κ, and usually a range of κ existed for which η_κ was essentially a constant, designated η_p. Whereas, for the isotropic solution, $\partial \ell n \eta/\partial T^{-1}$ is about equal to $\partial \ell n \eta_s w_c^3/\partial T^{-1}$, as expected with Eqn. 3.1, for the nematic solution, [6] $\partial \ell n \eta/\partial T^{-1}$ is smaller, and more nearly equal to $\partial \ell n \eta_s/\partial T^{-1}$. Consequently, data on η_p are reduced to 24°C using $\partial \ell n \eta_s/\partial T^{-1}$.

The decrease of "the viscosity" on the formation of the nematic phase is one of the most characteristic features noted for mesogenic polymer solutions [33]. The value of η_κ usually referred to in this correlation is η_p defined above. Values of η_p determined here are given in Fig. 1b.

With nematic solutions, the rheological properties are inherently anisotropic. According to the constitutive equation of Leslie [34] and Erickson [35], the steady-state shear viscosity η_0 (determined for flow in a wide gap in a region for which the stress tensor is a linear function of the velocity gradient tensor) may be expressed in the form

$$\eta_0 = \eta_b + \frac{1}{2}(1 - \lambda^{-1})[(\eta_c - \eta_b) + \frac{1}{2}\alpha_1(1 + \lambda^{-1})] \tag{3.6a}$$

$$2\eta_b = \alpha_3 + \alpha_4 + \alpha_6 \tag{3.6b}$$

$$\eta_c - \eta_b = -\alpha_2 - \alpha_3 \tag{3.6c}$$

$$\lambda^{-1} = (\alpha_2 - \alpha_3)/(\alpha_2 + \alpha_3) \tag{3.6d}$$

where λ and the α_i are related to the order parameter S for the quiescent nematic fluid, see below. Stable simple shear obtains only if $\lambda > 1$, with the rodlike chains at angle $(\arccos \lambda^{-1})/2$ with the flow direction. In theoretical treatments [36,37] based on the diffusion equation of Doi and Edwards [22], λ and the Leslie-coefficient α_i are all functions of the order parameter S for the quiescent material

$$S = \frac{1}{2}(3 \langle\cos^2\theta\rangle - 1) \tag{3.7}$$

where θ is the angle between the rodlike molecules with the average calculated using the equilibrium distribution of θ, and S depends on c/c_c, increasing toward unity with increasing c/c_c. For example, for the treatment in reference 36, $4S = 1 + 3(1 - 8C_c/9C)^{1/2}$. With the α_i and λ given in reference 36 (see Table 4),

$$\eta_0 = \eta_b \frac{(1 + 2S)(2 + 3S)}{(2 + S)} \tag{3.8a}$$

$$\eta_b = \eta_0^{ISO} \frac{(1 - S)^4}{(1 + S/2)} \tag{3.8b}$$

where η_0^{ISO} is given by Eqn. (3.1) with $B = 0$, so that η_0 decreases as S increases toward unity. With the α_i and λ given in reference 37 (see Table 4), $\lambda < 1$ and simple shear flow is predicted to be unstable. Since the calculation for the α_i is delicate, the significance of this result is unclear for the observed behavior, which results in stable steady-state shear stress at all κ studied. With the α_i given in reference 37,

$$\eta_b = \eta_0^{ISO} \frac{(1 - S)^4}{105} \{4(4S + 7) + 35(5S - 2)^{-1}\} \tag{3.9}$$

which is smaller (about 20%) than η_b given by Eqn. 3.6b for given S. (Similarly, the elongational viscosity obtained with the α_i in reference 36 is about twice that for the α_i in reference 37 for comparable S). Values of η_p plotted in Fig. 1b are much smaller than η_0 extrapolated for the isotropic fluid, in some cases being in the range expected for an isotropic fluid with $cL/M_L = \alpha^*$.

Flow birefringence data indicate a substantial degree of orientation for shear flows with $\eta_p R_0^{(S)} \kappa > 1$. For example, at such κ, the strong turbidity characteristic of the quiescent nematic fluid is lost, and the fluid in flow is much like that for a well oriented isotropic solution, with similar characteristics for the transmission of polarized light. For smaller κ (e.g., $\eta_p R_0^{(S)} \kappa < 0.1$), the transmitted intensity is smaller than expected, and tends to fluctuate. In particular, with $\eta_p R_0^{(S)} \kappa > 1$, the sum $I_+ + I_{\parallel}$ of the intensities I_+ and $I_{\parallel}$ transmitted between crossed and parallel polars, respectively, is smaller than that for the quiescent fluid, and the overall field appears optically homogeneous (albeit birefringent and oriented). For $\eta_p R_0^{(S)} \kappa < 0.1$, the sum $I_+ + I_{\parallel}$ is markedly depressed and the overall field is mottled in appearance.

In addition to the complicated flow birefringence behavior at small κ, it is found that η_κ determined in stress-growth with shear rate $\kappa = \partial\gamma_{ss}/\partial t$ increases with decreasing shear rate $\partial\gamma_{ss}/\partial t$, and for small κ is larger than η_p, whereas $\eta_\sigma = (\partial J_\sigma/\partial t)^{-1}$ is about equal to η_p. This unusual behavior may be related to the flow birefringence behavior reported above. In other respects, the rheological behavior is similar to that given by Eqns. 1.16-1.17. Thus, as shown in Figure 9, the $R_\sigma(t)$ determined after steady-state flow with $\eta_p R_\kappa^{(S)} \kappa \approx 1$ is similar to $R_0(t)$ for the nematic solutions. Values of the (apparent) λ_i and R_i calculated with Eqn. 1.12 and $R_\sigma(t)$ so determined lead to (apparent) τ_i and η_i that reproduce η_κ/η_p reasonably well for $\eta_p R_\kappa^{(S)} \kappa > 1$ using Eqn. 3.6, see Figure 10.

The behavior summarized above is not consistent with Eqn. 3.6 (and the α_i in reference 36), and may suggest some kind of flow instability, perhaps similar to the effect predicted in connection with α_i given in reference 37. Alternatively, the effects observed at small κ may be caused by the effects of chains strongly anchored to the surface of the platens, with the chain axis in the plane of the platen. The anchored chains should be parallel locally, but chains on different platens, or far removed from each other on a given platen need not be parallel. Consequently, the quiescent fluid exhibits texture imposed by the mostly smooth transitions of the director orientation throughout the volume to accommodate the variable orientations at the platens. When subjected to a steady shear deformation, this texture is maintained in some degree until κ is large enough that $\eta_p R_0^{(S)} \kappa \approx 1$, with preferred orientation in the flow direction replacing the texture for $\eta_p R_0^{(S)} \kappa > 1$.

Acknowledgement:

It is a pleasure to acknowledge partial support for the studies reported above from the Polymers Program Division of Material Research, National Science Foundation, and the Materials Laboratory, Wright-Patterson Air Force Base.

Table 1: Polymer Solutions Used[4-6]

Polymer	$[\eta]$/mLg^{-1}	L_η/nm[a]	Range w/gkg^{-1}	Range T/K
PBT 62	2650	170	15.0−82	272−353
PBT 72	1770	135	25.4−29.4	297−331
PBT 72R	1100	100	29.6	296−331
PBT 53	1400	120	255−42.7	283−333
PBT 53	1400	118	1.49−32.3	283−333
PBT 43	900	95	29.4−31.5	288−297

a) $L_\eta = M_\eta/M_L$, where M_L = 220 dalton nm^{-1} and $M_\eta = ([\eta]/K_\eta)^{1/1.8}$ with $M_L^{1.8}K_\eta$ = 0.26 mLg^{-1}.[10]

Table 2: Critical Concentrations for Solutions of Rodlike Polymers[5]

Polymer	T/K	w_c/gkg^{-1} [a]	w^*B^{-1}/gkg^{-1} [b]	Bw_c/w^*
PBT−72	296	29.7	27.5	1.08
	313	32.2	28.9	1.11
	333	34.9	30.4	1.15
PBT−72R	296	--	36.5	
	313	--	38.3	
	333	--	40.3	
PBT−53	296	29.9	31.5	0.95
	313	31.7	32.9	0.96
	333	34.9	34.9	1.00
PBT−43	296	32.2	38.8	0.83
	313	34.9	40.7	0.86
	333	37.9	42.9	0.88

(a) The concentration for formation of the ordered phases interpolated using the relation $w_c = k \exp - 433/T$, with k equal to 128.4, 128.8 and 139.2 for PBT 72, 53, and 43, respectively.

(b) Obtained by a fit of Eqn. 3.1 with data on η_0, as described in the text.

Table 3: Retardation and Relaxation Spectra[5,6]

λ_i/τ_c [a]	$R_i/R_0^{(S)}$	τ_i/τ_c [a]	η_i/η_0
	PBT−53; w = 32.3 g kg^{-1} (NEMATIC)[b]		
8.30	0.262	8.59	0.037
2.65	0.104	2.79	0.065
1.04	0.312	1.40	0.266
0.193	0.170	0.353	0.323
		0.086	0.149
0.042	0.111	0.0075	0.160
	PBT−53; w = 29.4 g kg^{-1} (ISOTROPIC)		
2.580	0.580	3.251	0.231
0.210	0.280	0.494	0.480
0.0175	0.130	0.0615	0.092
		0.0010	0.092
	PBT−53; w = 25.5 g kg^{-1} (ISOTROPIC)		
5.248	0.430	5.732	0.094
1.022	0.290	1.354	0.258
0.240	0.174	0.381	0.253
0.080_5	0.035	0.098_8	0.084
		0.025_2	0.312

(a) $\tau_c = \eta_0 R_0^{(S)}$

(b) $R_0^{(S)}$, η_0 and τ_c replaced by values obtained from creep with shear stress σ = 14.6 Pa, see text

Table 4: Leslie Coefficients for Rodlike Polymers

Parameter[a]	Ref 36	Ref 37[b]
α_1/k	$-S^2$	$-rS^2$
α_2/k	$-S(1+2S)/(2+S)$	$-(S/2)[3(r-1)+2(1+2S)]/[3(r-1)+2+S]=-S(4S-1)/(5S-2)$
α_3/k	$-S(1-S)/(2+S)$	$-(S/2)[3(r-1)+2(1-S)]/[3(r-1)+2+S]=S(1-S)/(5S-2)$
α_4/k	$(1-S)/3$	$(7-5S-2rS^2)/35=(1-S)[1-(2/35)(1-S)(7+4S)]/3$
α_5/k	S	$S(5+2rS)/7=S[1-(2/21)(1-S)(3-4S)]$
α_6/k	0	$-2S(1-rS)/7=-2S(1-S)(3+4S)/21$.
λS[c]	$1-(1-S)/3$	$r-(1-S)/3=-1-5(1-S)/3$
r	---	$1-4(1-S)/3$

(a) $k=\eta_0^{ISO}(1-S)^2$; η_0^{ISO} given by Eqn. (3.1) with B=0.
Since $\alpha_2+\alpha_3 = \alpha_6-\alpha_5$, only five of the α_i are independent.

(b) Based on asymptotic behavior for α_i given in ref. 37 and approximate relations for λ and r.

(c) $\lambda = (\alpha_2+\alpha_3)/(\alpha_2-\alpha_3)$.

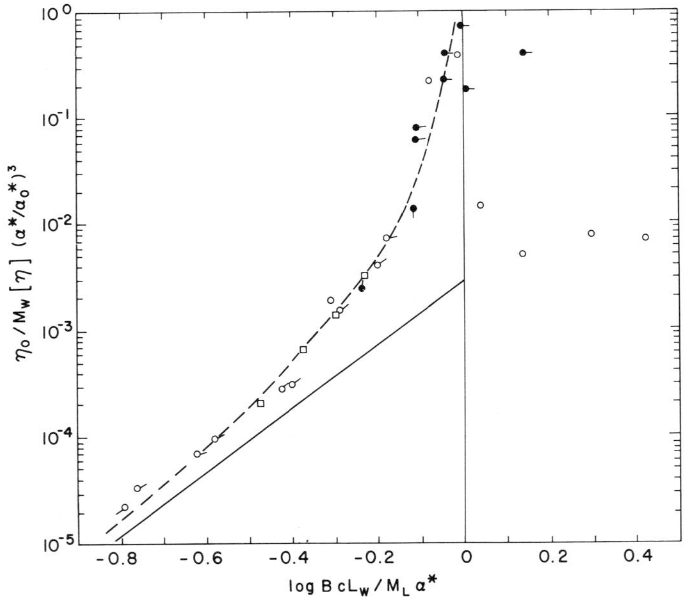

Figure 1:

$\eta_0/M_w[\eta](\alpha^*/\alpha_0^*)^3$ versus $BcL_w/M_L\alpha^*$ for isotropic (left) and nematic (right) solutions of PBT-53 and PBT-62. With the nematic solutions, η_0 is replaced by η_p (α_0^* is a constant). From reference 6.

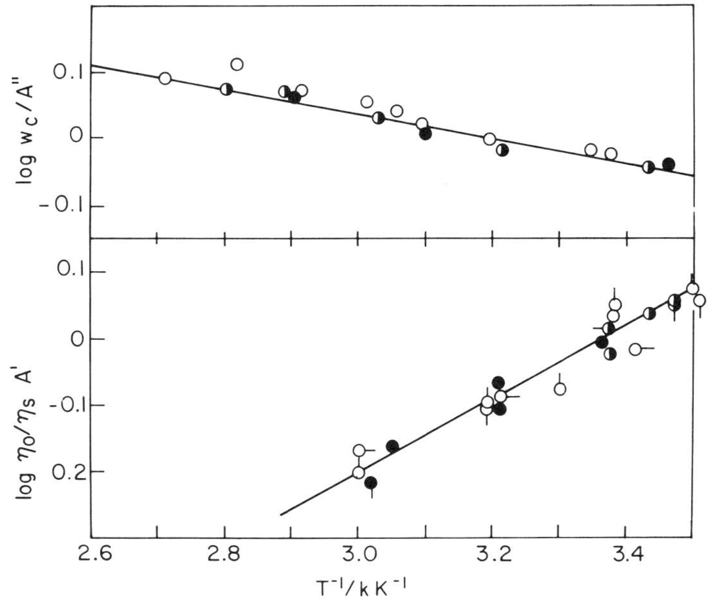

Figure 2:
Temperature dependence of the viscosity η_0 relative to that η_s of methane sulfonic acid, and the concentration w_c for the onset of the nematic phase. In the figures, A' and A" are the values of η_0/η_s and w_c, respectively, for T = 313K. The symbols denote PBT-53, O; PBT-72, • and PBT-43, ◐. From reference 5."

Figure 3:
The reciprocal of the critical strains γ^* and γ^+ versus weight percent 100w for several PBT-53 solutions in MSA. From reference 6.

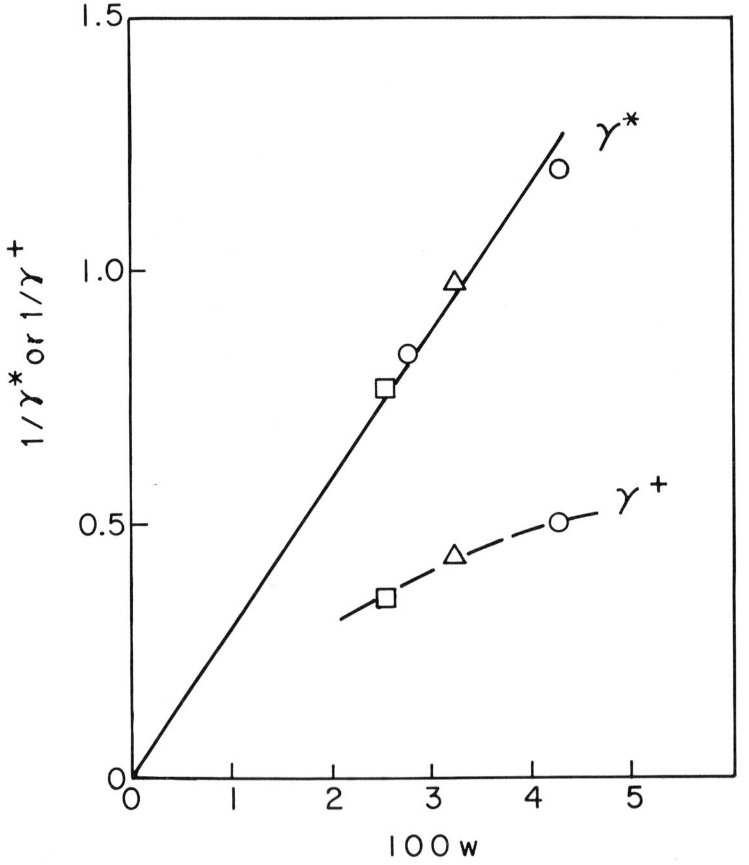

Figure 4:

The reduced creep compliance $J_\sigma(t)/R_0^{(S)}$ (open circles) and recoverable compliance $R_0(t)/R_0^{(S)}$ (filled circles) versus the reduced time t/τ_c for a solution of PBT−53 in methane sulfonic acid, w = 25.5 gkg^{-1}. The symbols O(•), Q, O-, Ȯ, -O denote stress σ/Pa equal to 2.8, 15.2, 68.3, 78.7 and 145, respectively. The data for O(•) are at 23 C; all others at 13 C. From reference 5.

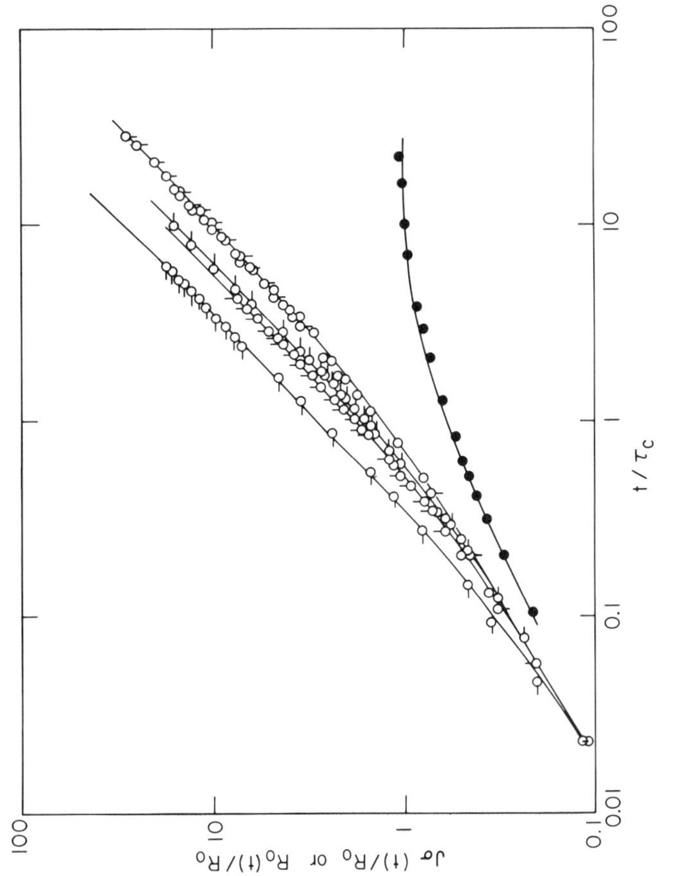

Figure 5:
η_κ/η_0 and $R_\kappa^{(S)}/R_0^{(S)}$ versus $R_0^{(S)}\eta_0\kappa$ for isotropic solutions of PBT−53 (w = 25.5g/kg, with symbols for different temperatures in the range specified in Table 1). The curves for η_κ/η_0 and $R_\kappa^{(S)}/R_0^{(S)}$ represent Eqns. 1.18 and 1.19, respectively, with the τ_i and η_i in Table 3. From reference 5.

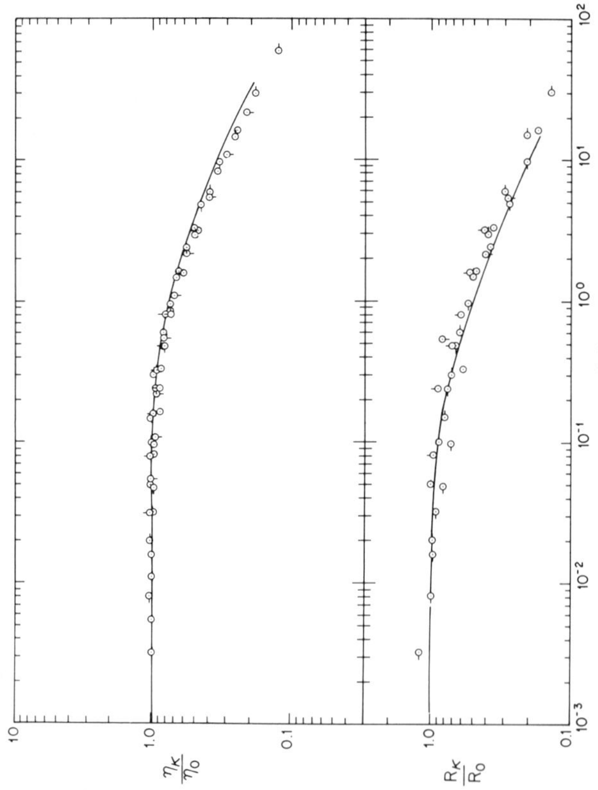

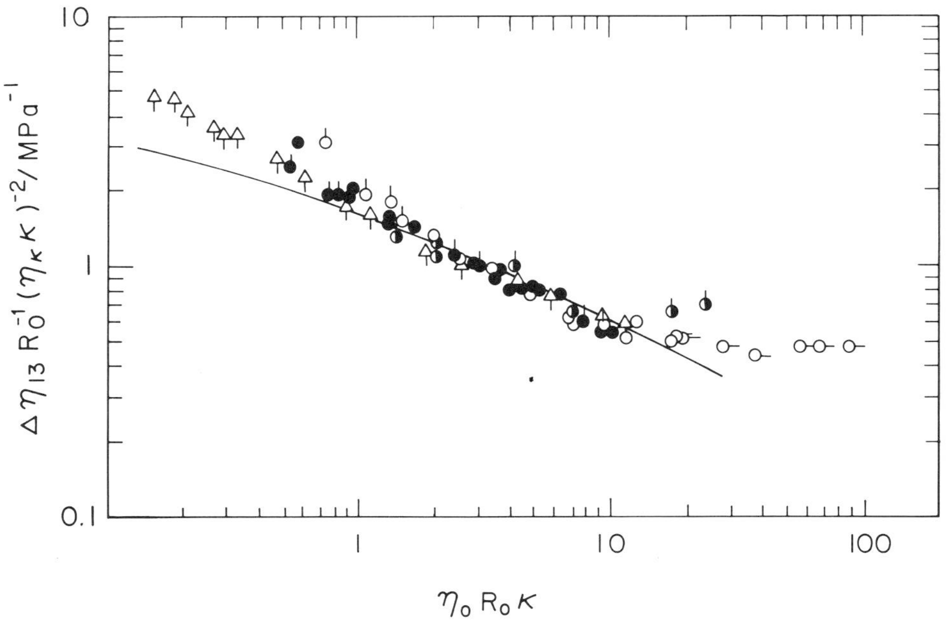

Figure 6:
The steady-state flow birefringence versus the reduced shear rate $\eta_0 R_0^{(S)} \kappa$ for an isotropic solution of PBT-53 (0.0255 weight fraction) at several temperatures. The curve is calculated with Eqns. 1.14 and 1.20 using experimentally determined values of τ_i and η_i. From reference 5.

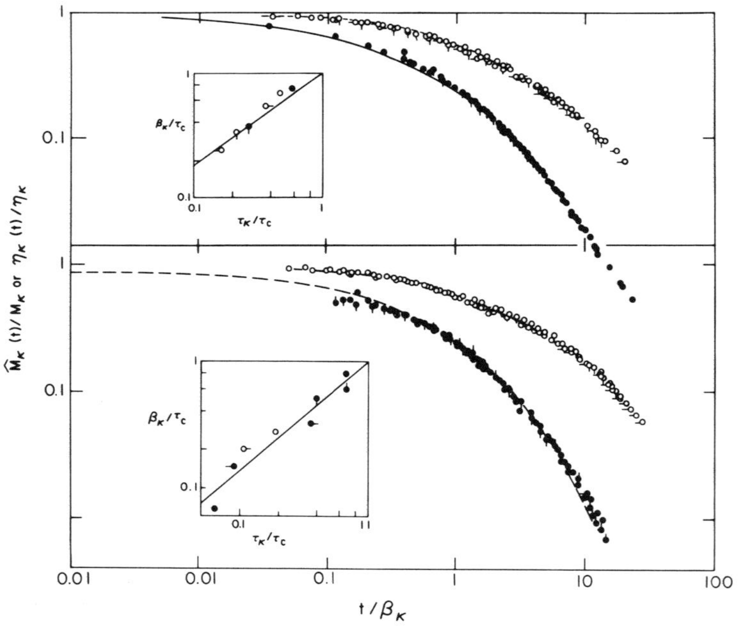

Figure 7:
The reduced flow birefringence relaxation function $\hat{M}_\kappa(t)/M_\kappa$ (filled circle) and stress relaxation function $\hat{\eta}_\kappa(t)/\eta_\kappa$ (open circles) for PBT-72, 25.4 gkg^{-1} (top) and PBT-53, 29.4 gkg^{-1} (bottom) in methane sulfonic acid. For PBT-72, κ/s^{-1} is 0.0208 (O), 0.0358 (O-), 0.0726 (Q), 0.100 (-O), 0.0126 (●), 0.0052 (●), all at 39° C. For PBT-53, κ/s^{-1} is 0.0227 (O), 0.0455 (O-), 0.0851 (Q), 0.186 (-O), 0.0050 (●), 0.0126 (●), 0.0502 (●), all at 21° C, and 0.0050 (●), 0.0252 (●), both at 40° C. From reference 5.

Figure 8:
The steady-state viscosity η_κ and recoverable compliance $R_\kappa^{(S)}$ for solutions of PBT-62 with $T \approx 60°$ C. The symbols indicate 100 W equal to 1.5, 2.5, 3.0, 3.4, 4.7, 6.1 and 8.2 for Q, O, O, Q, O and -O, respectively. From reference 6.

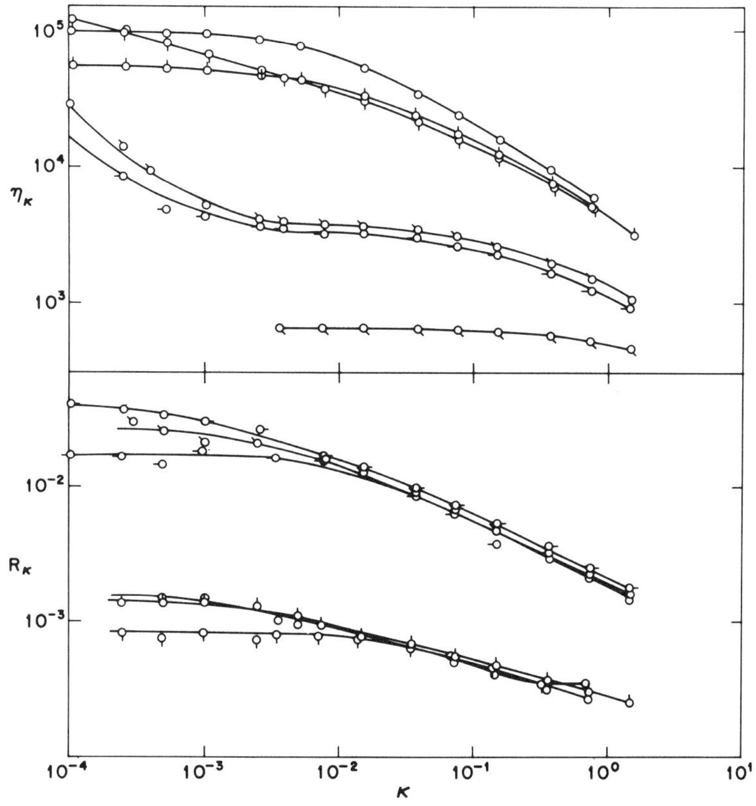

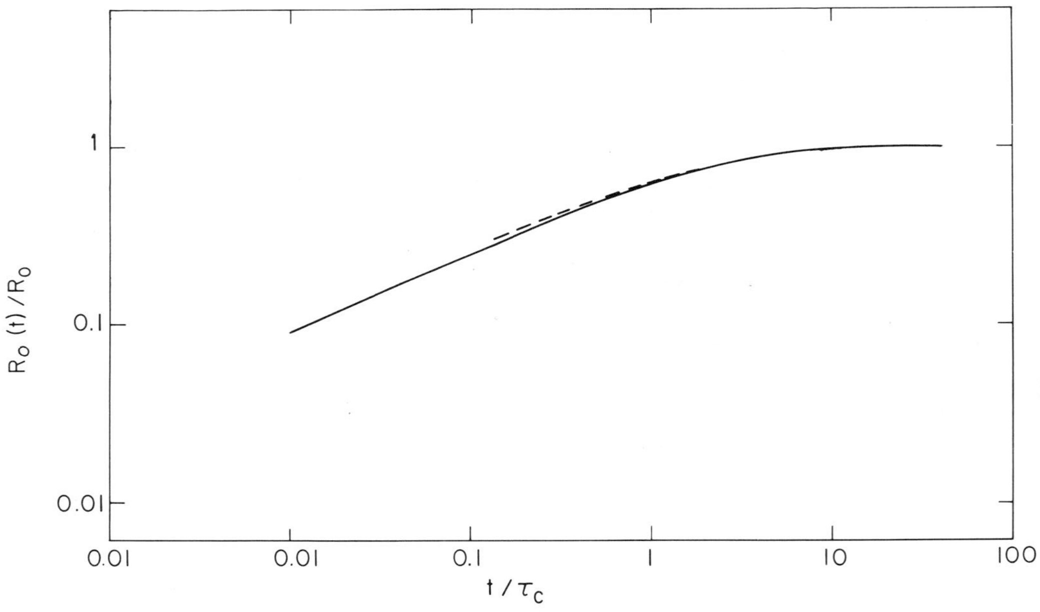

Figure 9:
$R_\sigma(t)/R_0$ versus t/τ_c for two isotropic solutions of PBT-53 (w/g kg^{-1} equal to 29.4 and a nematic solution (w^c = 32.3 g/kg, T = 19.5° C, solid curve). For the latter, τ_c is replaced by $\eta_\kappa R_\kappa^{(S)}$ and $R_0^{(S)}$ is replaced by $R_\kappa^{(S)}$. With the isotropic solutions, $R_\sigma(t) = \hat{R}_0(t)$, and the nematic solution, σ = 14.6 Pa. From reference 6.

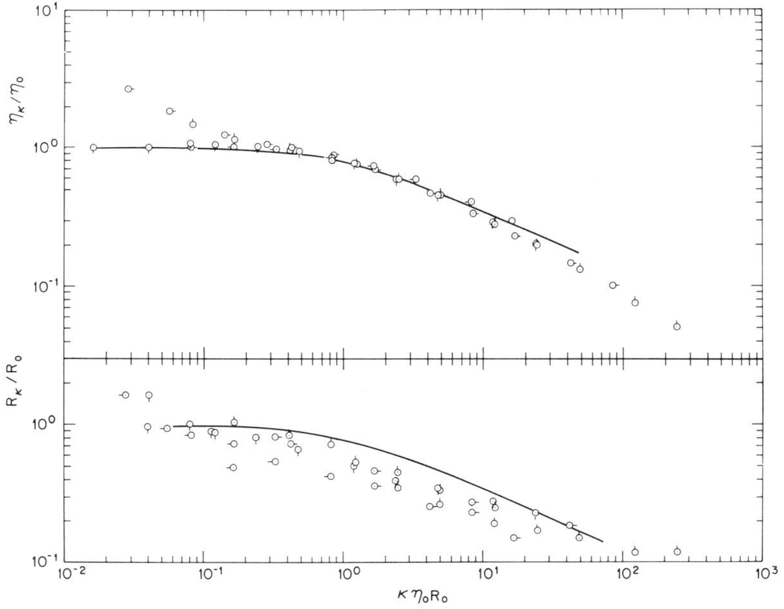

Figure 10:
Rheological data for isotropic (–○, ○) and nematic (○–, ○) solutions of PBT-53 (0.0323 weight fraction polymer. With the latter, η_0 is replaced by η_p, see text. The curves are calculated with Eqns. 1.18 and 1.19 using experimentally determined values of τ_i and η_i. From reference 6.

References

1. C.-P. Wong and G.C. Berry, Polymers 1979, 20, 229.

2. G.C. Berry, P. Metzger, D.B. Cotts, and S.-G. Chu, British Polym. J. 1980, 16, 947.

3. C.-P. Wong, H. Ohnuma, and G.C. Berry, J. Polym. Sci., Symp. 1978, 65 173.

4. S.-G. Chu, S. Venkatraman, G.C. Berry, and Y. Einaga, Macromolecules 1981, 14, 939.

5. S. Venkatraman, G.C. Berry, and Y. Einaga, J. Polym. Sci., Polym. Phys. Ed., in press.

6. Y. Einaga, G.C. Berry, and S.-G. Chu, Polymer J., 1985, 17, 239.

7. G.C. Berry, B.L. Hager, and C.-P. Wong, Macromolecules 1977, 10, 361.

8. K. Nakamura, G.C. Berry and C.-P. Wong, J. Polym. Sci., Polymer Phys., in press.

9. H. Markovitz and G.C. Berry, Ind. Eng. Chem. Prod. Res. Div. 1978, 17(2), 95.

10. H. Markovitz, in Am. Inst. Phys. 50th Anniversary Physics Vade Mecum, H.L. Anderson, ed., Am. Inst. Phys., NY, (1981), Chapter 19.

11. J.D. Ferry, Viscoelastic Properties of Polymers, Wiley, NY, 3rd Ed., (1980).

12. R. Sips, J. Polym. Sci. 1951, 7, 191.

13. H. Janeschitz-Kriegl, Adv. Polym. Sci. 1969, 6, 170.

14. B.D. Coleman, E.H. Dill, and R.A. Toupin, Arch. Ration. Mech. Anal. 1970, 39, 358.

15. B. Bernstein, E.A. Kearsley, and L.J Zapas, Trans. Soc. Rheol. 1963, 7, 391.

16. H.M. Laun, Rheol. Acta 1978, 17 1.

17. G.C. Berry and C.-P. Wong, J. Polym. Sci., Polym. Phys. Ed. 1975, 13, 1761.

18. M. Doi, J. Phys. (Paris) 1975, 36, 607.

19. J.A. Odell, E.D.T. Atkins, A. Keller, J. Polymer Sci., Polymer Lett. Ed. 1983, 21, 289.

20. A.V. Tobolsky, Properties and Structure of Polymers, Wiley, NY, (1960) p. 188.

21. W.W. Graessley, Adv. Polym. Sci. 1974, 20, 229.

22. M. Doi and S.F. Edwards, J. Chem. Soc., Faraday Trans. 2 1978, 74, 560.

23. N.Y Kuzuu and M. Doi, Polymer J. 1980, 12, 883.

24. G. Marrucci and N. Grizzuti, J. Non-Newtonian Fluid Mech. 1984, 14, 103.

25. K. Osaki, N. Bessho, T. Kojimoto, and M. Kurata, J. Rheology 1979, 23, 457.

26. L. Onsager, Am. N.Y. Acad. Sci. 1949, 51, 627.

27. P.J. Flory, J. Proc. R. Soc. London, Ser. A. 1956, 234, 73.

28. P.J. Flory and R.S. Frost, Macromolecules 1978, 11, 1126.

29. J.K. Moscicki and G. Williams, Polymers 1982, 23, 588.

30. H. Yamakawa, Modern Theory of Polymer Solutions, Harper and Row, NY 1971, p. 180.

31. C.C. Lee, S.-G. Chu, and G.C. Berry, J. Polym. Sci., Polym. Phys. Ed. 1983, 21, 1573.

32. G.C. Berry and D.B. Cotts, Macromolecules 1981, 14, 930.

33. K.F. Wissbrun, J. Rheology 1981, 25, 619.

34. F.M. Leslie, Archs. Ration. Mech. Anal. 1968, 28, 265.

35. J.L. Ericksen, Arch. Ration. Mech. Anal. 1960, 4, 231.

36. G. Marrucci, Mol. Cryst. Liq. Cryst. (Lett.) 1982, 72, 153.

37. N. Kuzuu and M. Doi, J. Phys. Soc. Japan 1983, 52, 3486; 1984, 53, 1031.

LIQUID CRYSTALS AND ENERGY ESTIMATES FOR S^2- VALUED MAPS

Haïm Brezis

Département de Mathématiques
Université Paris VI
4 Pl. Jussieu
75252 Paris Cedex 05

This report summarizes results obtained in collaboration with J.M. Coron and E. Lieb (see [3] and [4]); it answers some questions raised by J. Ericksen and D. Kinderlehrer. The original motivation comes from the theory of liquid crystals (see [7], [8], [10]), and is well explained in other contributions to this volume.

We deal with maps ϕ from a domain $\Omega \subset R^3$ with values into S^2 which admit a finite number of singularities. We consider two different kinds of problems. In the first type of problem the location and the degree of the singularities is prescribed; the main result is an explicit formula, when $\Omega = \mathbb{R}^3$, for the minimum value of the deformation energy. In the second type of problem the number, the location and the degree of the singularities are "free"; our main result asserts that if ϕ is a minimizer then all its singularities have degree ± 1, moreover, the first order expansion shows that ϕ (or $-\phi$) acts like a rotation near every singularity - a fact which agrees with experimental and numerical evidence (see [5] and [6]).

1. Prescribed Singularities

Fix N points $a_1, a_2, \dots a_N$ in $\mathbb{R}^3$ (the desired location of the singularities). Consider maps ϕ which are smooth on $\mathbb{R}^3 \setminus \bigcup_{i=1}^{N} \{a_i\}$, with values in S^2, and with finite energy, i.e.

$$E(\phi) = \int_{\mathbb{R}^3} |\nabla\phi|^2 < \infty$$

[The most general energy of interest in the theory of liquid crystals is

$$\tilde{E}(\phi) = K_1 \int (\mathrm{div}\phi)^2 + K_2 \int |\phi \cdot \mathrm{curl}\,\phi|^2 + K_3 \int |\phi \wedge \mathrm{curl}\,\phi|^2$$

which is equivalent to $E(\phi)$ when $K_1 = K_2 = K_3 = 1$; it is an interesting open problem to extend our results to $\tilde{E}$].

The fact that $E(\phi) < \infty$ does not imply that ϕ is continuous at the points a_i. A typical example of a ϕ with a singularity at $x = 0$ and locally finite energy is $\phi(x) = x/|x|$.

The degree of ϕ at a_i, $\deg(\phi,a_i)$, is defined to be the (Brouwer) degree of ϕ restricted to any small sphere around a_i. The class of admissible maps consists of

$$\mathcal{E} = \{\phi \in C^1(R^3 \setminus \bigcup_{i=1}^{N} \{a_i\}; S^2) \mid \int_{R^3} |\nabla\phi|^2 < \infty \text{ and } \deg(\phi,a_i) = d_i\}$$

where the d_i' s are given integers (positive or negative). [Experimental evidence shows that the only observed degrees are ± 1 and the reason will be explained in Section 2 ; but, a priori, it makes mathematical sense to consider all possible integers].

Note that if the class $\mathcal{E}$ of admissible maps is not empty, then we must have

$$(1) \qquad \sum_{i=1}^{N} d_i = 0$$

because the assumption $\int_{\mathbb{R}^3} |\nabla\phi|^2 < \infty$ implies that, in some weak sense, ϕ tends to a constant at infinity and therefore the total degree must be zero. Conversely, if (1) holds, then $\mathcal{E}$ is not empty (this follows from the construction below).

Our purpose is to investigate the least deformation energy E needed to produce singularities of assigned degree at a prescribed location, namely,

$$(2) \qquad E = \inf_{\phi \in \mathcal{E}} \int |\nabla\phi|^2 \quad .$$

The main results of Section 1 are the following:

Theorem 1 $E = 8\pi L$

where L is the length of a minimal connection (a notion which will defined later).

Theorem 2 The infimum in (2) is <u>not</u> achieved. If (ϕ_n) denotes a minimizing sequence, then there is a subsequence (ϕ_{n_k}) which converges to a constant a.e. and such that $|\nabla\phi_{n_k}|^2$ converges in the sense of measures to $8\pi\ \delta_C$ where C is some minimal connection and δ_C is the one-dimensional Hausdorff measure uniformly distributed over C.

In order to explain the concept of a minimal connection it is convenient to consider first some simple cases:

Example 1 The system consists only of two points a_1, a_2 with degrees $+1$ and -1. This basic example will be called a <u>dipole</u>. Here, $L = |a_1 - a_2|$ is the distance between the two points and δ_C is the uniform one-dimensional Hausdorff measure of the segment $[a_1, a_2]$. It is not surprising, from dimensional analysis, that E has the homogeneity of a length.

Example 2 The system consists of many points (a_i) and all the degrees d_i's are ± 1. Because of assumption (1) there are as many pluses as minuses. We relabel the points (a_i) by distinguishing the positive points $p_1, p_2 \dots p_k$ and the negative points $n_1, n_2 \dots n_k$.

Here,

(3) $$L = \operatorname*{Min}_{\sigma} \sum_{i=1} |p_i - n_{\sigma(i)}|$$

where the minimum is taken over the set of permutations σ of the integers $\{1,2,..k\}$.

A minimal connection is, by definition, a union of segments

$$C = \bigcup_{i=1}^{k} [p_i, n_{\sigma(i)}]$$

where σ is a minimizing permutation in (3). There may be several minimal connections.

Theorem 2 says that if (ϕ_n) is a minimizing sequence, then all its energy tends to "concentrate" near some minimal connection.

Example 3 In the general case where the d_i's are any integers one proceeds as in Example 2 except that the points a_i are repeated according to their multiplicity $|d_i|$.

Remark 1 There are variants of Theorem 1 when R^3 is replaced by a domain Ω and the class of admissible maps consist either of

$$\mathcal{E}_1 = \{\phi \in C^1(\Omega \setminus \bigcup_{i=1}^{k} \{a_i\}; S^2) \mid \int_\Omega |\nabla\phi|^2 < \infty\ ,\ \deg(\phi, a_i) = d_i\}$$

or of

$$_2 = \{\phi \in C^1(\bar{\Omega} \setminus \bigcup_{i=1}^{k} \{a_i\}; S^2) \mid \int_\Omega |\nabla\phi|^2 < \infty\ ,\ \deg(\phi, a_i) = d_i \quad \text{and} \quad \phi \text{ is constant on } \partial\Omega\}.$$

For example, in the latter case the formula is

$$E_2 = \operatorname*{Inf}_{\rho \varepsilon\ 2} \int |\nabla\phi|^2 = 8\pi L_2$$

where

$$L_2 = \operatorname*{Min}_{\sigma} \sum_{i=1}^{k} d_\Omega\, (p_i\ ,\ n_{\sigma(i)})$$

and $d_\Omega(p,n)$ denotes the geodesic distance between p and n <u>within</u> Ω (see [4]).

Remark 2 One may conceive of other problems where the <u>energy has the homogeneity of an area</u>. Consider, for example, a fixed Jordan curve Γ in R^3. The class of admissible maps consists of

$$\mathcal{E} = \{\phi \in C^1(\mathbb{R}^3 \setminus \Gamma; S^1) \mid \int_{\mathbb{R}^3} |\nabla\phi| < \infty \quad \text{and} \quad \deg(\phi, \Gamma) = 1\},$$

where $\deg(\phi,\Gamma)$ is the circulation of ϕ around Γ i.e. the degree of ϕ restricted to any circle which links with Γ. The energy

$$E(\phi) = \int_{\mathbb{R}^3} |\nabla\phi|$$

has the dimension of an area (instead of a length).

We conjecture that

$$(4) \qquad E = \operatorname*{Inf}_{\phi \varepsilon \mathcal{E}} \int |\nabla\phi| = 2\pi A$$

where A is the area of an area - minimizing surface spanned by Γ. (Formula (4) is established in the case where Γ is a planar curve, see [4]). This is just the analogue of the dipole formula. One may imagine a similar problem for a collection of oriented curves (Γ_i), the class of admissible maps being

$$E = \{\phi \,\varepsilon\, C^1(R^3 \setminus \bigcup_{i=1}^{k} \Gamma_i ; S^1) | \int |\nabla\phi| < \infty \text{ and } \deg(\phi,\Gamma_i) = d_i\}.$$

Sketch of the proof of Theorem 1 The proof is divided into two distinct parts:

Part A : The upper bound $E \leq 8\pi L$

Part B : The lower bound $E \geq 8\pi L$.

Part A: The upper bound $E \leq 8\pi L$.

The main ingredient in the proof is a basic dipole construction summarized in

Lemma 1 Consider a dipole $\{p,n\}$. Then, for every $\varepsilon > 0$ there is a map $\phi_\varepsilon \,\varepsilon\, \mathcal{E}$ (relative to the dipole), i.e.,

$$\phi_\varepsilon \subset C^1(\mathbb{R}^3 \setminus \{p,n\};S^2) , \deg(\phi_\varepsilon , p) = +1 , \deg(\phi_\varepsilon,n) = -1$$

such that

$$(5) \qquad \int |\nabla\phi_\varepsilon|^2 \leq 8\pi |p - n| + \varepsilon$$

and

(6) ϕ_ε is constant outside an ε - neighborhood of the segment $[p,n]$.

Proof Without loss of generality we may assume that $p = (0,0,1)$ and $n = (0,0,-1)$. Let $\Pi: R^2 \to S^2$ be the inverse of stereographic projection from the north pole N. It is easy to check that

$$\int_{\mathbb{R}^2} |\nabla \Pi|^2 = 8\pi .$$

By a small modification of Π near infinity we obtain a smooth map $\omega_\varepsilon : \mathbb{R}^2 \to S^2$ such that

$$\begin{cases} \int |\nabla \omega_\varepsilon|^2 < 8\pi + \varepsilon \\ \omega_\varepsilon \text{ is constant } (=N) \text{ far out} \\ \deg \omega_\varepsilon = 1 \quad (\mathbb{R}^2 \cup \{\infty\} \text{ is identified with } S^2). \end{cases}$$

After a dilation we may further assume that ω_ε is constant outside the unit disc (note that $\int |\nabla \omega_\varepsilon|^2$ is invariant under dilations).
Next, consider the map $\phi : \mathbb{R}^3 \to S^2$ defined by

$$\phi(x,y,z) = \begin{cases} N & \text{if } |z| > 1 \\ \omega_\varepsilon \left(\frac{x}{1-z^2}, \frac{y}{1-z^2}\right) & \text{if } |z| < 1 \end{cases}$$

and the sequence of maps $\phi_n : \mathbb{R}^3 \to S^2$ defined by

$$\phi_n(x,y,z) = \phi(nx,ny,z) .$$

Note that $\phi_n \in \mathcal{E}$ and moreover ϕ_n is constant (= N) outside the region

$$V_n = \{(x,y,z) \mid z^2 + n\sqrt{x^2 + y^2} < 1\}$$

which is a small neighborhood of the segment $[p,n]$.

We have

$$\frac{\partial \phi_n}{\partial x} = \frac{n}{1-z^2} \frac{\partial \omega_\varepsilon}{\partial x} \left(\frac{nx}{1-z^2}, \frac{ny}{1-z^2}\right)$$

$$\frac{\partial \phi_n}{\partial y} = \frac{n}{1-z^2} \frac{\partial \omega_\varepsilon}{\partial y} \left(\frac{nx}{1-z^2}, \frac{ny}{1-z^2}\right)$$

$$\frac{\partial \phi_n}{\partial z} = \frac{2nz}{(1-z^2)^2} \quad x \frac{\partial \omega_\varepsilon}{\partial x} \left(\frac{nx}{1-z^2}, \frac{ny}{1-z^2}\right) + y \frac{\partial \omega_\varepsilon}{\partial y} \left(\frac{nx}{1-z^2}, \frac{ny}{1-z^2}\right) .$$

So that

$$\frac{|\partial \phi_n|}{\partial z} \leq \frac{2nz}{(1-z^2)^2} \sqrt{x^2+y^2} \; C_\varepsilon \leq \frac{2z}{1-z^2} C_\varepsilon \qquad \text{in } V_n ,$$

where $C_\varepsilon = \text{Max } |\nabla \omega_\varepsilon|$. It follows that

$$\int_{R^3} |\nabla \phi_n|^2 \leq 2 \int_{R^2} |\nabla \omega_\varepsilon|^2 + 4 C_\varepsilon^2 \int_{V_n} \frac{z^2}{(1-z^2)^2} \, dxdydz$$

(in order to compute the first two integrals one uses the change of variable $\zeta = \frac{nx}{1-z^2}$, $\eta = \frac{ny}{1-z^2}$) .

Therefore we obtain

$$\int_{\mathbb{R}^3} |\nabla \phi_n|^2 \leq 2\,(8\pi + \varepsilon) + \frac{8\pi}{3} C_\varepsilon^2 \frac{1}{n^2}$$

and the conclusion follows by choosing n large enough.

In the general case, let $C = \bigcup_{i=1}^{k} [p_i , n_{\sigma(i)}]$ be any minimal connection. On each segment $[p_i, n_{\sigma(i)}]$ consider the basic dipole construction as above and then glue these objects. Note that they glue well since ϕ_ε is constant (=N) outside a small neighborhood of [p,n] and also since two segments have no self-intersection because C is a minimal connection. [Two segments may overlap or intersect at their end points but these cases are easy to handle].

Part B: The Lower Bound $E \geq 8\pi L$.

We have to prove that

$$(7) \qquad \int |\nabla\phi|^2 \geq 8\pi L \qquad \forall\ \phi \in \mathcal{E}.$$

For this purpose it is extremely convenient to associate with every map $\phi \in \mathcal{E}$ a vector field D (a kind of electric field) defined by its coordinates

$$(8) \qquad D = (\phi\cdot\phi_y \wedge \phi_z \ , \ \phi\cdot\phi_z \wedge \phi_x \ , \ \phi\cdot\phi_x \wedge \phi_y) \ .$$

The vector field D has some remarkable properties. First, we have

$$(9) \qquad |D| \leq \frac{1}{2} |\nabla\phi|^2 \qquad \text{on } \mathbb{R}^3 \ .$$

Indeed, choose a coordinate system so that

$$\phi = (0,\ 0,\ 1)$$

and then, since $|\phi| = 1$, we may write

$$\phi_x = (a_1 \ , \ b_1 \ , \ 0)$$
$$\phi_y = (a_2 \ , \ b_2 \ , \ 0)$$
$$\phi_z = (a_3 \ , \ b_3 \ , \ 0) \ .$$

Therefore we find

$$D = a \wedge b$$

with $a = (a_1, a_2, a_3)$ and $b = (b_1 , b_2, b_3)$.

It follows that

$$|D| \leq |a| \ |b| \leq \frac{1}{2} \ (|a|^2 + |b|^2) = \frac{1}{2} \ |\nabla\phi|^2 \qquad .$$

Next, we have

(10) $$\operatorname{div} D = 4\pi \sum_{i=1}^{N} d_i \, \delta_{a_i} \qquad \text{in } \mathcal{D}'(R^3) .$$

[Note that $D \in L^1$ since $\int|\nabla\phi|^2 < \infty$, and thus (10) makes sense in $\mathcal{D}'$].

Indeed, it is easy to check that

$$\operatorname{div} D = 0 \qquad \text{in } \mathbb{R}^3 \setminus \bigcup_{i=1}^{N} \{a_i\} .$$

In order to prove (10) it suffices to observe that if Σ is any smooth closed surface in $R^3 \setminus \bigcup_{i=1}^{N} \{a_i\}$, then the flux of D across Σ is given by

$$\int_\Sigma D \cdot \nu \, d\sigma = \int_\Sigma J_\phi \, d\sigma$$

where ν is the normal to Σ and J_ϕ is the Jacobian determinant of ϕ restricted to Σ ; on the other hand the degree of ϕ (considered as a map from Σ to S^2) is given by an analytic formula (see e.g. [13])

$$\deg \phi_{|\Sigma} = \frac{1}{4\pi} \int_\Sigma J_\phi \, d\sigma .$$

It is a surprising fact that we may now ignore the map ϕ and work only with the vector field D. More precisely, we claim

(11) $$\int |D| \geq 4\pi L$$

for every $D \in L^1(\mathbb{R}^3, \mathbb{R}^3)$ such that $\operatorname{div} D = 4\pi \sum_{i=1}^{N} d_i \, \delta_{a_i}$.

Note that, in view of (9) and (10), (11) implies (7). Let $\zeta : \mathbb{R}^3 \to \mathbb{R}$ be any function with $\|\zeta\|_{Lip} \leq 1$, so that $\|\nabla\zeta\|_{L^\infty} \leq 1$. We have

$$\int |D| \geq -\int D \cdot \nabla\zeta = 4\pi \sum_{i=1}^{N} d_i \, \zeta(a_i) .$$

Relabelling the points (a_i) as positive and negative points and taking into account their multiplicity we may write

$$\sum_{i=1}^{N} d_i \, \zeta(a_i) = \sum_{i=1}^{k} (\zeta(p_i) - \zeta(n_i)) .$$

Claim (11) is a consequence of the following general Lemma:

Lemma 2 Let M be a metric space and let $p_1, p_2, \dots p_k$ and $n_1, n_2, \dots n_k$ be $2k$ points in M .

Then

$$\max_{\substack{\zeta: M \to \mathbb{R} \\ \|\zeta\|_{Lip} \leq 1}} \left\{ \sum_{i=1}^{k} (\zeta(p_i) - \zeta(n_i)) \right\} = L \tag{12}$$

where

$$\|\zeta\|_{Lip} = \sup_{x \neq y} |\zeta(x) - \zeta(y)| \,/\, d(x,y)$$

and

$$L = \min_{\sigma} \sum_{i=1}^{k} d(p_i, n_{\sigma(i)}) .$$

Proof of Lemma 2 It is clear

$$\sum (\zeta(p_i) - \zeta(n_i)) \leq \sum_{i=1}^{k} d(p_i \, , n_{\sigma(i)}) .$$

for every function ζ with $\|\zeta\|_{Lip} \leq 1$ and every permutation σ . It follows that

$$\sup_{\|\zeta\|_{Lip} \leq 1} \left\{ \sum(\zeta(p_i) - \zeta(n_i)) \right\} \leq L .$$

In order to prove equality it suffices to construct a function ζ defined <u>only</u> on the set $Q = (\bigcup_{i=1}^{k} \{p_i\}) \cup (\bigcup_{i=1}^{k} \{n_i\})$ with $\|\zeta\|_{Lip} \leq 1$ on Q and such that

$$\sum_{i=1}^{k} (\zeta(p_i) - \zeta(n_i)) = L$$

[Because such a function ζ may be extended to all of M by letting

$$\tilde{\zeta}(x) = \operatorname*{Inf}_{y \in Q} \{ \zeta(y) + d(x,y) \}$$

which has all the required properties].

The existence of ζ is a consequence of two facts:

a) A min-max equality of Kantorovich [12] (see also [14]) which - in our special situation - says that

$$\operatorname*{Max}_{\substack{\zeta:\, Q \to R \\ \|\zeta\|_{Lip} \leq 1}} \{ \sum(\zeta(p_i) - \zeta(n_i)) \} = \operatorname*{Inf}_{(a_{ij}) \in \mathcal{A}} \sum_{i,j=1}^{k} a_{ij}\, d(p_i, n_j)$$

where $\mathcal{A}$ denotes the (convex) set of doubly stochastic matrices, i.e.

$$a_{ij} \geq 0 \quad \forall\, i,j \quad , \quad \sum_{i=1}^{k} a_{ij} = 1 \quad \forall j \quad \text{and} \quad \sum_{j=1}^{k} a_{ij} = 1 \quad \forall i \quad .$$

b) A classical result of Birkhoff which asserts that the extreme points of $\mathcal{A}$ are the permutation matrices.

For the convenience of the reader we present a direct elementary argument. After relabelling the points (n_i) we may always assume that L is given by

$$L = \sum_{i=1}^{k} d(p_i, n_i) \ .$$

Set $d_i = d(p_i, n_i)$ and consider $\lambda_i = \zeta(n_i)$, $1 \leq i \leq k$, as being the unknowns so that $\zeta(p_i) = \lambda_i + d_i$. We are led to the following system of inequalities which expresses that $\|\zeta\|_{Lip} \leq 1$ on Q :

$$(13_1) \qquad | \lambda_i - \lambda_j | \leq d\,(n_i, n_j) \qquad \forall i,j$$

$$(13_2) \qquad | (\lambda_i + d_i) - (\lambda_j + d_j) | \leq d(p_i, p_j) \qquad \forall i,j$$

(13_3) $$|(\lambda_i + d_i) - \lambda_j| \leq d(p_i, n_j) \qquad \forall i,j \quad ,$$

which in turn is equivalent to

(14) $$\lambda_i + d_i - \lambda_j \leq d(p_i, n_j) \qquad \forall i,j$$

[All the other inequalities in (13) are consequences of (14) and of the triangle inequality]. In other words, we have to find a solution (λ_i) for a linear programming system of the form

(15) $$\lambda_i - \lambda_j \leq b_{ij} \qquad \forall i,j = 1,2,\ldots k$$

where $$b_{ij} = d(p_i, n_j) - d_i \quad .$$

Such a system has a solution if and only if the matrix (b_{ij}) satisfies the condition

(16) $$\begin{cases} b_{ii} \geq 0 & \text{for every } \quad i = 1,2,\ldots k \\ \sum_{i=1}^{k} b_{i,\sigma(i)} \geq 0 & \text{for every permutation } \sigma \quad , \end{cases}$$

which in our case, is precisely the assumption that L is the length of a minimal connection.

Indeed, assume that (16) holds. We shall construct a solution of (15) by using essentially the method of [1]. By a chain K we mean any finite sequence of elements (not necessarily distinct) taken from $\{1,2,\ldots k\}$; we write

$$K = \{n_1, n_2, \ldots n_\ell\}$$

where $\ell \geq 2$ can be any integer. We say that a chain is a <u>loop</u> if $n_1 = n_\ell$ and we say that the chain K <u>connects</u> i to j if $n_1 = i$ and $n_\ell = j$. Given a chain K we set

$$S_K = b_{n_1 n_2} + b_{n_2 n_3} + \ldots b_{n_{\ell-1} n_\ell} \quad .$$

It follows from assumption (16) that $S_K \geq 0$ for every loop K. This is obvious if K is a simple loop (i.e. all elements are distinct except the two end points) because we may apply (16) to the permutation $\sigma : n_1 \to n_2 , n_2 \to n_3 , \dots n_{\ell-1} \to n_1$ with all other integers being invariant. If K is a general loop we may split it as the union of simple loops.

For every integer $i = 1, 2, \dots k$, set

$\lambda_i = \text{Inf} \{ S_K \mid K \text{ is a chain connecting } i \text{ to } 1 \}$.

Note that λ_i is well defined ($\lambda_i > -\infty$) since for every chain K connecting i to 1 , we have $S_K \geq - b_{1i}$ (because $\{1, K\}$ is a loop) . It is clear that (λ_i) satisfies (15). Indeed if K is any chain connecting j to 1 , then $\{i, K\}$ is a chain connecting i to 1 and so

$$\lambda_i \leq b_{ij} + S_K ,$$

which implies that $\lambda_i \leq b_{ij} + \lambda_j$

The proof of Theorem 2 is more delicate (see [4].) I will only give a brief indication in the case of a dipole $\{a_1, a_2\}$. First, note that if B is a ball of radius R centered at a and $\phi \in C^1 (B \setminus \{a\}; S^2)$ with $\deg(\phi,a) = 1$, then,

$$\int_B |\nabla\phi|^2 \geq 8\pi R . \tag{17}$$

Indeed, consider the D field associated with ϕ .

We have

$$\int_B |\nabla\phi|^2 \geq 2 \int_B |D| \geq - 2 \int_B D \cdot \nabla\zeta = 8\pi \, \zeta(0)$$

for every function ζ such that $\|\nabla\zeta\|_{L^\infty} \leq 1$ and $\zeta = 0$ on ∂B ; then, choose ζ to be the distance to $\partial\Omega$. Assume now, by contradiction, that the least energy E is achieved for the dipole by a map ϕ . Let B_1 (respectively B_2) be a ball centered at a_1 (respectively a_2) with radius R_1 (respectively R_2) such that $R_1 + R_2 = |a_1 - a_2| = L$. By (17) we have

$$\int_{B_1} |\nabla\phi|^2 \geq 8\pi R_1 \quad \text{and} \quad \int_{B_2} |\phi|^2 \geq 8\pi R_2$$

and thus

$$\int_{B_1 \cup B_2} |\phi|^2 \geq 8\pi(R_1 + R_2) = 8\pi L \ .$$

Since, on the other hand,

$$\int_{R^3} |\nabla\phi|^2 = 8\pi L \ ,$$

we conclude that $\nabla\phi = 0$ outside $B_1 \cup B_2$. By varying R_1 and R_2 we find that $\nabla\phi = 0$ outside the segment $[a_1,a_2]$, so that ϕ is constant on R^3 - which is absurd. In fact this argument shows that if (ϕ_n) is a minimizing sequence then

$$\int_K |\nabla\phi_n|^2 \to 0$$

for every compact set K such that $K \quad [a_1,a_2] = \emptyset$. It follows that $|\nabla\phi_{n_k}|^2$ converges to a measure μ concentrated on the segment $[a_1,a_2]$. A similar argument shows that μ is uniformly distributed on the segment $[a_1,a_2]$

2. Free Singularities

Let $\Omega \quad R^3$ be a (smooth) bounded domain. Let $g: \partial\Omega \to S^2$ be a given boundary data. We consider now the problem of minimizing the energy in the class

$$\mathcal{E} = \{ \phi \in H^1(\Omega;S^2) \mid \phi = g \quad \text{on} \quad \partial\Omega \}$$

where $H^1(\Omega;S^2) = \{\phi \in H^1 \ (\Omega; R^3) \mid |\phi| = 1$ a.e. on $\Omega \}$. It is clear, by a standard lower semicontinuity argument, that

$$E = \min_{\phi \in \mathcal{E}} \int |\nabla\phi|^2$$

is achieved. Moreover, every minimizer satisfies the Euler equation i.e. the equation of harmonic maps

$$- \Delta\phi = \phi|\nabla\phi|^2 \qquad \text{on} \quad \Omega \quad .$$

[The Lagrange multiplier $|\nabla\phi|^2$ comes from the constraint $|\phi| = 1$] . It is known (see [15], [16]) that every minimizer is smooth, except at a finite number of points. In contrast with Section 1, the number and the location of the singularities is not prescribed and in fact, it would be interesting to estimate the number of singularities. Here, singularities are free to appear wherever they want as long as they help to lower the energy. A natural question is whether singularities really appear. The answer is yes and there are two reasons:

1) If $\deg(g,\partial\Omega) \neq 0$, there is a topological obstruction since g can not be extended smoothly inside Ω ; every map in the class $\mathcal{E}$ must have at least one singularity.

2) If $\deg(g,\partial\Omega) = 0$, there is no topological obstruction : g can be extended smoothly inside Ω . A very interesting example of Hardt-Lin [11] shows that there may still be singularities. In other words, the system is not forced (topologically) to have singularities, but it pays for the system to create singularities in order to lower its energy. Here is an alternative simple example of a map g from $\partial\Omega$ to S^2 , of degree zero, such that

$$(18) \qquad E = \inf_{\substack{\phi \in H^1(\Omega;S^2) \\ \phi = g \text{ on } \partial\Omega}} \int |\nabla\phi|^2 \sim \varepsilon$$

while

$$(19) \qquad E_{reg} = \inf_{\substack{\phi \in C^1(\overline{\Omega};S^2) \\ \phi = g \text{ on } \partial\Omega}} \int |\nabla\phi|^2 \sim 16\pi$$

(with ε arbitrarily small). Let Ω be the unit ball with north pole N and south pole S . Along the NS axis we place two dipoles with the same orientation: $\{p_1,n_1\}$ is centered at N and $\{p_2, n_2\}$ is centered at S (see Fig. 1) .

Fig. 1

We assume that $| p_1 - n_1 | = | p_2 - n_2 | = \varepsilon$ is small. Using the construction of Lemma 1 we obtain a map ϕ_ε which is smooth except at the points $\{p_1, n_1, p_2, n_2\}$, which is constant except on $B(N, \varepsilon/2)$ and $B(S, \varepsilon/2)$ and such that

$$\int |\nabla\phi_\varepsilon|^2 < 16\pi\varepsilon + 2\varepsilon .$$

Define g to be the restriction of ϕ_ε to $\partial\Omega$, so that g is smooth and g has degree zero. Clearly we have $E < 16\pi\varepsilon + 2\varepsilon$ (since we may use ϕ_ε as an admissible map). For the proof of (19) it is convenient to use the D field associated with ϕ ; we find

$$\int_\Omega |\nabla\phi|^2 \geq 2 \int_\Omega |D| \geq 2 \int_\Omega D \cdot \nabla\zeta = 2 \int_{\partial\Omega} (D \cdot n)\zeta \, d\sigma$$

(since $\operatorname{div} D = 0$ because ϕ is smooth), for every function ζ such that $\|\nabla\zeta\|_{L^\infty} \leq 1$. Choosing a function ζ such that $\zeta \equiv 0$ in $B(S, \varepsilon/2)$ and $\zeta \equiv 2-\varepsilon$ in $B(N, \varepsilon/2)$ we obtain

$$\int |\nabla\phi|^2 \geq 2\,(2-\varepsilon) \int_{\partial\Omega \cap B(N,\varepsilon/2)} (D \cdot n) .$$

But $D \cdot n = \operatorname{Jac} g$ is the Jacobian determinant of g , which vanishes except near N and S , and thus

$$\frac{1}{4\pi} \int_{\partial\Omega \cap B(N,\varepsilon/2)} (D \cdot n) = \deg(\phi_\varepsilon, p_i) = 1 .$$

Remark 3 This gap phenomenon ($E < E_{reg}$) raises many interesting questions:

a) Is E_{reg} achieved?

b) It implies that smooth maps from B^3 into S^2 are not dense in $H^1(B^3;S^2)$ - a fact already pointed out in [16]. More generally, one may ask whether smooth maps from B^k to S^ℓ are dense in the Sobolev space $W^{1,p}(B^k; S^\ell)$, $1 \leq p < \infty$. Some surprising partial results have been obtained by F. Bethuel and X. Zheng [2]. Assume for example $k = 3$:

if $\ell = 1$, smooth maps are dense iff $p \in [2,\infty)$

if $\ell = 2$ smooth maps are dense iff $p \in [1,2) \cup [3,\infty)$

if $\ell \geq 3$ smooth maps are dense for all $p \in [1,\infty)$.

The main results of Section 2 are the following

Theorem 3 Assume Ω is the unit ball and $g(x) = x$ is the identity map on $\partial\Omega$. Then $\phi(x) = x/|x|$ is a minimizer for E .

Theorem 4 Assume Ω is the unit ball and $g : \partial\Omega \to S^2$ is arbitrary. Then the homogeneous extension $\phi(x) = g(x/|x|)$ is <u>not</u> a minimizer for E unless g is an isometry or a constant.

Remark 4 By contrast, if we ask the question whether $\phi(x) = g(x/|x|)$ is a critical point, i.e. a solution of $-\Delta\phi = \phi|\nabla\phi|^2$, then there are many more g's (all harmonic maps from S^2 to S^2) .

These results have an interesting consequence:

Corollary 5 Assume Ω is any domain and g is any map. Let ϕ be a minimizer for E, then all its singularities have degree ± 1. Moreover, for every singularity x_0, there is a rotation R such that

$$\phi(x) \sim \pm R\left(\frac{x - x_0}{|x-x_0|}\right) \quad \text{as} \quad x \to x_0 .$$

Corollary 5 is derived from Theorem 4 by a standard blow-up procedure. Assume for example $x_0 = 0$; as $\varepsilon \to 0$, $\phi(\varepsilon x) \to \psi(x)$ (see [15] and [17]) which is a minimizing harmonic map and which depends only on the direction $x/|x|$. It follows from Theorem 4 that $\psi(x) = \pm Rx / |x|$.

Sketch of the Proof of Theorem 3 Our proof is rather indirect and it would be interesting to find a different argument. An obvious calculation shows that the energy of $x/|x|$ is 8π. Therefore, we have only to prove that

$$\int |\nabla\phi|^2 \geq 8\pi \qquad \forall \phi \in H^1(\Omega;S^2) \, , \ \phi(x) = x \ \text{on} \ \partial\Omega . \tag{20}$$

It suffices to establish (20) for ϕ's which are smooth except at a finite number of points. The reason is that, by [15], every minimizer has that property; alternatively one may also invoke a result of [2] which asserts that such ϕ's are dense in H^1. Consider such a ϕ and its D field. We have

$$\int_\Omega |\nabla\phi|^2 \geq 2\int_\Omega |D| \geq 2\int_\Omega D \cdot \nabla\zeta = 2 \int_{\partial\Omega} (D \cdot n)\zeta - 2\int_\Omega (\operatorname{div} D))\zeta$$

for every ζ such that $\|\nabla\zeta\|_{L^\infty} \leq 1$.

But $D \cdot n = \operatorname{Jac}(\phi_{|\partial\Omega}) = 1$ (since $\phi(x) = x$ on $\partial\Omega$) and $\operatorname{div} D = \sum_{i=1}^N d_i \, \delta_{a_i}$ with $d_i \in \mathbb{Z}$ and $\sum_{i-1}^N d_i = 1$.

Therefore, we have

$$\frac{1}{8\pi} \int_\Omega |\nabla\phi|^2 \geq \frac{1}{4\pi} \int_{\partial\Omega} \zeta \, d\sigma - \sum_{i=1}^N d_i \, \zeta(a_i) .$$

Lemma 3 below (applied with $M = \Omega$ and $d\mu = \frac{1}{4\pi} d\sigma$) shows that

$$\frac{1}{8\pi} \int_\Omega |\nabla\phi|^2 \geq \underset{y \in \overline{\Omega}}{\text{Min}} \frac{1}{4\pi} \int_{\partial\Omega} |y-\sigma| d\sigma = 1 \ .$$

Lemma 3 Let M be a compact metric space and let μ be a fixed probability measure on M .

Then

$$\underset{\nu \in \mathcal{A}}{\text{Inf}} \ \underset{\substack{\zeta : M \to \mathbb{R} \\ \|\zeta\|_{Lip} \leq 1}}{\text{Max}} \left\{ \int \zeta d\mu - \int \zeta d\nu \right\} = \underset{y \in M}{\text{Min}} \int d(x,y) d\mu(x)$$

where the infimum is taken over the class $\mathcal{A}$ of all measures ν of the form $\nu = \sum_{\text{finite}} d_i \delta_{a_i}$, with $d_i \in \mathbb{Z}$ and $\sum d_i = 1$.

Sketch of the Proof of Lemma 3 It is clear that $\underset{\nu}{\text{Inf}} \ \underset{\zeta}{\text{Max}} \leq \underset{y}{\text{Min}}$. Indeed, if we choose $\nu = \delta_y$ we obtain

$$\int \zeta d\mu - \int \zeta d\nu = \int (\zeta(x) - \zeta(y)) d\mu(x) \leq \int d(x,y) d\mu(x) \ .$$

For the reverse inequality, it suffices - by density - to consider the case where μ is a discrete measure with rational coefficients, which we may always write as

$$\mu = \frac{1}{m} \sum_{i=1}^{m} \delta_{c_i}$$

(the points c_i need not be distinct).

Fix a measure $\nu \in \mathcal{A}$; relabelling the points (a_i) as positive and negative points and taking into account their multiplicity we may write

$$\nu = \sum_{j=1}^{k} \delta_{p_j} - \sum_{j=1}^{k-1} \delta_{n_j} \ .$$

We have to prove that $A \geq B$ where

$$A = \operatorname*{Max}_{\|\nabla\zeta\|_{Lip} \leq 1} \left\{ \int \left(m \sum_{j=1}^{k} \delta_{p_j} - m \sum_{j=1}^{k-1} \delta_{n_j} - \sum_{i=1}^{m} \delta_{c_i}\right)\zeta \right\}$$

and

$$B = \operatorname*{Min}_{y \in M} \sum_{i=1}^{m} d(c_i, y) .$$

It follows from Lemma 2 that $A = L$, the length of a minimal connection of a system which consists of mk positive points and mk negative points. The positive points are the points $(p_j)_{1 \leq j \leq k}$ counted with multiplicity m . The negative points are the points $(n_j)_{1 \leq j \leq k-1}$ counted with multiplicity m together with the points $(c_i)_{1 \leq i \leq m}$ counted with multiplicity one. Finally we invoke the following Lemma from Graph Theory (whose statement has been conjectured by us and proved by Hamidoune-Las Vergnas [9])

Lemma 4 Consider a family of k boys $B_1, B_2 \ldots B_k$ and k girls $G_1, G_2 \ldots G_k$. Assume $\mathcal{G}$ is a graph connecting the boys and the girls such that, in $\mathcal{G}$, every boy is joined exactly to m girls and every girl is joined exactly to m boys. Then, given any girl G there is some boy B joined to G by m disjoint paths in $\mathcal{G}$.

Proof of Lemma 3 completed

The boys are the points $p_1, p_2, \ldots p_k$; the girls $G_1, G_2, \ldots G_{k-1}$ are the points $n_1, n_2, \ldots n_{k-1}$, while G_k consists of $\bigcup_{i=1}^{m} \{c_i\}$. The graph $\mathcal{G}$ is any minimal connection.

It follows from Lemma 4, that given the girl $G = G_k$, there is some boy, say p_ℓ , such that $\mathcal{G}$ contains m disjoint paths joining p_ℓ to all the points $(c_i)_{1 \leq i \leq m}$. We conclude that

$$L \geq \sum_{i=1}^{m} d(c_i, p_\ell) \geq \operatorname*{Min}_{y \in M} \sum_{i=1}^{m} d(c_i, y) = B .$$

The proof of Theorem 4 is quite involved and I will not discuss it here (see [3]). Rougly speaking, there are two steps:

Step 1 If $|\deg g| > 1$ one constructs a map ϕ with more than one singularity and with energy lower than $g(x/|x|)$

Step 2 If $|\deg g| = 1$ and g is not an isometry, one can lower the energy by "moving the singularity" towards the center of mass of $|\nabla g|^2$ i.e. $\int_{\partial\Omega} |\nabla g|^2 \sigma d\sigma$.

References

[1] S.N. Afriat, The system of inequalities $a_{rs} > X_r - X_s$, Proc. Camb. Phil. Soc. 59 (1963) p. 125-133.

[2] F. Bethuel - X. Zheng, Sur la densité des fonctions régulières entre deux variétés dans des espaces de Sobolev, C.R. Acad. Sc. Paris 303 (1986) p. 447-449.

[3] H. Brezis - J.M. Coron - E. Lieb, Estimations d'energie pour des applications de R^3 a valeurs dans S^2, C.R. Acad. Sc. Paris 303 (1986), p. 207-210.

[4] H. Brezis - J.M. Coron - E. Lieb, Harmonic maps with defects, Comm. Math Phys. (to appear). IMA preprint 253.

[5] W.F. Brinkman - R.E. Cladis, Defects in liquid crystals, Physics Today, May 1982, p. 48-54.

[6] R. Cohen - R. Hardt - D. Kinderlehrer - S.Y. Lin - M. Luskin, Minimum energy configurations for liquid crystals: computational results, in this Volume

[7] P.G. De Gennes, The Physics of Liquid Crystals, Clarendon Press, Oxford (1974).

[8] J.L. Ericksen, Equilibrium Theory of Liquid Crystals, in Advances in Liquid Crystals 2, Brown G.H. ed., Acad. Press, New York (1976), p. 233-299.

[9] Y.O. Hamidoune - M. Las Vergnas, Local edge-connectivity in regular bipartite graphs (to appear).

[10] R. Hardt - D. Kinderlehrer, Mathematical questions of liquid crystal theory, in this Volume.

[11] R. Hardt - F.H. Lin, A remark on H^1 mappings, Manuscripta Math. 56 (1986) p. 1-10.

[12] L.V. Kantorovich, On the transfer of masses, Dokl. Akad. Nauk SSSR 37 (1942) p. 227-229.

[13] L. Nirenberg, Topics in Nonlinear Functional Analysis, New York University Lecture Notes, New York (1974)

[14] S.T. Rachev, The Monge - Kantorovich mass transference problem and its stochastic applications, Theory of Prob. and Appl. 29 (1985), p. 647-676.

[15] R. Schoen - K. Uhlenbeck, A regularity theory for harmonic maps, J.Diff. Geom. 17 (1982), p. 307-335.

[16] R. Schoen - K. Uhlenbeck, Boundary regularity and the Dirichlet problem for harmonic maps, J. Diff. Geom. 18 (1983), p. 253-268.

[17] L. Simon, Asymptotics for a class of nonlinear evolution equations with applications to geometric problems, Ann. of Math. 118 (1983), p. 525-571.

ON VIRTUAL INERTIA EFFECTS DURING DIFFUSION OF A DISPERSED MEDIUM IN A SUSPENSION

G. Capriz and P. Giovine

Istituto Matematico
Via F. Buonarotti 2
Universita degli Studi
56100 Pisa, Italy

1. Introduction

The equations of motion for a mixture of two immiscible fluids (or a suspension of solid particles in a fluid) are still a matter for discussion. Indeed the fundamental choice of an appropriate expression for the inertia terms is itself not trivial, because virtual mass effects have an important rôle during diffusion and are not easily modelled within the scheme of a continuum.

It is usually claimed that the additional inertia force f due to those effects is proportional to a relative acceleration; however, the volume fraction itself may vary in time and from place to place and these changes are not irrelevant for the correct evaluation of f. Some authors believe that f should be objective (see, for instance, [1]), but such requirement does not appear to be mandatory (an up to date discussion is promised in [2]).

It seems easier, and perhaps less controversial, to start with the proposal of an expression for the virtual kinetic energy, following suggestions from classical analyses of the motion of a sphere (or of a disc) in a fluid, and then to formulate an appropriate variational principle. If one chooses a variational principle of material type, however, one must, perhaps implicitly, assume a mechanism for the transport of the virtual kinetic energy (see, for instance, [4]); thus, in a sense, one begs at least part of the question. We prefer to adopt here a spatial variational principle, along lines followed within classical contexts in [3].

Because we want to concentrate attention on virtual mass effects, we examine only the most elementary type of two-phase flow that implies them: a suspension of incompressible particles, or bubbles, in an incompressible perfect fluid. After a brief justification of the choice of the expression for the total kinetic energy density, we introduce the variational principle. We give some details of the developments that lead to the balance equations of momentum. We check the consistency of those equations with Truesdell's 'metaphysical' principles for mixtures and we conclude with the study of a special problem of sedimentation.

The equations of motion we obtain seem consistent with those that have been derived within an extremely general context in the paper already quoted by Bedford and Drumheller [4]. In view of the simplicity of the situation envisaged here, the details of the derivation can be followed explicitly; a constant effort is also made to assure physical insight into the analytical developments.

2. The Variational Principle

In this section we propose a continuum model of a two-phase system where the particles of the dispersed phase are large enough so that virtual inertia effects cannot be disregarded. As is usual in theories of continua with microstructure, the idea is to obtain hints for an appropriate expression of the relevant densities from results of analyses of simple motions in classical continua.

One can imagine an element of our continuum as a spherical drop of an incompressible perfect fluid containing a concentric spherical inclusion (the dispersed phase), the latter moving with an instantaneous purely translational relative speed z with respect to the drop. If υ is the volume of the element, β the ratio of the volume of the inclusion to υ, v_1 the absolute speed of the mass centre of the fluid, $v_2 = v_1 + z$ the absolute speed of the inclusion, ρ_1 the density of the fluid and ρ_2 the density of the inclusion, then the total kinetic energy of the element turns out to be (see [5], Sect. 93) υ times the expression

$$\frac{1}{2}\left(\rho_1(1-\beta)v_1^2 + \rho_2\beta v_2^2 + \rho_1\psi(\beta)z^2\right), \tag{2.1}$$

where

$$\psi(\beta) = \frac{1}{2}\,\beta\,\frac{1+2\beta}{1-\beta} \quad . \tag{2.2}$$

The expression (2.1) will be taken below as the density of kinetic energy per unit volume of the mixture; the special choice (2.2) for the function ψ, presumably adequate only when β is sufficiently small, need not be made.

Remark. The model suggested above for an elementary drop of the mixture could be sharpened to include the effects of relative rotation of the inclusion. One could also, perhaps more appropriately in certain situations, attribute to the suspended particle an ellipsoidal shape, perhaps even the shape of a disc. But then the analysis would become much more complex, because the suspended phase would have to be modelled through a continuum with microstructure, where the unit vector of the axis of symmetry of the ellipsoid would be the microstructural variable. Also, the effects of the interference between micromotions in neighbouring elements could be brought to bear; then a non-local expression for the kinetic energy density would be required (a very interesting notion, in principle), or at least an expression involving the gradients of β and z; for instance, formulae (16) of Sect. 98 and (7) of Sect. 99 of [5] hint at expressions of the latter type. Finally the idea, (embodied in (2.2) and which conforms to the spirit of the paper) that the flow between the two spheres is that pertaining to a perfect fluid, could be forfeited in favour of a similar hypothesis but involving a viscous fluid; but then (as far as concerns formula (2.1)) only the dependence of ψ on β would change.

To reduce developments to essentials, we consider only the case when the mixture is contained in a fixed rigid vessel with impermeable walls; we call B the region of space delimited by the vessel and ∂B the boundary. Kinematic compatibility imposes the conditions

$$v_1 \cdot n = 0 \quad , \quad v_2 \cdot n = 0 \ , \quad \text{on} \quad \partial B \ , \tag{2.3}$$

if n is the unit exterior vector normal to ∂B .

Again to simplify matters radically, both phases are assumed to be incompressible; ρ_1 and ρ_2 are then constant. Also mass exchanges between phases are excluded. Thus conservation of mass requires that

$$\left.\begin{aligned} &\frac{\partial\,((1-\beta)\rho_1)}{\partial\tau} + \operatorname{div}\,((1-\beta)\rho_1 v_1) = 0 \\ &\frac{\partial(\beta\rho_2)}{\partial\tau} + \operatorname{div}\,(\beta\rho_2 v_2) = 0\,, \end{aligned}\right\} \quad \text{in } B;$$

or

$$\left.\begin{aligned} &-\frac{\partial\beta}{\partial\tau} + \operatorname{div}\,((1-\beta)v_1\ ) = 0\,, \\ &\frac{\partial\beta}{\partial\tau} + \operatorname{div}\,(\beta v_2) = 0\,, \end{aligned}\right\} \quad \text{in } B\,. \qquad (2.4)$$

As for the external body forces, they are taken here to be conservative with a potential energy ω per unit mass. The potential energy of internal actions (σ, per unit volume) is taken to depend at most on β.

A mechanical process of duration $\bar{\tau}$ in the mixture is portrayed through the assignment at each instant τ in $[0,\bar{\tau}]$ of the fields v_1, v_2 and β over B, subject to (2.3). Among all these virtual processes the natural process is distinguished because it satisfies a variational principle and the balance conditions (2.4). To state the principle, the usual minor technical definitions are required.

Let $\{\hat{v}_1(x,\tau),\hat{v}_2(x,\tau),\ \hat{\beta}(x,\tau)\}$ be the natural process and $\{v_1(x,\tau,\varepsilon),v_2(x,\tau,\varepsilon),\ \beta(x,\tau,\varepsilon)\}$ a family of virtual processes depending smoothly on a parameter ε, for ε in a neighbourhood N_ε of the origin, and such that

$$\{v_1(x,\tau,0),v_2(x,\tau,0),\beta(x,\tau,0)\} \equiv \{\hat{v}_1(x,\tau),\hat{v}_2(x,\tau),\hat{\beta}(x,\tau)\},\quad \forall x \in B,\ \forall\tau\in[0,\bar{\tau}],$$

and

$$\{v_1(x,0,\varepsilon),v_2(x,0,\varepsilon),\ \beta(x,0,\varepsilon)\} \equiv \{\hat{v}_1(x,0),\hat{v}_2(x,0),\ \hat{\beta}(x,0)\},$$
$$\{v_1(x,\bar{\tau},\varepsilon),v_2(x,\bar{\tau},\varepsilon),\ \beta(x,\bar{\tau},\varepsilon) \equiv \{\hat{v}_1(x,\bar{\tau}),\hat{v}_2(x,\tau),\ \hat{\beta}(x,\bar{\tau})\},\quad \forall\varepsilon \in N_\varepsilon\ ,\ \forall x\in B\,.$$

The variation $\delta\Gamma$ of any quantity Γ defined on a process class is given by

$$\delta\Gamma := \frac{\partial\Gamma}{\partial\varepsilon}\Big|_{\varepsilon=0} \quad . \tag{2.5}$$

The conditions imposed upon $v_1(x,\tau,\varepsilon), v_2(x,\tau,\varepsilon), \beta(x,\tau,\varepsilon)$ assure us that

$$\delta v_1 = 0 \ , \quad \delta v_2 = 0 \ , \ \delta\beta = 0, \text{ for } \quad \tau = 0, \ \tau = \bar{\tau} \ , \tag{2.6}$$

whereas

$$\delta v_1 \cdot n = \delta v_2 \cdot n = 0 \quad \text{in} \quad \partial B. \tag{2.7}$$

The variational principle asserts that, during the natural motion of the body, the equality

$$\delta \int_0^{\bar{\tau}} d\tau \int_B \Big(\tfrac{1}{2}\,(\rho_1(1-\beta)v_1^2 + \rho_2\beta v_2^2 + \rho_1\psi(\beta)z^2) - (\rho_1(1-\beta) + \rho_2\,\beta)\omega +$$

$$+ \quad \sigma + \phi(\frac{\partial\beta}{\partial\tau} + \operatorname{div}(\beta v_2)) + \gamma(-\frac{\partial\beta}{\partial\tau} + \operatorname{div}\,((1-\beta)v_1))\Big)\, dB = 0$$

holds for all virtual processes; here ϕ and γ are Lagrange multipliers of the constraints imposed upon the natural process by mass balance.

3. The Balance Equations for Momentum

In view of the restrictive hypothesis made on B, reflected in the relations (2.3), the transport theorem leads, for any Γ , to the equalities

$$\delta \int_0^{\bar{\tau}} d\tau \int_B \Gamma\, dB = \int_0^{\bar{\tau}} d\tau \int_B \delta\Gamma\, dB \ ,$$

$$\int_0^{\bar{\tau}} dt \int_B \frac{\partial\Gamma}{\partial\tau}\, dB = \Big[\int_B \Gamma\, dB\Big]_0^{\bar{\tau}} \ ,$$

and these, plus repeated recourse to integration by parts in the usual manner, lead to the following consequences of (2.8)

$$\frac{1}{2}\,(\rho_2 v_2^2 - \rho_1 v_2^2) + \frac{1}{2}\,\rho_1\frac{d\psi}{d\beta}\, z^2 + (\rho_1 - \rho_2)\,\omega - \frac{d\sigma}{d\beta} =$$

$$= \frac{\partial\phi}{\partial\tau} + v_2 \cdot \operatorname{grad}\,\phi - \frac{\partial\gamma}{\partial\tau} - v_1 \cdot \operatorname{grad}\,\gamma \ , \tag{3.1}$$

$$\rho_1 v_1 - \rho_1 \frac{\psi z}{(1-\beta)} = \text{grad}\ \gamma \ , \tag{3.2}$$

$$\rho_2 v_2 + \rho_1 \frac{\psi z}{\beta} = \text{grad}\ \phi \quad .$$

The similarity of these consequences with those obtained from a variational principle in classic cases is obvious (see [3]); one can also proceed similarly to eliminate ϕ and γ by cross-differentiation and obtain

$$\begin{aligned}
&\text{grad}((\rho_1 - \rho_2)\omega - \frac{1}{2}\rho_1 \frac{d\psi}{d\beta} z^2 - \frac{d\sigma}{d\beta}) = \rho_2 a_2 - \rho_1 a_1 + \\
&+ \frac{\rho_1}{\beta}(\frac{\partial(\psi z)}{\partial\tau} + (\text{grad } v_2)^T \psi z) + \frac{\rho_1}{1-\beta}(\frac{\partial(\psi z)}{\partial\tau} + (\text{grad } v_1)^T \psi z) + \\
&+ \{\rho_1 \psi z \frac{\partial}{\partial\tau}(\frac{1}{\beta}) + (\text{grad}(\frac{\rho_1 \psi z}{\beta}))^T v_2 - v_2 \times \text{rot}(\frac{\rho_1 \psi z}{\beta})\} + \\
&+ \{\rho_1 \psi z \frac{\partial}{\partial\tau}(\frac{1}{1-\beta}) + (\text{grad}(\frac{\rho_1 \psi z}{1-\beta}))^T v_1 - v_1 \times \text{rot}(\frac{\rho_1\ \psi z}{1-\beta})\} ,
\end{aligned} \tag{3.3}$$

where the peculiar accelerations a_1, a_2 have been introduced

$$a_i = \frac{\partial v_i}{\partial\tau} + \text{grad}\frac{v_i}{2} - v_i \times \text{rot } v_i \ , \ i = 1,2$$

and obvious consequences of (3.2) have been exploited, i.e.,

$$\text{rot } \rho_1 v_1 = \text{rot } \rho_1 \frac{\psi z}{(1-\beta)} \quad , \quad \text{rot } \rho_2 v_2 = - \text{rot } \rho_2 \frac{\psi z}{\beta} \ . \tag{3.4}$$

Now one can use (2.4) to show that the two terms between curly brackets in (3.4) are equal respectively to

$$\frac{1}{\beta}\text{div}(\rho_1 \psi z \otimes v_2) \quad \text{and} \quad \frac{1}{1-\beta}\text{div}(\rho_1 \psi z \otimes v_1)$$

with the conclusion that

$$\begin{aligned}
&\beta(\rho_1(1-\beta)(a_1 + \text{grad}(\omega + \frac{1}{2}\frac{d\psi}{d\beta} z^2) - \rho_1(\frac{\partial\psi z}{\partial\tau} + (\text{grad } v_1)^T \psi z) - \rho_1 \text{div}(\psi z \times v_1)) = \\
&= (1-\beta)(\beta(\rho_2 a_2 + \rho_2 \text{grad}\omega + \text{grad}\frac{d\sigma}{d\beta}) + \rho_1(\frac{\partial\psi z}{\partial\tau} + (\text{grad} v_1)^T \psi z) + \rho_1 \psi \text{ grad}\frac{z^2}{2} +
\end{aligned}$$

$$+ \rho_1 \operatorname{div}(\psi z \otimes v_2)) . \tag{3.5}$$

If one calls $\beta(1-\beta)\, u$ the common value of right and left-hand side of (3.5), one discovers, by way of (2.4) and (3.4), that the vector $u + \beta \operatorname{grad} \frac{d\sigma}{d\beta}$ is irrotational . Then, introducing a scalar π (the physical meaning of which will be clear in the sequel), such that

$$u + \beta \operatorname{grad} \frac{d\sigma}{d\beta} = \operatorname{grad} \left(\pi - \frac{1}{2} \rho_1 z^2 \left(\psi + (1-\beta) \frac{d\psi}{d\beta}\right)\right) , \tag{3.6}$$

one is led to the balance equations of momentum for the two constituents of the mixture

$$\begin{aligned} \rho_1(1-\beta)a_1 &= -\rho_1(1-\beta) \operatorname{grad} \omega - (1-\beta) \operatorname{grad} \pi - \operatorname{div} (\rho_1 \psi z \otimes z) + f , \\ \rho_2 \beta a_2 &= -\rho_2 \beta \operatorname{grad} \omega - \beta \operatorname{grad} \pi - f , \end{aligned} \tag{3.7}$$

where: π has the obvious meaning of a pressure, and the terms which involve it express the Archimedean buoyancy forces; a term involves the Reynolds stress $\rho\psi z \times z$ due to relative motion; the interaction force f between the two phases has the expression:

$$\begin{aligned} f = (1-\beta) \operatorname{grad} \left(\frac{1}{2} \rho_1 z^2 \left(\psi - \beta \frac{d\psi}{d\beta}\right)\right) + \rho_1 \left(\frac{\partial(\psi z)}{\partial\tau} + (\operatorname{grad} v_1)^T (\psi z)\right) + \\ + \operatorname{div} (\rho_1 \psi z \otimes v_2) + \beta(1-\beta) \operatorname{grad} \frac{d\sigma}{d\beta} . \end{aligned} \tag{3.8}$$

Thus our analysis leads to an expression for the interaction force which is not objective, contrary to what some authors would prefer it to be.

There are many other forms into which eqns (3.7), (3.8) can be put; one variant shows that they obey the 'metaphysical' principles set by Clifford Truesdell as the basis of any theory of mixtures [6]:

$$\begin{aligned} \rho_1(1-\beta)a_1 &= - \rho_1(1-\beta) \operatorname{grad} \omega + \operatorname{div} T_1 + h , \\ \rho_2 \beta a_2 &= - \rho_2 \beta \operatorname{grad} \omega + \operatorname{div} T_2 - h , \end{aligned} \tag{3.9}$$

with

$$T_1 = \rho_1 \psi z \otimes v_1 - (1-\beta)(\pi - \frac{1}{2}\rho_1 z^2(\psi - \beta\frac{d\psi}{d\beta}))1 ,$$
$$T_2 = -\rho_1 \psi z \otimes v_2 - (\beta\pi + \frac{1}{2}\rho(1-\beta)z^2(\psi - \beta\frac{d\psi}{d\beta}))1 , \tag{3.10}$$

and

$$h = f - \text{div}\,(\rho\psi z \otimes v_2 + \frac{1}{2}(1-\beta)\rho_1 z^2(\psi - \beta\frac{d\psi}{d\rho})1) - \tag{3.11}$$

$$- \pi \text{ grad } \beta .$$

Thus each constituent satisfies the balance equation of momentum of an ordinary continuum; besides, if one sums up term by term either (3.7) or (3.9), introduces mixture density ρ , velocity v and acceleration a

$$\rho = \rho_1(1-\beta) + \rho_2\beta ,$$
$$\rho v = \rho_1(1-\beta)v_1 + \rho_2\beta v_2 ,$$
$$a = \frac{\partial v}{\partial \tau} + (\text{grad } v)v ,$$

and remembers the kinematic relation ([6] , formula (5.16))

$$\rho a = \rho_1(1-\beta)a_1 + \rho_2\beta a_2 - \text{div}(\rho_1(1-\beta)v_1 \otimes v_1 + \rho_2\beta\, v_2 \otimes v_2 - \rho v \otimes v), \tag{3.13}$$

one is led to the balance equation for the mixture

$$\rho a = - \rho \text{ grad } \omega + \text{div } T, \tag{3.14}$$

where

$$T = \rho v \otimes v - \rho_1(1-\beta)v_1 \otimes v_1 - \rho_2\beta v_2 \otimes v_2 - \rho_1 \psi z \times z - \pi 1. \tag{3.15}$$

Again here Truesdell's general requirement is satisfied.

4. A Problem of Sedimentation

As we have repeatedly declared, our aim was the derivation, from a single (and perhaps less disputable) principle, of an expression for the interaction force due to virtual mass effects. For a discussion of practical problems, however, our balance equations are far too special. In particular the absence of terms accounting for viscosity is unrealistic; of course, such terms could easily be added, but the process of adaptation of (3.7) to a more concrete form will not be pursued.

Here, just for the sake of showing what can be expected from our analysis, we treat a particular and, as already admitted, rather artificial sedimentation problem. We consider a mixture occupying the half-space $\zeta>0$, under the effect of gravity ($\omega = -|g|\zeta$; g , acceleration due to gravity). We look for steady-state solutions of (3.7), (3.8) where: (i) all unknown functions depend only on ζ ; (ii) more specifically

$$v_1 = \alpha(\zeta)c \quad , \; v_2 = \gamma(\zeta)c \; , \tag{4.1}$$

(c, unit vector of ζ-axis); (iii) the mean velocity vanishes.

Conservation of mass (i.e., eqns (2.4)) requires

$$\begin{aligned} \alpha &= (1-\beta_0)\alpha_0(1-\beta)^{-1} \; , \\ \gamma &= \beta_0\gamma_0\beta^{-1} \; , \end{aligned} \tag{4.2}$$

if $\beta_0, \alpha_0, \gamma_0$ are values of β , α , γ at $\zeta = 0$, for instance. Hypothesis (iii) imposes the relation

$$\varepsilon(1-\beta_0)\alpha_0 = -\beta_0\gamma_0 \; , \tag{4.3}$$

if $\varepsilon = \rho_1/\rho_2$.

By introducing (4.2) in (3.7) we obtain a system of two ordinary differential equations of the first order in $\beta(\zeta)$ and $\pi(\zeta)$. Actually π can be easily eliminated; in fact one could refer to (3.5) and obtain directly a single equation in β .

We spare the reader the details of the algebraic manipulations and register below results valid for small β for the case when ψ is taken to be zero

$$\beta = \beta_0 \left(1 + \frac{2(1-\varepsilon)|g|\zeta}{\gamma_0^2} \right)^{-1/2}$$

and for the case when ψ is given by (2.2)

$$\beta = \beta_0\left(1 + \frac{4(1-\varepsilon)|g|\zeta}{(2+\varepsilon)\gamma_0^2}\right)^{-1/2} .$$

References

[1] D. Drew, L. Cheng and R.T. Lahey, The analysis of virtual mass effects in two-phase flow. Int. J. Multiphase Flow, 5 (1979), 233-242.

[2] S.L. Passman, Invariance of the virtual mass. Society of Rheology Meeting, Blacksburg (1985).

[3] G. Capriz, Spatial variational principles in continuum mechanics. Arch. Rat. Mech. Analysis, 85 (1984), 99-109.

[4] D.S. Drumheller and A. Bedford, A thermomechanical theory for reacting immiscible mixtures. Arch. Rat. Mech. Analysis, 73 (1980), 257-284.

[5] H. Lamb, Hydrodynamics. Cambridge University Press (1932).

[6] C.A. Truesdell, Rational thermodynamics, McGraw Hill, New York (1969).

DEGENERATE HARMONIC MAPS AND LIQUID CRYSTALS

Hyeong In Choi

Department of Mathematics
University of Illinois, Chicago
Chicago, Illinois 60680

The equilibrium theory of nematic liquid crystals can be formulated as a variational problem: The equilibrium state of a nematic liquid crystal is represented by a map $u: \Omega \quad R^3 \to S^2$ which minimizes the energy integral

$$\int_\Omega k_1(\operatorname{div} n)^2 + k_2(n \cdot \operatorname{curl} n)^2 + k_3\|n \times \operatorname{curl} n\|^2$$

$$+ (k_2 + k_4)[\operatorname{tr}(\nabla n)^2 - (\nabla \cdot n)^2].$$

In the special case when $k_1 = k_2 = k_3$ and $k_4 = 0$, it turns out to be a harmonic map from $\Omega \subset R^3$ into S^2. (See [Ha] for more details)

Harmonic maps are non-linear generalizations of harmonic functions. In the past twenty years harmonic maps have been studied extensively by many geometers and analysts, and many important applications have been found. The present day understanding of harmonic maps is substantial: Important existence theorems are established ([ES], [H], [HKW]), and good partial regularity results are available [SU1,2] to name a few.

One of the important properties of harmonic functions is the so-called Liouville theorem which says that every positive harmonic function on R^n is a constant function. There are many generalizations of this fact to harmomic maps. But the gist of all these results is that if the image of a smooth harmonic map is sufficiently small, it must degenerate to a point map. There are corresponding local versions which imply the global ones.

Translated into the language of liquid crystals, these Liouville type theorems suggest that if the angles the director field of a nematic liquid crystal makes with a fixed direction is smaller than a certain fixed angle at every point in a very large region, or if the total energy is very small in a very large region ,for, in either case, the regularity results [S] tell us that the harmonic map is smooth, then the director field must align in almost one direction. This assertion is true for harmonic maps, therefore it is reasonable to believe it for general nematic liquid crystals. It is an interesting problem to quantify this assertion in a sharp fashion.

The purpose of this paper is to point out a few important Liouville type theorems for harmonic maps which may be relevant to the study of liquid crystals. We also discuss some differential geometric formalisms.

1. Basic Differential Geometric Formulas

Let (M^m,g) and (N^n,h) be Riemannian manifolds of dimension m and n respectively. Let $x^1,\dots,x^m$ be local co-ordinates of M, and let the Riemannian metric of M be $g = g_{\alpha\beta}dx^\alpha dx^\beta$. Denote the Riemannian metric of h of N by $h_{ij}dy^i dy^j$, where $y^1,\dots,y^n$ are local co-ordinates of N. Throughout this paper the summation convention is used. All repeated Greek indices are summed from 1 to m, and all repeated Latin indices are summed from 1 to n. Let $u: M \to N$ be a map, and let $u^i = y^i(u)$. The energy density of u at the point x is given by

$$e(u)(x) = TR_g(u^*h) = g^{\alpha\beta}(x)h_{ij}(u(x))\,\frac{\partial u^i}{\partial x^\alpha}(x)\,\frac{\partial u^j}{\partial x^\beta}(x)$$

where $(g^{\alpha\beta})$ is the inverse matrix of $(g_{\alpha\beta})$. The energy functional is defined to be

$$E(u) = \int_M e(u)dvol.$$

A critical point of this functional is called a harmonic map. The Euler-Lagrange equation of $E(u)$ is

$$\Delta_M u^i + \Gamma^i_{jk}\, g^{\alpha\beta}\,\frac{\partial u^j}{\partial x^\alpha}\,\frac{\partial u^k}{\partial x^\beta} = 0 \qquad \text{for } i = 1,2,\dots,n, \tag{1.1}$$

where Γ^i_{jk} is the Christoffel symbol of N.

When N is flat, all Christoffel symbols vanish, and Equation (1.1) becomes decoupled, so that a harmonic map is locally an n-tuple of harmonic functions. In particular, a harmonic map from M into the real line R is a harmonic function on M. When N is not flat, as is the case in the liquid crystals, Eq. (1.1) is a system of nonlinear second order elliptic equations. In other words, the nonlinearity is due to the presence of the curvature in the target manifold N.

It is more convenient to do differential geometric computations in orthonormal frame. (See [CG] for example.) Let $\{f_\alpha\}_{1\leq\alpha\leq m}$ be local orthonormal vector fields on M and $\{\theta^\alpha\}_{1\leq\alpha\leq m}$ be the dual 1-forms. Then the metric of M is

$$g = (\theta)^2 + (\theta^2)^2 + \ldots + (\theta^m)^2.$$

Let $\{e_i\}_{1\leq i\leq n}$ be local orthonormal vector fields in N and $\{\omega^i\}_{1\leq i\leq n}$ be the dual 1-forms. Again the metric of N is

$$h = (\omega^1)^2 + (\omega^2)^2 + \ldots + (\omega^n)^2.$$

The connection forms $\{\theta^\alpha_\beta\}$ of M and $\{\omega^i_j\}$ of N are defined by the following equations

$$d\theta^\alpha = -\theta^\alpha_\beta \wedge \theta^\beta$$

$$\theta^\alpha_\beta + \theta^\beta_\alpha = 0$$

$$d\omega^i = -\omega^i_j \wedge \omega^j$$

$$\omega^i_j + \omega^j_i = 0$$

Let $u\colon M \to N$ be a map. In terms of the above local frames, one can define u^i_α by $u^i_\alpha\theta^\alpha = u^*\omega^i$. In fact $u^i_\alpha\theta^\alpha \otimes e_i$ is an invariantly defined quantity, and is a u^*TN valued 1-form. It is easy to see that the energy density of u is then

$$e(u) = \sum_{i,\alpha} (u^i_\alpha)^2.$$

In terms of the above orthonormal frame, the covariant derivatives g_i, g_{ij}, g_{ijk} of a function g on N are defined by

$$dg = g_i \omega^i$$

$$dg_i = g_{ij} w_i^j + g_j \omega_i^j$$

$$dg_{ij} = g_{ijk} \omega^k + g_{ik} \omega_j^k + g_{kj} \omega_i^k .$$

The second covariant derivatives define the Hessian tensor. The Hessian tensor $D^2 g$ at a point q of a function g is defined as follows. For vectors X and Y at q, define $D^2 g(X,Y) = X(Yg) - (D_X Y)g$, where X and Y are arbitrarily extended to vector fields in a neighborhood of q, and $D_X Y$ is the usual covariant derivative. It is easy to see that the above definition does not depend on the extension of X and Y. If $\{f_i\}$ is a local orthonormal frame near q, $g_{ij} = D^2 g(f_i, f_j)$, and $\Delta g = \sum_i g_{ij}$.

When u is a map from M to N, the covariant derivatives u_α^i, $u_{\alpha\beta\gamma}^i$ of u_α^i involve both connection forms of M and N, and they are defined by

$$du_\alpha^i = u_{\alpha\beta}^i \theta^\beta + u_\beta^i \theta_\alpha^\beta - u_\alpha^k u^* \omega_k^i$$

$$du_{\alpha\beta}^i = u_{\alpha\beta\gamma}^i \theta^\gamma + u_{\alpha\gamma}^i \theta_\beta^\gamma + u_{\gamma\beta}^i \theta_\alpha^\gamma - u_{\alpha\beta}^k u^* \omega_k^i$$

It is well known [CG] that Eq. (1.1) is equivalent to

$$\sum_\alpha u_{\alpha\alpha}^i = 0, \qquad \text{for } i = 1,\dots,n. \tag{1.2}$$

Since $de(u) = \sum_{i,\alpha,\beta} 2u_\alpha^i u_{\alpha\beta}^i \theta^\beta$, we have by the Cauchy-Schwarz inequality

$$|de(u)|^2 = 4 \sum_\beta \Big(\sum_{i,\alpha} u_\alpha^i u_{\alpha\beta}^i \Big)^2 \le 4 \sum_\beta \Big[\Big(\sum_{i,\alpha} (u_\alpha^i)^2 \Big) \Big(\sum_{i,\alpha} (u_{\alpha\beta}^i)^2 \Big) \Big] \tag{1.3}$$

$$= 4\, e(u) \sum_{i,\alpha,\beta} (u_{\alpha\beta}^i)^2$$

In the differential geometric computations the curvature terms are the obstruction to commuting indices as the following formulas show.

$$g_{ij} = g_{ji}$$

$$g_{ijk} = g_{ikj} + R^N_{\ell ijk} g_\ell$$

$$u^i_{\alpha\beta} = u^i_{\beta\alpha}$$

$$u^i_{\alpha\beta\gamma} = u^i_{\alpha\gamma\beta} - R^N_{ijk\ell} u^j_\alpha u^k_\beta u^\ell_\gamma + R^M_{\alpha\delta\gamma\beta} u^i_\delta$$

Now it is not hard to derive the following formula from the above definitions and the commutation relations

(1.4) $$\frac{1}{2} \Delta e(u) = \sum_{i,\alpha,\beta} (u^i_{\alpha\beta}) - R^N_{ijk\ell} u^i_\alpha u^j_\beta u^k_\alpha u^\ell_\beta + R^M_{\alpha\beta} u^i_\alpha u^i_\beta$$

where $R^N_{ijk\ell}$ is the curvature tensor of N, and $R^M_{\alpha\beta}$ is the Ricci tensor of M. Formula (1.4) is called the Bochner formula for the energy of a harmonic map. If N has nonpositive sectional curvature, the Bochner formula is generally strong enough to derive many interesting results. However, when N has positive sectional curvature, the problem usually gets more delicate.

2. Interior Energy Estimate by Maximum Principle

In [Co], the following is proved.

Theorem 2.1

Suppose M and N are complete Riemannian manifold: M with the Ricci curvature bounded below by $-A, A \geq 0$, N with the sectional curvature bounded above by a positive constant $K > 0$. Let $u: M \to N$ be a harmonic map whose image lies in the ball $B_R(q)$ of radius R centered at a point q. If the ball lies within the cut locus of q, and $R < \pi/2\sqrt{K}$, then $e(u) \leq CA$, where C is a constant depending only on A,K and R. Therefore, if $A = 0$, u is a constant map.

This type of Liouville theorem was first proved by S.T. Yau [Y] for harmonic functions, and later generalized by S.Y. Cheng [Ce] and the author [Co]. In the context of liquid crystals, Theorem 2.1 implies the following

Corollary 2.2

Let $u: R^3 \to S^2$ be a harmonic map whose image lies strictly in the upper hemisphere of S^2, then u is a constant map.

The method of proof of this theorem is the use of the maximum principle as developed by Cheng and Yau [Co]. The crucial part of this method is a good choice of the function, such as (2.2), to differentiate. Here we simply indicate some essential ideas of the proof of Corollary 2.2.

Let r be the distance function from the origin in R^3. So if $p = (x^1,x^2,x^3)$, then $r(p) = [(x^1)^2 + (x^2)^2 + (x^3)^2]^{1/2}$. Let ρ be the intrinsic (spherical) distance function in S^2 from the north pole. The Riemannian metric of S^2 in terms of the polar co-ordinates (r,θ) in S^2 is $ds^2 = d\rho^2 + \sin^2\rho d\theta^2$. It is easy to check (e.g., see [GW]) that $D^2\rho = \sin\rho\cos\rho d\theta^2$. Since $D^2g(\rho) = g'(\rho)D^2\rho + g''(\rho)d\rho^2$, we have $D^2\phi = \cos\rho ds^2$, where $\phi = 1 - \cos\rho$. Let $u: R^3 \to S^2$ be a harmonic map. Then

$$
\begin{aligned}
(2.1) \qquad \Delta(\phi \circ u) &= \textstyle\sum_{i,\alpha}(\phi_i u^i_\alpha)_\alpha = \sum_{i,j,\alpha} \phi_{ij}\, u^i_\alpha u^j_\alpha + \sum_{i,\alpha}\phi_i u^i_{\alpha\alpha} \\
&= \textstyle\sum_{i,j,\alpha}\phi_{ij}u^i_\alpha u^j_\alpha \\
&= \cos\rho\, e(u)
\end{aligned}
$$

Our assumption is that $u(R^3)$ $B_R(q)$, where q is the north pole, and $R < \pi/2$. Choose b such that $\phi \circ u \leq 1 - \cos R < b < 1$. Consider the function

$$
(2.2) \qquad \Phi = \frac{(a^2 - r^2)^2\, e(u)}{(b - \phi \circ u)^2}
$$

defined on the ball $B_a(0) \subset R^3$. Since Φ vanishes on the boundary of $B_a(0)$, Φ attains its maximum at an interior point $p \in B_a(0)$. Then clearly at p, $d \log \Phi = 0$ and $\Delta \log \Phi \leq 0$. These together with (1.3), (1.4) and (2.1) imply the following inequality

$$
(2.3) \qquad 0 \geq \frac{12a^2 + 4r^2}{(a^2 - r^2)^2} - \frac{8r\, \text{sni}\, \rho \sqrt{e(u)}}{(a^2 - r^2)(b - \Phi \cdot u)} + \Big(\frac{2\cos\rho}{b - \phi \cdot u} - 2\Big)e(u)
$$

which is valid at q. (See [Co].) Since $\phi \circ u \leq 1 - \cos R < b < 1$, there exists a constant C_1 such that

$$\frac{2\cos\rho}{b - \phi \circ u} - 2 \geq C_1 > 0 \tag{2.4}$$

It is easy to check that if $ax^2 - bx - c \leq 0$, where a,b and c are positive constants, then $x \leq \max\{2b/a, 2\sqrt{c}/a\}$. Therefore (2.3) and (2.4) imply that, at p,

$$e(u) \leq 4 \max\left\{ \frac{64r^2\sin^2\rho}{c^2(a^2 - r^2)^2(b - \phi \circ u)^2}, \frac{12a^2 + 4r^2}{c_1(a^2 - r^2)^2} \right\} \tag{2.5}$$

Let R.H.S. denote the right hand side of (2.5). Then clearly at p

$$\phi \leq \frac{(a^2 - r^2)^2 \text{R.H.S.}}{(b - \phi \circ u)^2} . \tag{2.6}$$

Letting $a \to \infty$ in (2.5), we obtain $e(u) \equiv 0$.

3. Other Liouville Theorems

There are many generalizations of the Liouville theorem for harmonic maps, and interesting related results. Many of them are nicely surveyed in [Hi]. Among them, the following theorem due to Hildebrandt, Jost and Widman [HJW] seems most relevant for our purpose.

Theorem 3.1

Let M be a Riemannian manifold diffeomorphic to R^n with the Riemannian metric $ds^2 = g_{\alpha\beta}dx^\alpha dx^\beta$. Suppose that there are constant C_1 and C_2 such that $C_1|\xi|^2 \leq g_{\alpha\beta}\xi^\alpha\xi^\beta \leq C_2|\xi|^2$, for any $\xi = (\xi^1, \xi^2, \ldots, \xi^m)$. Then any harmonic map from M into N whose image lies in $B_R(y)$ is a constant map, where N and $B_R(y)$ satisfy the same condition as in Theorem 2.1.

This theorem again implies Corollary 2.2. The essential idea of the proof of Theorem 3.1 is the growth estimate of the energy integral by potential estimates and the Moser iteration procedure. The growth estimate then implies by Morrey's result a Hölder estimate, which again implies the Liouville theorem.

Another condition of "smallness" of a map is the smallness of the total energy. The following theorem is proved in [S].

Theorem 3.2

Let N be an arbitrary Riemannian manifold. Suppose $B_R \subset R^m$ is equipped with the Riemannian metric $ds^2 = g_{\alpha\beta}dx^\alpha dx^\beta$. Assume that there is a constant Λ such that $\Lambda^{-1}(\delta_{ij}) \leq (g_{ij}) \leq \Lambda(\delta_{ij})$ and $|\nabla g_{ij}| \leq \Lambda R^{-1}$. Then there exist constants $\varepsilon > 0$ and $C > 0$ depending only on Λ, m and N such that if $u: B_R \to N$ is a harmonic map satisfying

$$R^{2-m} \int_{B_R} e(u) \leq \varepsilon$$

then

$$\sup_{B_{R/2}} e(u) \leq CR^{-m} \int_{B_R} e(u).$$

Note that there is no restriction on the image, and the smallness of the total energy is phrased in terms of $R^{2-m} \int_{B_R} e(u)$, which is a scale invariant quantity. It may be interesting to find a good estimate for ε. The proof of this theorem uses the technique developed by Schoen and Uhlenbeck utilizing the monotonicity formula. The monotonicity formula for the harmonic map states that

$$R^{2-m} \int_{B_R} e(u)$$

is a nondecreasing function of R. Therefore $\lim_{R\to\infty} R^{2-m} \int_{B_R} e(u)$ exists. Letting $R \to \infty$, we obtain the following

Corollary 3.3

Let $u: R^m \to N$ be a harmonic map such that $\lim_{R\to\infty} R^{2-m} \int_{B_R} e(u) \leq \varepsilon$. Then u is a constant map.

In particular, we have the following result proved essentially by Garber, Ruijsenaars, Seiler and Burns [GRSB].

Corollary 3.4

Let $u: R^m \to N$ be a harmonic map such that the total energy $\int_{R^m} e(u)$ is finite. Then u is a constant map.

References

[Ce] S.Y. Cheng: Liouville theorem for harmonic maps, Proc. Sympos. Pure Math., vol. 36, 1980, 147-151.

[CG] S.S. Chern and S. Goldberg: On the volume decreasing property of a class of real harmonic mappings, Amer. J. Math., vol. 97, 1975, 133-147.

[Co] H.I. Choi: On the Liouville theorem for harmonic maps, Proc. Amer. Math. Soc., vol. 85, 1982, 91-94.

[ES] J. Eells and J. Sampson: Harmonic mappings of Riemannian manifolds, vol. 86, 1964, 109-160.

[GRSB] W.-D. Garber, S. Ruigsenaars, E. Seiler and D. Burns: On the finite action solution of the nonlinear σ-model, Ann. of Physics, vol. 119, 1979, 305-325.

[GW] R. Greene and H. Wu, Function theory of manifolds which possess a pole, Lecture notes in Math. #699, 1979, Springer-Verlag.

[H] R. Hamilton: Harmonic maps of manifolds with boundary, Lecture Notes in Math. #471, 1975, Springer-Verlag.

[Ha] R. Hardt and D. Kinderlehrer: Mathematical questions of liquid crystal theory, this volume.

[Hi] S. Hildebrandt: Liouville theorem for harmonic mappings and an approach to Bernstein theorems, Seminar on differential geometry edited by S.T. Yau, Annals of Math. Studies #102, Princeton U. Press, 1982, 107-131.

[HJW] S. Hildebrandt, J. Jost, and K.-O. Widman: An existence theorem for harmonic mappings of Riemannian manifolds, Acta Math., vol. 138, 1977, 1-16.

[HJW] S. Hildebrandt, J. Jost, and K.-O. Widman: Harmonic mappings and minimal submanifolds, Inv. Math., vol. 62, 1980, 269-298.

[S] R. Schoen: Analytic aspects of the harmonic map problem, Seminar on Nonlinear Partial Diff. Eqs. edited by S.S. Chern, Math. Sci. Res. Institute Publications #2, 1984, Springer-Verlag, 321-358.

[SU1] R. Schoen and K. Uhlenbeck: A regularity theory for harmonic maps, J. Diff. Geom., vol. 17, 1982, 307-335.

[SU2] R. Schoen and K. Uhlenbeck: Boundary regularity and the Dirichlet problem for harmonic maps, J. Diff. Geom., vol. 18, 1983, 253-268.

[Y] S.-T. Yau: Harmonic functions on complete Riemannian manifolds, Comm. Pure & Appl. Math., vol. 28, 1975, 201-228.

A REVIEW OF CHOLESTERIC BLUE PHASES

P.E. Cladis

AT&T Bell Laboratories
Murray Hill, New Jersey 07974

Abstract

Blue phases (BP) challenge our understanding of phase transitions since they exhibit three dimensional translational order yet occur between phases of higher symmetry: the cholesteric phase which has translational order in only one dimension and the isotropic phase (or the amorphous blue fog phase) which has no translational order at all. Materials exhibiting blue phases are necessarily chiral; that is to say, on a microscopic scale they lack reflection symmetry. In liquid crystal systems, a macroscopic expression of this microscopic property is often a helicoidal structure. For example, in cholesteric liquid crystals the director, or the direction along which the long axis of the molecules tends to align, rotates with a constant pitch along a single direction, the z axis, say, this structural feature distinguishes the z direction from the other two directions and a cholesteric liquid crystal is optically anisotropic. In addition, cholesterics are optically active: the plane of polarization of light rotates as it travels along the cholesteric twist axis but is unchanged traveling in directions perpendicular to it. Blue phases share the optical activity of cholesteric liquid crystals but they are optically isotropic.

There are three blue phases, BPI, BPII and BPIII which appear to be periodic extremals to the Landau - de Gennes Free Energy functional for cholesterics. Experimental evidence has been found that BPI and BPII are soft solids with large unit cells. They are believed to be composed of defect arrays and stabilized by the proximity of the isotropic liquid phase. It has been suggested, and there is some evidence to support this idea, that the BPI-BPII transition is a percolation threshold where the continuous phase changes from being cholesteric to isotropic. On the other hand, that BPII may be an example of a bicontinuous phase cannot be excluded. Another suggestion is that 3-dimensional bond orientational order is associated with BPIII.

Because of their large unit cells, crystals composed of only 20^3 or so unit cells of BPI or BPII are visible in the optical microscope providing us with a rare opportunity to observe directly mechanisms and dynamics of crystal growth.

Two remarks based on observations of defects in cholesteric systems are:

1. the most frequently observed defects appear to have strength (topological index) $S = 2$, perhaps because the $S = 2$ structure is homotopic to the ground state.
2. the director orientation on the boundary, $\partial\Omega$, plays a crucial role in determining the director configuration in Ω.

It is emphasized that our understanding of blue phases is limited owing to the absence of a complete theory of cholesteric defects.

The outline is as follows:

I. Introduction

II. Historical Survey of Blue Phases

A. From the beginning of liquid crystal time (~1890) to the mid 1970's.

1. Textures and Stability of Blue Phases
2. Spatial Periodicity and Fluctuations
3. Heuristic Demonstration that Periodicity Counts

B. From the mid 1970's to the present.

1. Thermodynamic Properties of Blue Phases
2. Blue Phase Structure
3. Electric Fields and Blue Phases
4. Viscosity of Blue Phases
5. Elastic Measurements
6. Blue Phase Paradigms of Crystal Growth

III. Particularly Fascinating Aspects of Blue Phases

A. Double Twist and Negative Surface Energies

B. Why Three Blue Phases?

C. Remarks on Cholesteric Defects

D. Orienting Effect of the Cholesteric - Isotropic Interface

IV. Conclusions

I. Introduction

The term liquid crystal refers to intermediate phases that occur between an ordered solid and the isotropic liquid phase characterized by long range orientational order [1]. The direction along which the system chooses to order is denoted by a unit vector, **n**, called the director. There is a loss in symmetry as one passes from the isotropic liquid to the nematic liquid crystal phase, characterized only by orientational order, then to the smectic A phase where there is a one-dimensional density wave along **n** (i.e. the system is now layered), to finally the solid phase characterized by three dimensional positional order.

The most well-known liquid crystal phase, the nematic, belongs to a larger class of intermediate phases called "cholesteric liquid crystals" in which there is a twist axis, **t**, perpendicular to **n** around which **n** turns with constant pitch, $p_0 = (2\pi/q_0)$. The nematic is the special case $q_0 = 0$ - the cholesteric phase with infinite pitch. On scales very small compared to the pitch, a cholesteric is nematic: **n** is defined but **t** is not. On scales comparable to the pitch, **n** and **t** are defined and the system is locally biaxial with the complicating feature of one dimensional periodicity in the direction of **t**. On scales <u>very</u> large compared to the pitch, only **t** is defined and the cholesteric is uniaxial negative.

Cholesteric blue phases refer to three intermediate phases frequently observed between the cholesteric liquid crystal phase and the isotropic liquid when $p_0 \lesssim 2500Å$. Blue phases are optically isotropic, optically active (they rotate the plane of polarization of incoming light) and they Bragg reflect between two and four different wave lengths of light the same way a solid crystal Bragg reflects x-rays. The reflected wavelengths are in the ratio : $1/\sqrt{2}/\sqrt{3}/\ldots/$ leading to the conclusion that blue phases have some kind of cubic order. Thus, although the system is fluid-like on a scale small compared to the pitch, there is macroscopic three-dimensional crystalline ordering with a lattice constant $p_0 \cong 2500Å$. Despite the fact they occur at higher temperatures, translational order in three dimensions makes blue phases I and II more ordered than the cholesteric which only has translational order in one dimension.

Fig. 1 shows an organizational chart of two of the many liquid crystal systems known to date: smectics and cholesterics. The distinction between smectic and cholesteric is determined by the scale of positional ordering: when comparable to a molecular length, a ≅ 30Å, it is called smectic, when much larger than a, it is called cholesteric. Smectic layering is determined by the molecular length whereas the helicoidal structure responsible for cholesteric translational ordering, is the macroscopic expression of the absence of reflection symmetry at the molecular level. Chiral smectics are amusing compounds in which both lengths are present (see for example reference [1]).

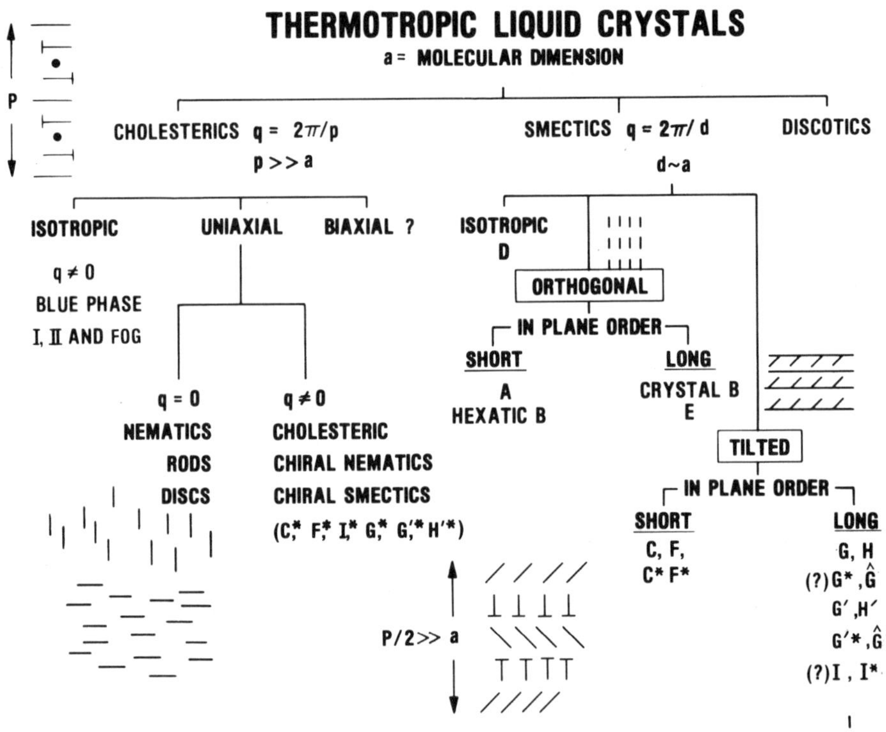

Blue phases have their smectic analogue in the isotropic smectic D phase as well as the cubic phases of lyotropic liquid crystals which occur between the lamellar neat soap phase and the middle soap phase where infinitely long cylindrical micelles are stacked in a hexagonal array. Middle soap has the same symmetry as the hexagonal discotic phases [2]. The unit cell of blue phases involves about 10^7 molecules and the cubic soap phases $\sim 10^3$ molecules.

The cholesteric is difficult to characterize properly because, in addition to being periodic, the reciprocal lattice vector q is a "free angle variable" - it can point anywhere in space. In the optical microscope, this freedom is manifested by a variety of "textures" - often beautiful to see but difficult to understand. Kleman [3] gives a review of the status of our understanding of liquid crystal defects up to the late 1970's, however, a complete defect theory of the cholesteric phase doesn't exist. For example, the homotopic description of cholesteric defects by Volovik and Mineyev [4], lucidly reviewed by Mermin [5], concentrates on the free angle aspects and ignores periodicity.

Our chances of completely understanding the structure of Blue Phases, which the available evidence indicates is a stationary, stable array of cholesteric defects, depends upon how effectively more recent and future theories consolidate these two faces of cholesteric liquid crystals. "Escaping into the fourth dimension" [6] may well be necessary to do this.

In the meantime, experimentalists build their intuition about the kinds of defects that exist by simple observations in the polarizing microscope. This has been our strategy [7-9]. It should also be mentioned that Bouligand [10] has prepared many beautiful drawings showing the geometric disposition of cholesteric defects and textures from all perspectives.

II. History and Generalities

In Fig. (2), the number of papers published on blue phases is shown versus decade. This graph demonstrates that interest in blue phases has intensified relatively recently. Up to the mid- nineteen seventies, blue phases were not studied and it was not really appreciated how common they were [11]. The most interesting result to emerge in the last decade concerning blue phases is that they are indeed crystals with giant unit cells.

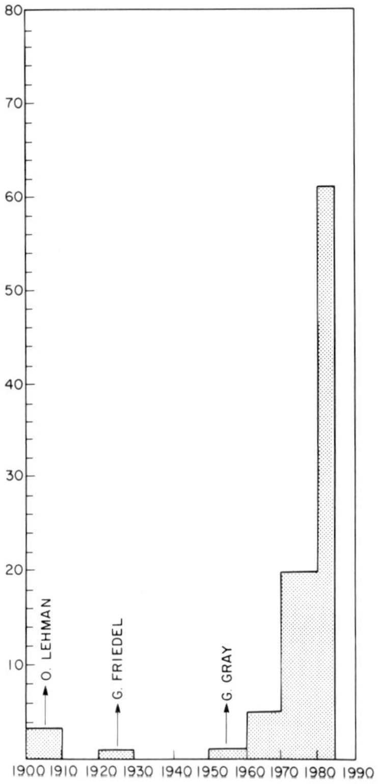

A. ~1890 - mid 1970's

1. Textures and Stability of Blue Phases

In their review article [12], Stegemeyer and Bergmann recount that blue phases were the first thermotropic liquid crystal phase ever to be observed by Reinitzer and Lehman ~1890 when they cooled cholesterol benzoate from the isotropic liquid phase.

In his great work [13], G. Friedel describes the appearance of the cholesteric phase from the isotropic liquid as a "veil" whose structure cannot be resolved in the polarizing microscope. He noted in exceptional cases the formation of cholesteric drops in the form of "batonnets" which looked and acted exactly like the "batonnets" formed by the smectic A phase when it condensed out of the isotropic phase. Friedel pointed out that it was unusual for the planar cholesteric phase and the uniform smectic A phase to form spontaneously from the isotropic liquid. This is easily understood because of the translational order of the two phases i.e. it is the same problem any crystal has in growing. However, he did note that a certain Stumpf believed he saw a "first phase" distinct from the "main phase" which was optically active and reflected colours. Stumpf even fixed the limits of the domain of stability of this first phase: between 102°C and 105°C. But then, Stumpf also observed that he could recover the "main phase" by shearing his preparation. This suggested that his blue phase was simply a question of two different textures of the same phase.

Similar objections to the stability of blue phases have been raised more recently by Chystiakov and Gusakova [14] who state that in a very pure sample of cholesterol propionate, they do not observe any blue phases. They obtain blue phases by adding just a tiny amount ($\lesssim$ 1%) of impurity. Their cholesterol propionate must have been exceptionally pure because the commonly available samples of this compound exhibit at least one blue phase [36,41].

2. Spatial Periodicity and Fluctuations

Quite apart from detailed considerations of blue phases, Friedel's observations of the similarities of the "petites plages grises" in the vicinity of both the smectic A - isotropic and cholesteric - isotropic transitions could be a manifestation of the large fluctuations typical for systems characterized by one dimensional translational order. We use Crooker's description from his review article on blue phases [15].

In terms of a Landau-de Gennes theory,in the absence of periodicity, the free energy, F, can be transformed to

$$F = \sum_k (a + bk^2)\phi_k^2, \tag{2}$$

where ϕ_k is the k-th fourier component of the order parameter, $a = a_0(T - T^*)$ and b is a constant. The expectation value of

$$\langle\phi(r)^2\rangle = \int \frac{dk_x dk_y dk_z}{(a + bk^2)} . \tag{3}$$

Since $dk_x dk_y dk_z = k^2 dk d\Omega$, in the limit of small k or large wavelength fluctuations, $\langle\phi(r)^2\rangle \rightarrow 0$ when $a > 0$. Long wavelength fluctuations remain small.

In the case of 1 - d periodicity, Eq. (2) is replaced by

$$F = \sum_k [a + b(k - k_0)^2]\phi_k^2 \tag{4a}$$

and

$$\langle\phi(r)^2\rangle = k^2 \int \frac{dk}{a + b(k - k_0)^2} . \tag{4b}$$

Instead of decaying to zero as it does in the absence of periodicity, (Eq. 2), small fluctuations lead to a degeneracy in the direction of k_0. Long wavelength fluctuations go to zero slowly compromising order parameter rigidity.

3. Heuristic Demonstration that Peridocity Counts

Early on, de Gennes and Shtrikman [16] appreciated that the cholesteric-isotropic transition had some interesting aspects to it when one considered a straight forward expansion of the free energy in terms of the cholesteric tensor order parameter [16]. Consider a cholesteric with twist axis in the z direction. The director is given by

$$\mathbf{n} = \begin{bmatrix} \cos(qz) \\ \sin(qz) \\ 0 \end{bmatrix} . \tag{5}$$

Constructing the tensor order parameter from the following recipe [16],

$$Q_{ij} = S\langle n_i n_j - \frac{1}{3} \delta_{ij} \rangle \tag{6}$$

one obtains the following expression,

$$Q = \frac{S}{6} \begin{bmatrix} 1 & 0 & 0 \\ 0 & 1 & 0 \\ 0 & 0 & -2 \end{bmatrix} + \frac{S}{2} \begin{bmatrix} \cos(2qz) & \sin(2qz) & 0 \\ \sin(2qz) & -\cos(2qz) & 0 \\ 0 & 0 & 0 \end{bmatrix} . \tag{7}$$

The first term in the above expression is uniaxial negative. It represents the twist axis perpendicular to the director. The second term is purely biaxial. In order to track the effect of the biaxial term on the cholesteric - isotropic phase transition, replace the second S in Eq. (7) with B. The free energy is then:

$$F = \frac{a}{6} S^2 + [\frac{a}{2} - q_0^2]B^2 + \frac{bS}{4} [B^2 - \frac{S^2}{9}] + c[\frac{S^4}{72} + \frac{S^2B^2}{9} + \frac{B^4}{8}] . \tag{8}$$

a. Nematic. $S = B$ and $q = 0$.

Eq. (7) then becomes

$$Q = \frac{S}{3} \begin{bmatrix} 2 & 0 & 0 \\ 0 & -1 & 0 \\ 0 & 0 & -1 \end{bmatrix} , \tag{7a}$$

the usual uniaxial positive order parameter for rod-like nematics. In this case the free energy is

$$F = \frac{2}{3} aS^2 + \frac{2}{9} bS^3 + \frac{2}{9} cS^4 \tag{9}$$

and a first order transition to the isotropic phase (determined by the conditions

$F = 0$ and $\frac{\partial F}{\partial S} = 0$) takes place at a temperature given by

$$T_N = T^* + \frac{3b^2}{8a_0c} . \tag{10}$$

b. **Cholesteric.** $S = B$.

A first order transition takes place at the temperature

$$T_{Ch} = T_N + \frac{3}{2}\frac{K}{a_0} q_0^2, \tag{11}$$

where T_N is given by Eq. (10). K is a Frank constant $\approx 2 \times 10^{-7}$ dynes. Thus, compared to its racemate, the transition temperature to the isotropic phase of a cholesteric is higher by an amount proportional to its inverse pitch square; twist enhances orientational order.

c. **Blue Phases?** $B = S/3$.

Then,

$$\mathbf{Q} = \frac{S}{3}\begin{bmatrix} \cos^2(qz) & \sin(qz)\cos(qz) & 0 \\ \sin(z)\cos(qz) & \sin^2(qz) & 0 \\ 0 & 0 & -1 \end{bmatrix} . \tag{12}$$

Q is purely biaxial. The free energy F becomes

$$F = S^2[\frac{2}{9}a - Kq_0^2] + \frac{c}{36}S^4 + 0(S^6). \tag{13}$$

There is no longer a cubic term and the fourth order term being positive, the biaxial transition to the isotropic phase is second order. The transition temperature to the isotropic phase occurs when the coefficient of the S^2 term is zero, at T_B, say, where

$$T_B = T^* + \frac{9}{2}\frac{K}{a_0} q_0^2. \tag{14}$$

Comparing the q_0 dependence of the transition temperatures given by Eq. (11) and Eq. (14), it is seen that even though $T^* < T_N$, the greater slope of T_B

overcomes T_{Ch} at some positive value of q. The second order transition will always win... which we know doesn't happen. A more general description is needed and it shows that the second order transition at T_B is preempted by a first order transition: - the blue phases [17-20].

B. Mid 1970's - 1984

1. **Thermodynamic Properties of Blue Phases**

In the mid -70's, Armitage and Price [21] were the first to make volume and heats of transition measurements on a variety of cholesterol esters which seemed to indicate that blue phases were autonomous liquid crystal phases. They found a small latent heat and a small volume change at the cholesteric to blue phase transition - it had not yet been established that there were three blue phases [22], and a large change in these quantities at the blue phase to isotropic transition. Similar results have been obtained by Stegemeyer et al. [12,22] on still more cholesterol esters and even some chiral compounds (which I refer to as chiral nematics) not derived from cholesterol [23]. Stegemeyer's group was the first to point out that there was more than one blue phase.

It is important to note that cholesterol is a complicated molecule usually not synthesized from scratch but extracted from even more complex dairy systems like butter. Consequently, the purity of cholesterol derivatives is usually very low. Generally, purity is not measured, but in one instance it was, and found to be typically 75% [24] for a nominally "very pure" sample of cholesterol nonanoate. Cholesterol nonanoate has a blue phase range of heating of about 0.5°C. A 99% pure chiral nematic [24] exhibited a blue phase range of less than .05°C.

Very recently, high resolution heat capacity measurements [25] have confirmed the small heat capacity at the cholesteric to blue phase transition and the order of magnitude larger heat capacity of the BPIII to isotropic transition suggesting that BPIII is a distinct thermodynamic phase.

2. Blue Phase Structure

There have been many theoretical contributions to the structure of blue phases [6,18-20,26-30]. Most of these have been reviewed in a balanced but concise fashion in the thesis of Wright [31,32] and I will not discuss them.

From an experimental point of view, blue phases reflect several wavelengths and all reflections are circularly polarized. As many as five different wavelengths have been observed in BPI and generally only two are observed in BPII [33-36]. The relative intensities and polarizations of these reflections give information about the structure of blue phases [28,37]. The consensus is that BPI is a body centered cubic (bcc) arrangement of helical elements and BPII is a simple cubic (sc) structure.

Tanimoto and Crooker [39] found one mixture with a BPII phase exhibiting the bcc structure. They deduced this by observing a rare, for BPII, third line. They reasoned that in a sc structure this line would be indexed as [111]. Hornreich and Shtrikman [28] had shown, quite generally, that the [111] reflection is never cicularly polarized. Since their third line was circularly polarized, Tanimoto and Crooker were forced to conclude - in the absence of a better alternative - that their BPII phase was bcc not sc.

3. Electric Fields and Blue Phases

When the dielectric constant parallel to **n**, $\varepsilon_{\parallel}$, is greater than $\varepsilon_{\perp}$, the dielectric constant perpendicular to **n**, an electric field, E_c, can untwist a cholesteric so that it becomes nematic [1]. The blue phases of cholesteric mixtures consisting of a chiral nematic added to a non-chiral nematic, both of which possess large dielectric anisotropies, $\varepsilon_a = \varepsilon_{\parallel} - \varepsilon_{\perp} > 10$, have been untwisted by an electric field [39-43].

From the magnitude of the field, one can estimate the difference in energy between a nematic and blue phases: $\delta F \cong \varepsilon_a E^2 \cong$ (the blue phase → chol. latent heat). This energy difference is very small. When $\varepsilon_a \cong 10$ and $E \cong 25$ volts/12μm, $\delta F \cong 4.8 \times 10^4$ ergs/cm^3. The corresponding latent heat, one millicalorie per gram, is one order of magnitude smaller than the measured values

With increasing field, the observed phase sequence can be BPI or II→ cholesteric or nematic [41,42] or BPII→ BPI→ cholesteric or nematic [42].

4. Viscosity of Blue Phases

A large increase in viscosity is observed in the vicinity of the isotropic phase in cholesteric mixtures and pure compounds exhibiting blue phases, whereas, in the absence of blue phases, no anomaly is observed [44-46]. Stegemeyer and Pollman [46] report more than one order of magnitude increase in viscosity at the BPI-isotropic transition.

Interestingly, a more detailed measurement of the BP shear viscosity of two cholesterol esters shows a viscosity minimum in the BPI phase just before the BPII or isotropic transition [25,47]. The viscosity minimum is ~1.5 poise independent of frequency. In these studies [25], the viscosity only increases by ~20% at ~100 hz, and not at all at 1000 hz. At ~10 hz, the increase is about a factor or 3 [47].

5. Elastic Measurements

The shear modulus of BPI, measured by exciting normal shear models was found to be on the order of 200 dynes/cm^2 in a chiral nematic mixture [48] and 2000 dynes/cm^2 in a cholesterol ester [47]. The order of magnitude of the shear modulus is similar to colloidal crystals but the temperature dependence of the BPI modulus is dramatic ... it changed by nearly two orders of magnitude in a 5°C temperature interval.

In a percolation scenario supposed to account for this [48], blue phase helical elements are envisaged to fill space and isotropic "holes", formed by the cores of defects, are located on a cubic lattice. The spring constant is inversely proportional to the number of "holes". Increasing the temperature increases the number of holes and the elastic modulus declines dramatically.

A low frequency torsional oscillator operating between 50 and 1000 hertz [25], gave data suggesting that BPI and BPII are viscoelastic solids (i.e. shear modulus → constant as frequency → 0) over the frequency range studied. In order to prove they are not viscoelastic liquids, (i.e. shear modulus → 0 as frequency → 0) it is necessary to supplement this data with measurements made at lower frequencies. However, the data of Clark et al. [47] taken on a similar compound in the ≈2 - 10 hertz range seems to bear out the viscoelastic solid conclusion - at least for BPI. An increase in shear modulus on going from BPI - BPII is interpreted as resulting from the greater line tension of the BPII defects [25].

6. Blue Phase Paradigms of Crystal Growth

There are three blue phases; BPI, BPII and BPIII. BPIII is closest to the isotropic phase and BPI is closest to the cholesteric phase. BPIII, unlike BPI and BPII has no positional order. BPIII is also known as the "blue fog" [12,49]. The light intensity reflected by BPIII is very weak. Of all the blue phases, it is the most difficult to observe in the optical microscope despite having the largest heat of transition to the isotropic phase. Sometimes an interface is observed between the isotropic liquid and BPIII [41,49].

Blue phase I and II are soft solids with large lattice constants. Small crystals of only 20 unit cells are big enough to be observed in the optical microscope (Fig. (3)). Blue phase crystals thus offer a rare opportunity to study mechanisms and dynamics of crystal growth by direct observation.

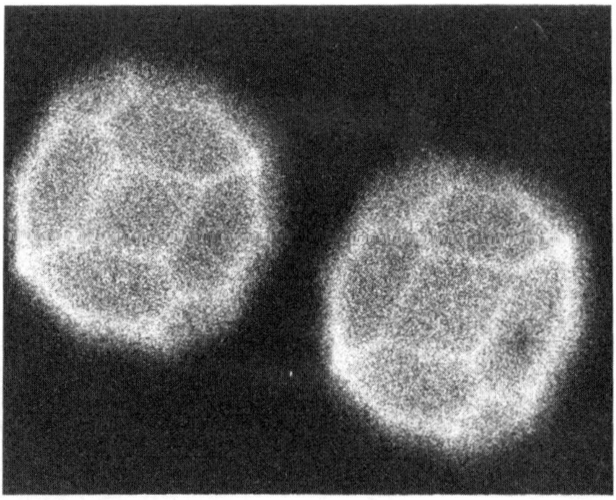

BPII frequently exhibits a square habit [50]. Since this presentation of BPII reflected the longest wavelength, Marcus [50] deduced that the structure of BPII was sc. At the BPII $\rightarrow$ BPI transition, the square crystals develop a cross-hatched pattern running diagonal to the sides of the square [36,50].

BPI has only recently been observed to condense out of the isotropic phase as large polyhedral shaped monocrystals with well defined facets [8,48,51-54]. Fig. (3) is a photomicrograph showing a typical BPI crystallite [48]. The photograph was taken using a microspectrometer [55] - a reflection microscope with monochromatic illumination, developed for the study of colloidal crystals. Although colloidal crystals, like blue phases, also have three dimensional crystalline order with large lattice constants, they do not exhibit facets.

Detailed studies of the BPI crystals have been made and the various facets have been indexed [53,54]. Steps have also been observed on the facets. As the crystal grows, steps turn around screw dislocations as predicted by F.C. Frank [56] long ago. It was argued that the crystalline forms of BPI are compatible with a particular bcc structure known as $I4_132$ [51,53].

Crystallites resembling ones shown in Fig. (3), have also been observed in the cubic phases of the lyotropic liquid crystal [57] cousins of blue phases.

III. Some Interesting Aspects of Blue Phases

A. DOUBLE TWIST AND NEGATIVE SURFACE ENERGIES

An interesting idea to emerge recently is that negative surface energies of certain defects could stabilize blue phase structures [27] in much the same way the vortex lattice of type two superconductors is stabilized [58]. A surface term in the Frank free energy [59] of the form,

$$F_{surf} = (K_2 + K_{24})\int \mathrm{div}(\mathbf{n} \times \mathrm{curl}\,\mathbf{n} + \mathbf{n}\,\mathrm{div}\,\mathbf{n})dV, \tag{16}$$

usually neglected in calculations of bulk properties or calculations involving a liquid crystal interface with solid objects and which usually sums to zero for singular line defects, can have a negative contribution for some non-singular defects. These defects are known as double twist although they are actually stabilized by a divergence of the bend distortion (see Eq. (16)).

Singular defects have two surfaces, one far from the core and one around the core [60]. The contribution of Eq. (16) on each of these surfaces is equal but opposite in sign. In the case of non-singular defects, there is no surface around the core (because there is no core) so the magnitude of the far surface contribution is not canceled out.

In cylindrical co-ordinates (r,ϕ,z), the director for double twist is

$$\mathbf{n} = \begin{bmatrix} 0 \\ \sin(qr) \\ -\cos(qr) \end{bmatrix} \tag{17}$$

and the elastic energy for a cholesteric [59] is

$$F_{e\ell} = \int [K_1(\operatorname{div}\mathbf{n})^2 + K_2(\mathbf{n}\cdot\operatorname{curl}\mathbf{n} + q_0)^2 + K_3(\mathbf{n}\times\operatorname{curl}\mathbf{n})^2]dV. \tag{18}$$

As is obvious, Eq. (18) is positive definite but Eq. (16) is negative for any $(qr) < \pi$ for the configuration described by Eq. (17). The energy density associated with this defect, $(F_{e\ell} + F_{surf})$/unit volume is negative for $(qr) < \pi/4$. These objects constitute the main building block of the defect models of Meiboom et al. [27,33].

The negative energy of "double twist" is sapped by singular defects needed to fill space. Currently, it is not clear how the cores associated with the singular lines should be modelled.

An interesting footnote is that similar tube models, without the compelling arguments of "double twist" however, have been proposed (and rejected) for the smectic relative of blue phases - the cubic smectic phases [61].

B. WHY THREE BLUE PHASES?

The suggestion that it might be profitable to think of blue phases as stable defect arrays was first made by Saupe [26]. Finn and Cladis [41] point out that this idea might account for exactly three blue phases. They assume blue phases are two phase regions of cholesteric and isotropic liquid. In BPI, the continuous phase is cholesteric and the isotropic liquid, the discontinuous phase. Perhaps for reasons such as the negative surface energy associated with "double twist" [27], the isotropic regions do not coalesce but rather form a periodic array. With increasing temperature, the amount of isotropic liquid increases and eventually there is an inversion (or percolation threshold) at the BPI-BPII phase transition. In BPII, the continuous phase is isotropic and the discontinuous phase, cholesteric. Cholesteric regions are periodically arrayed in BPII but not BPIII.

Sir Charles Frank points out [62] that, unlike 2-dimensions, the percolation threshold in 3-dimensions need not be sharp. Thus, one cannot exclude the possibility of a temperature interval where both phases are continuous, a bicontinuous phase. A not unreasonable scenario, with all the advantages listed above, would then be that BPI is as above, BPII is the bicontinuous phase and in BPIII, the isotropic phase is continuous, as above.

Considering BPII as a bicontinuous phase, makes Blue Phases exactly similar to a current model for microemulsions [63] of, for example, oil and water. Microemulsions are stabilized by the competition between entropic terms and surface energies and form thermodynamically stable phases when "droplets" of one phase in the other are small enough, [63] i.e. the analogues of BPI and BPIII. Similar arguments can be made for the thermodynamic stability of the bicontinuous phase but experimental evidence for its existence is rather weak. So BPII may provide an interesting opportunity to study this elusive state of matter. A large dispersion in droplet sizes occurs in microemulsions, whereas, blue phase droplets being controlled by the pitch, are the same size. For this reason, they settle easily into a lattice, whereas microemulsions are not generally known to order except in the case of lyotropic liquid crystals, a subset of microemulsions.

Evidence supporting this picture is:

1. three blue phases are observed
2. the strong temperature dependence of the BPI modulus [47,48]

3. viscosity rise at the BPI-BPII transition [25]. The fact that the viscosity of BPII is larger than BPI in some systems would seem to favor the bicontinuous model for BPII.
4. the similarity in the magnitude of the field to unwind BPI compared to the field to unwind the cholesteric phase [40-43]
5. the 5°C blue phase range observed in the 13 component mixtures [48] compared to the 0.05°C blue phase range in a 99.9% pure compound [24].

The above picture accounts for many of the properties of BPI and BPII but one might still question the large heat of transition accompanying the transformation of BPIII to isotropic liquid. Recently, Anderson [64] suggested that 3-dimensional bond orientational order (BOO) may be associated with BPIII. For example, one can imagine a dense cluster of 13 spheres - each one filled with cholesteric, all with the same radius - composed of one sphere in the middle surrounded by a shell of 12 spheres to form an icosohedron. Mackay [65] has called this icosohedral shell packing (isp). He has shown that it is locally a denser packing for spheres than bcc but not cubic close packing to which it can easily transform. Next, one fills a space with such clusters, all oriented the same way but their centres of mass are not on a lattice. There is isotropic liquid between the clusters. At the BPIII - isotropic transition it is this orientational order, BOO, which is lost accounting for the large BPIII - isotropic heat of transition relative to the weaker binding energy of the BPI and BPII lattices.

Icosohedral packing has been observed in the complex crystal structures with large unit cells of certain complex transition metal alloys called Frank-Kasper compounds and "quasi-crystals", a solid phase also characterized by BOO, have recently been observed by Schechtman et al. [65].

Icosohedral symmetry arises from a dense, non-crystallographic packing of equal spheres. One outstanding question is, how do you put a cholesteric into a sphere? Several schemes [41,66,67], some not requiring a singularity in the volume of the sphere have been proposed. Next we consider some general aspects of defects in cholesterics.

C. Some Remarks on Cholesteric Defects

Assuming blue phases are stable arrays of cholesteric defects, it is natural to inquire what are the defects of cholesteric phases? As yet a complete theoretical understanding of cholesteric defects [3] eludes us but the recent homotopic classification of defects [4,5] tells us that, if periodicity does not count, there are 4 classes of cholesteric defects and the quaternions, [1,-1,i,j,k], are a one dimensional representation of these classes. The uniform state is represented by 1. "Double twist" is in the class -1. Cholesteric defects, previously proposed by Kleman and Friedel [3] on the basis of a Volterra construction, fit nicely into this picture†. The interesting feature of this classification is that the first defect homotopic to the ground state has strength, $S = 2$. This is in contrast to the nematic where $S = 1$ is homotopic to the uniform configuration and therefore "escapes into the third dimension". Indeed, many of the textures seen in cholesteric liquid crystals are topologically similar to the $S = 2$ class of defects but a single "double twist" has never been observed.

For example, Fig. (4) shows a lattice texture obtained by applying an electric field perpendicular to the plane of the figure [4]. The sample (cholesterol propionate) was originally uniformly oriented with the director in the plane of the glass plates and thus, with the twist axis perpendicular to the plane of the figure. In the middle of each of the spherical objects as well as around them, the director is parallel to the field direction. The simplest director pattern compatible with these boundary conditions, a toroidal loop of "double twist", is topologically similar to an $S = 2$. Another example is the Frank-Pryce sphere [66] in which an $S = 2$ defect is projected onto the surface of a sphere giving rise to a single radial defect. Bouligand [67] has shown that double

---

† Neither the homotopic classification nor the Volterra method predicts the most commonly observed defects of cholesteric and smectic phases, focal conics. These amusing objects, invented by the mathematician, Dupin, are composed of equidistant surfaces. The fractal nature of focal conics when filling polyhedral spatial domains has been pointed out by Bidaux et al. and commented upon by Mandelbrot in his book [68].

spirals observed frequently in cholesterics of large pitch are a natural feature when this construction is sliced by a plane.

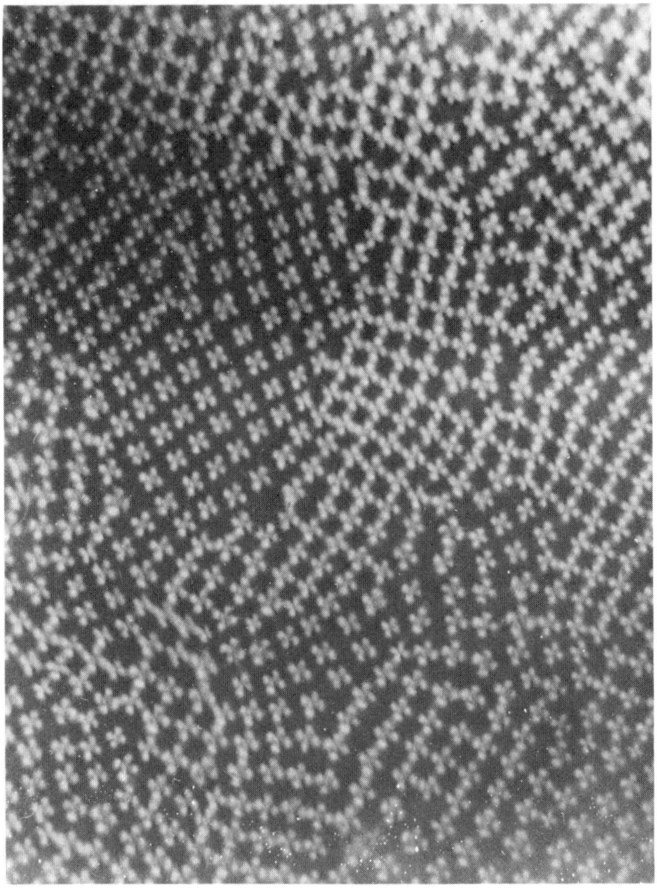

More examples can be found in the recent paper of Bouligand and Livolant [67], the paper by Cladis, White and Brinkman [7] as well as in the book by Kleman [3]. There is no energy calculation available to explain this fact nor has it been shown by a simple drawing that the S = 2 can be continuously deformed to produce the ground state.

Conclusion: the most frequently occurring defect in cholesterics is topologically similar to S = 2.

D. The Orienting Effect of the Cholesteric-Isotropic Interface

Presumably, to stabilize the interface [69], the director inclination at an isotropic-liquid crystal interface is constant. This is graphically seen in the photomicrographs of Volovik and Lavrentovich [70] shown to us by Fergason [71]. In these pictures, the homeotropic (perpendicular) director orientation at the nematic-isotropic liquid interface, changes at a point first. This squeezes the central point defect towards the drop surface where it eventually transforms to two point defects located at the poles of the sphere. The surface boundary condition drives the bulk configuration ... as is expected. In other words, the isotropic liquid orients the director in the interface in much the same way an electric or magnetic field does and this determines the equilibrium configuration inside the sphere.

Viewing blue phases as an ordered array of defects then makes them analogous to the vortex phase of type II superconductors which Abrikosov found as periodic solutions of the Ginzburg - Landau equations for superconductors [58]. In analogy, Brazovskii [19] found periodic blue phase solutions of de Gennes' free energy [16]. The analogue of the magnetic field in superconductors is the boundary condition imposed on the director at the isotropic - liquid crystal interface. Of course an electric field is an exact counterpart to a magnetic field so that the lattice texture shown in Fig. (3) is a simple analogue of the superconducting lattice except that the lattice is square rather than triangular as it is in the superconducting case. This agrees with the analysis of Delrieu [72].

Conclusion: it is important to know the boundary conditions when filling a sphere (or any space for that matter) with cholesteric.

IV. Conclusions

Although the first thermotropic liquid crystals ever observed, blue phases were virtually ignored until about ten years ago. They appear to be the periodic extremals of the Landau-de Gennes free energy for cholesterics, in much the same way the vortex lattice is a periodic extremal of the Ginzburg-Landau equations for superconductors.

The experimental evidence is that BPI and BPII are soft crystals with large unit cells. They occur at higher temperatures than the liquid cholesteric phase and are believed to be stable defect arrays. Finn and Cladis [41] have suggested that the BPI-BPII transition is a percolation threshold: in BPI the continuous phase is cholesteric whereas in BPII it is isotropic liquid. Frank [62] has pointed out that a bicontinuous phase cannot be excluded and BPII is a candidate for that. Recently Anderson [64] has suggested BPIII is a phase with 3-d bond orientational order which would account for the large heat of transition accompanying the transformation of BPIII to the isotropic liquid state. One outstanding question is: how do you fill a sphere with cholesteric? Pryce and Frank [66] provided us with one ingenious solution subject to a particular boundary condition. It seems likely that a better theory of defects in cholesterics is needed to sort through all the possibilities.

Because of their large unit cells, crystals composed of only 20^3 or so unit cells of BPI or BPII are visible in the optical microscope providing us with a rare opportunity to observe directly mechanisms and dynamics of crystal growth.

Acknowledgements

It is a pleasure to acknowledge stimulating discussion with my fellow workshop participants. The hospitality and support of the Institute for Mathematics and its Applications of the University of Minnesota is gratefully acknowledged. This work was partially supported by the NSF Industry/University Cooperative Research Activity DMR-8404942 Materials Research.

References

1. P.G. de Gennes, The Physics of Liquid Crystals, Oxford Univeristy Press (1974).

2. F.B. Rosevear, J. Amer. Oil Chem. Soc. 31. 68 (1954).

3. M. Kléman, Points, Lines and Walls, John Wiley and Sons, New York (1983).

4. G.E. Volovik and V.P. Mineyev, Sov. Phys. JETP Lett. 24, 561 (1976).

5. N.D. Mermin, Rev. Mod. Phys. 51, 591 (1979).

6. J.P. Sethna, Phys. Rev. Lett. 51, 2198 (1983); J.P. Sethna, D.C. Wright and N.D. Mermin, Phys. Rev. Lett. 51, 467 (1983).

7. P.E. Cladis, A.E. White and W.F. Brinkman, J. Physique. 40, 325 (1979).

8. R. Barbet-Massin, P.E. Cladis and P. Pieranski, La Recherche 154, 548 (1984).

9. W.F. Brinkman and P.E. Cladis, Physics Today 35, 48 (1982).

10. Y. Bouligand, B. Derrida, V. Poenaru, Y. Pomeau and G. Toulouse, J. Physique 39, 863 (1978).

11. D. Coates and G.W. Gray, Phys. Lett. 45A, 115 (1973).

12. H. Stegemeyer and K. Bergmann, in Liquid Crystals of One and Two Dimensional Order, edited by W. Helfrich and A. Heppke (Springer, Berlin, 1980), p. 161.

13. G. Friedel, Ann. Phys. 18, 273 (1922).

14. I.G. Chystiakov and L.A. Gusakova, Kristallografiya 14, 153 (1969).

15. P.P. Crooker, Mol. Cryst. Liq. Cryst. 98, 31 (1983).

16. P.G. de Gennes, Mol. Cryst. Liq. Cryst. 12, 193 (1971).

17. S. Shtrikman, Letter to D. Mukamel, Sept. 20, 1971. I thank Professor Shtrikman for sharing this information with me, P.G. de Gennes, European Conference on Smectics, Madonna di Campiglio (1976) (unpublished).

18. S.A. Brazovskii, Sov. Phys. JETP 41, 85 (1975).

19. S.A. Brazovskii and S.G. Dmitriev, Sov. Phys. JETP 41, 497 (1976); S.A. Brazovskii and V.M. Filev, Sov. Phys. JETP 48, 573 (1978).

20. H. Grebel, R.M. Hornreich and S. Shtrikman, Phys. Rev. A28, 1114 (1983).

21. D. Armitage and F.P. Price, J. Chem. Phys. 66, 3414 (1977); J. Physique. 36, C1-133 (1975).

22. K. Bergmann and H. Stegemeyer, Z. Naturforsch. 34A, 1031 (1979); 251 (199); K. Bergmann, P. Pollmann, G. Scherer and H. Stegemeyer, Z. Naturforsch. 34A 253 (1979); P. Pollmann and G. Scherer, Z. Naturforsch. 34A, 255 (1979).

23. H. Onusseit and H. Stegemeyer, Z. Naturforsch. 39A, 658 (1984).

24. J.W. Goodby and T.M. Leslie, (private communication).

25. R. Kleiman, D.J. Bishop, R. Pindak and P. Taborek, Phys. Rev. Lett. 53, 2137 (1984).

26. A. Saupe, Mol. Cryst. Liq. Cryst. 7, 59 (1969).

27. S. Meiboom, J.P. Sethna, P.W. Anderson and W.F. Brinkman, Phys. Rev. Lett. 46, 1216 (1981); S. Meiboom, M. Sammon and W.F. Brinkman, A27, 438 (1983).

28. M. Hornreich and S. Shtrikman, (NMR) Phys. Rev. 28A, 2544 (1983); (light scattering) Phys. Rev. 28A, 1791 (1983); (structure) Liquid Crystals of One and Two Dimensional Order, edited by W. Helfrich and G. Heppke (Springer, Berlin, 1980) p. 185; Phys. Rev. 24A, 635 (1981); Phys. Lett. 84A, 20 (1981); 82A, 345 (1981).

29. R.M. Hornreich, M. Kugler and S. Shtrikman, Phys. Rev. Lett. 48, 1404 (1982).

30. H. Kleinert and K. Maki, Fortschritte der Physik 29, 219 (1981).

31. David C. Wright, Cornell University PhD Thesis, Ithaca, New York (1983).

32. David C. Wright and N.D. Mermin, to appear in Rev. Mod. Phys.

33. S. Meiboom and M. Sammon, Phys. Rev. Lett. 44, 882 (1980); Phys. Rev. A24, 468 (1981); S. Meiboom, M. Sammon and D.W. Berreman, Phys. Rev. A28, 3553 (1983).

34. D.L. Johnson, J.H. Flack and P.P. Crooker, Phys. Rev. Lett. 45, 641 (1980).

35. J. Her, B.B. Rao and J.T. Ho, Phys. Rev. A24, 3272 (1981).

36. A.J. Nicastro and P.H. Keyes, Phys. Rev. A27, 431 (1983).

37. V.A. Kizel' and V.V. Prokhorov, Sov. Phys. JETP 38, 25 (1983).

38. V.A. Belyakov, V.E. Dmitrenko, S.M. Osadchii, Sov. Phys. JETP 56, 322 (1982). D. Bensimon, E. Domany and S. Shtrikman, Phys. Rev. 28A, 427 (1983); E.I. Demikov and V.K. Dolganov, Sov. Phys. JETP Lett. 38, 445 (1983); V.K. Dolganov, S.P. Krylov and V.M. Filev, Sov. Phys. JETP 57, 1177 (1980).

39. K. Tanimoto and P.P. Crooker, Phys. Rev. A29, 1566 (1984).

40. D. Armitage and R.J. Cox, Mol. Cryst. Liq. Cryst. 64, 41 (1980).

41. P.L. Finn and P.E. Cladis, Mol. Cryst. Liq. Cryst. 84, 159 (1982); Mol. Cryst. Liq. Cryst. Lett. 72, 107 (1981).

42. G. Heppke, M. Krumrey and F. Oestreicher, Mol. Cryst. Liq. Cryst. 99, 99 (1983).

43. H. Stegemeyer and F. Porsch, Phys. Rev. A30, 3369 (1984).

44. T. Yamada and E. Fukuda, Jap. J. Appl. Phys. 12, 68 (1973).

45. P.H. Keyes and D.B. Ajgaonkar, Phys. Lett. 64A, 298 (1977).

46. H. Stegemeyer and P. Pollmann, Mol. Crystl. Liq. Cryst. 82, 123 (1982).

47. N.A. Clark, S.T. Vora and M.A. Handschy, Phys. Rev. Lett. 52, 57 (1984).

48. P.E. Cladis, P. Pieranski and M. Joanicot, Phys. Rev. Lett. 52, 542 (1984).

49. M. Marcus, J. Physique 42, 61 (1981).

50. H. Onusseit and H. Stegemeyer, Z. Naturforsch 36A, 1083 (1981); 38a, 1114 (1983); J. Cryst. Growth 61, 409 (1983); M. Marcus, Phys. Rev. 25A, 2272 (1982).

51. Th. Blumel and H. Stegemeyer, J. Cryst. Growth 66, 163 (1984).

52. R. Barbet-Massin and P. Pieranski, J. Physique Lett. 45, L799 (1984).

53. R. Barbet-Massin, P.E. Cladis and P. Pieranski, Phys. Rev. A30, 1161 (1984).

54. P. Pieranski, R. Barbet-Massin and P.E. Cladis, Phys. Rev. A. 31, 3912 (1985).

55. P. Pieranski and B. Pansu, J. de Physique Coll. 46, (C3-28) (1985).

56. F.C. Frank, Disc. Faraday Soc., 5, 48 (1949).

57. P.A. Windsor in Liquid Crystals and Plastic Crystals Vol. I, edited by G.W. Gray and P.A. Winsor, John Wiley and Sons, New York (1974) p. 199.

58. P.G. deGennes, Superconductivity, W.A. Benjamin Press, New York (1966) p. 55.

59. F.C. Frank, Disc. Faraday Soc. 25, 19 (1958).

60. J.L. Ericksen, in Liquid Crystals and Ordered Fluids, Plenum Press (1970) p. 181.

61. M. O'Keefe and S. Andersson, Acta. Cryst. A33, 914 (1977); P. Saludjian, F. Reiss-Husson, PNAS (USA) 77, 6991 (1980); V. Luzzatti, A. Tardieu and T. Gulik-Krzywicki, PNAS (USA)78, 4683 (1981).

62. F.C. Frank, private communication.

63. L.E. Scriven, P.G. deGennes, Microemulsions.

64. P.W. Anderson (unpublished). For references on BOO see P.J Steinhardt, D.R. Nelson and M. Rouchetti, Phys. Rev. B28, 784 (1983); D. Levine and P.J. Steinhardt Phys. Rev. Lett. 53, 2477 (1984).

65. A.L. Mackay, Acta Cryst. 15, 916 (1962), Izvj. Jugosl. centr. Krist: (Zagreb) 10, 15 (1975); F.C. Frank and J.S. Kasper, Acta Cryst. 11, 184 (1958); ibid. 12, 483 (1959); D. Shechtman, I. Bleech, D. Gratias and J.W. Cahn, Phys. Rev. Lett. 53, 1951 (1984).

66. M. Pryce and F.C. Frank in C. Robinson and J.C. Beevers article, Disc. Faraday Soc. 25, 29 (1958).

67. Y. Bouligand (unpublished); Y. Bouligand and F. Livolant, J. de Physique 45, 1899 (1984).

68. see for example: J. Sethna and M. Kleman, Phys. Rev. A26, 3037 (1982); R. Bidaux, N. Boccara, G. Sarma, L. Seze, P.G. deGennes and O. Parodi, J. de Physique 34, 661 (1973); B.B. Mandelbrot, The Fractal Geometry of Nature, W.H Freeman and Company, New York (1983) p. 176. See also E.N. Gilbert, Ann. Math. Statistics, 33, 958 (1962) and Canadian J. Math 16, 286 (1964).

69. J.L. Ericksen, Arch. Ratl. Mech. Anal. 4, 231 (1960).

70. G.E. Volovik and O.D. Lavrentovich, JETP 85, 12 (1983).

71. J. Fergason, this workshop proceedings.

72. J.M. Delrieu, J. Chemical Physics 60, 1081 (1974).

MINIMUM ENERGY CONFIGURATIONS FOR LIQUID CRYSTALS: COMPUTATIONAL RESULTS

Robert Cohen[1], Robert Hardt[2],

David Kinderlehrer[3], San-Yih Lin[1], and Mitchell Luskin[1]

School of Mathematics
University of Minnesota
Minneapolis, MN 55455

Abstract

Two numerical algorithms which have been successfully employed to compute minimum energy configurations for liquid crystals are given. The results of computational experiments using these algorithms show that several critical points of the energy functional are not local minima.

1. Introduction

In the Ericksen-Leslie continuum theory for nematic liquid crystals, the bulk energy is given by [3]

$$W(n) = \int_\Omega W(\nabla n, n)dx \qquad (1.1)$$

where $n = n(x)$ is the unit vector which describes the optical axis at the point $x \in \Omega$, $\Omega \quad R^3$ is the volume occupied by the material, and $W(\nabla n,n)$ is the Oseen-Frank energy density. This energy density is given by

$$2W(\nabla n, n) = \kappa_1(\text{div } n)^2 + \kappa_2(n \cdot \text{curl } n)^2 + \kappa_3|n \wedge \text{curl } n|^2 + (\kappa_2 + \kappa_4)(\text{tr}(\nabla n)^2 - (\text{div } n)^2),$$

where the constants κ_i, $i = 1,2,3,4$ are usually assumed to satisfy

$$\kappa_1 > 0, \ \kappa_2 > 0, \ \kappa_3 > 0, \ \kappa_2 \geq |\kappa_4|.$$

[1]Supported by the NSF, Grant DMS 83-51080.
[2]Supported by the NSF, Grant DMS 85-11357.
[3]Supported by the NSF, Grant MCS 83-01345.
[4]Supported by the NSF, Grant DMS 83-01575.

The static equilibrium configurations which we observe are presumably configurations which at least locally minimize the bulk energy subject to the imposed boundary conditions, such as

$$n(x) = u(x), \qquad x \in \partial\Omega \, , \tag{1.2}$$

where u is a given unit vector defined on $\partial\Omega$. Both the theoretical and the numerical analysis of this problem are difficult because the admissible configurations, $n(x)$, must satisfy the non-convex constraint

$$|n(x)| = \sqrt{n_1^2 + n_2^2 + n_3^2} = 1, \tag{1.3}$$

for all $x \in \Omega$. Although solutions to elliptic variational problems are typically smooth, critical points and minima of (1.1) subject to the constraint (1.3) typically have singularities (see below). Further difficuluties are due to the fact that although the energy density is quadratic in the first derivatives of n, the coefficients depend on n. For simplicity of exposition, we shall restrict our focus in this paper to the energy density for $\kappa_1 = \kappa_2 = \kappa_3 = 1$ and $\kappa_4 = 0$. In this special case,

$$W(\nabla n, n) = \frac{1}{2} |\nabla n|^2, \tag{1.4}$$

and (1.4) is the integrand for a harmonic mapping of Ω into S^2. Although the choice of constants which gives (1.4) is not likely to occur in material systems, (1.4) is a useful model to use to discover qualitative features about minima for realistic constants. In what follows, we shall assume that

$$W(\nabla n, n) = \frac{1}{2} |\nabla n|^2$$

unless specific reference is made to the general Oseen-Frank energy density.

A complete classification of solutions to the Euler-Lagrange equations for critical points of (1.1) subject to the constraint (1.3) which are homogeneous of degree zero, i.e.,

$$n(x) = n\left(\frac{x}{|x|} \right)$$

is known. These solutions are

$$n(x) = \pi^{-1} \circ r \circ \pi\left(\frac{x}{|x|}\right) \qquad (1.5)$$

where π: $S^2 - \{(0,0,1)\} \to \mathbb{C}$ is the stereographic projection and r is a rational function of z or $\bar{z}$, i.e.,

$$r(z) = p(z)/q(z)$$

or

$$r(z) = p(\bar{z})/q(\bar{z})$$

for polynomials $p(z)$, $q(z)$.

At the time of the commencement of this work it was not known which of these critical points were in fact minima. It had recently been shown by Hardt, Kinderlehrer, and Lin, c.f. [5] that there was a maximum degree of $p(z)$ and $q(z)$ for which $n(x)$ of (1.5) could be a minimum. The computations that we present in this paper demonstrated that $r(z) = z$ was a local minimum, but that $r(z) = z^2$, $r(z) = \bar{z}^2$, $r(z) = z^3$, and $r(z) = \frac{z}{2}$ were not local minima. Partially motivated by our computational results, Brezis, Coron, and Lieb [2] have now been able to give a complete classification of the singularities of minima of (1.1) subject to the constraint (1.3), cf. Brezis [1].

In section 2 we shall describe a relaxation method for computing local minima, and in section 3 we shall describe a fractional step method. Finally, in section 4, we shall present and describe the computational experiments that we have done on the Cray-2 at the University of Minnesota.

2. Relaxation Method

For simplicity of exposition, let $\Omega = (0,1) \times (0,1) \times (0,1)$ be a unit cube. Set $h = 1/(M+1)$ and define

$$x_{ijk} = (ih, jh, kh)$$

for integers $0 \leq i,j,k \leq M+1$. Denote our approximation to $\underline{n}(\underline{x})$ by

$$n_{ijk} \sim n(x_{ijk}) .$$

We approximate the bulk energy function $W(n)$ by

$$\tilde{W}(n) = \sum_{i=0}^{M} \sum_{j=0}^{M} \sum_{k=0}^{M} \left(\left| \frac{n_{i+1,jk} - n_{ijk}}{h} \right|^2 h + \left| \frac{n_{i,j+1,k} - n_{ijk}}{h} \right|^2 h + \left| \frac{n_{ij,k+1} - n_{ijk}}{h} \right|^2 h \right)$$

We are searching for local minima of $\tilde{W}(n)$ subject to the non-convex constraint that

$$|n_{ijk}| = 1 \qquad \text{for} \qquad 0 \leq i,j,k \leq M+1 \tag{2.1}$$

and the boundary constraint that

$$n_{ijk} = u(x_{ijk}) \qquad \text{for} \qquad x_{ijk} \in \partial\Omega. \tag{2.2}$$

We can use the method of Lagrange multipliers to remove the constraint (2.1).

Local minima of $\tilde{W}(n)$ subject to the constraints (2.1) and (2.2) are critical points of

$$\tilde{W}(n, \lambda) = W(\tilde{n}) + L(n,\lambda) \tag{2.3}$$

subject to the constraint (2.2) where

$$L(n,\lambda) = \sum_{i=1}^{M} \sum_{j=1}^{M} \sum_{k=1}^{M} \lambda_{ijk} \left(|n_{ijk}|^2 - 1 \right) h .$$

The Euler-Lagrange equations for the functional (2.3) are

$$-\tilde{\Delta}\, n_{ijk} = \lambda_{ijk}\, n_{ijk}, \qquad i,j,k=1,\ldots, M,$$

$$|n_{ijk}|=1, \qquad i,j,k=1,\ldots, M, \tag{2.4}$$

$$n_{ijk} = u(x_{ijk}), \qquad x_{ijk} \varepsilon\ \partial\Omega,$$

where

$$-h^2\tilde{\Delta}\, n_{ijk} = (-n_{i+1,jk} + 2n_{ijk} - n_{i-1,jk})$$

$$+ (-n_{i,j+1,k} + 2n_{ijk} - n_{i,j-1,k}) + (-n_{ij,k+1} + 2n_{ijk} - n_{ij,k-1}).$$

We now describe the Gauss-Seidel relaxation method. Let n_{ijk}^{ℓ} be the ℓth iterate. If n_{ijk}^{ℓ} has been computed, we compute $n_{ijk}^{\ell+1}$, $\lambda_{ijk}^{\ell+1}$ from

$$(-n_{i+1,jk}^{\ell} + 2n_{ijk}^{\ell+1} - n_{i-1,jk}^{\ell+1}) + (-n_{i,j+1,k}^{\ell} + 2n_{ijk}^{\ell+1} - n_{i,j-1,k}^{\ell+1})$$

$$+ (-n_{ij,k+1}^{\ell} + 2n_{ijk}^{\ell+1} - n_{ij,k-1}^{\ell+1})$$

$$= h^2\, \lambda_{ijk}^{\ell+1}\, n_{ijk}^{\ell+1}, \qquad i,j,k=1,\ldots,M,$$

$$|n_{ijk}^{\ell+1}|=1, \qquad i,j,k=1,\ldots M, \tag{2.5}$$

$$n_{ijk}^{\ell+1} = n_{ijk}^{\ell} = u\,(x_{ijk}), \qquad x_{ijk}\ \varepsilon\ \partial\Omega.$$

We can eliminate $\lambda_{ijk}^{\ell+1}$ from (2.5) to obtain the algorithm

$$m_{ijk}^{\ell+1} = \frac{1}{6}\,(n_{i-1,jk}^{\ell+1} + n_{i,j-1,k}^{\ell+1} + n_{ij,k-1}^{\ell+1} + n_{i+1,jk}^{\ell}$$

$$+ n_{i,j+1,k}^{\ell} + n_{ij,k+1}^{\ell}), \quad i,j,k=1,\ldots,M,$$

$$n_{ijk}^{\ell} = n_{ijk}^{\ell+1} = u(x_{ijk}), \qquad x_{ijk}\ \varepsilon\ \partial\Omega, \tag{2.6}$$

$$n_{ijk}^{\ell+1} = \frac{m_{ijk}^{\ell+1}}{|m_{ijk}^{\ell+1}|}, \qquad i,j,k=1,\ \ldots,M\ .$$

We note that the first step of our method is the classical, unconstrained Gauss-Seidel method and the second step is a normalization which imposes the constraint. Our Gauss-Seidel method also shares with the unconstrained Gauss-Seidel method the energy decreasing inequality

$$\tilde{W}(n^{\ell+1}) \leq \tilde{W}(n^{\ell}) \quad . \tag{2.7}$$

The inequality (2.7) guarantees the stability of our method.

We also need to make some comments on the possible singularity of our algorithm when $m_{ijk}^{\ell+1} = 0$.

When

$$m_{ijk}^{\ell+1} = \frac{1}{6} \left(n_{i-1,jk}^{\ell+1} + n_{i,j-1,k}^{\ell+1} + n_{ij,k-1}^{\ell+1}\right.$$

$$\left. + n_{i+1,jk}^{\ell} + n_{i,j+1,k}^{\ell} + n_{ij,k+1}^{\ell} \right) = 0,$$

the approximate solution must be highly oscillatory. It is reasonable to assume that there is a singularity at x_{ijk} and that therefore we can assign $n_{ijk}^{\ell+1}$ an arbitrary normalized value.

It is well-known that for unconstrained minimization problems an order of magnitude improvement in the convergence rate can be obtained by over-relaxation. We have used over-relaxation for our constrained problem to also obtain a significant improvement in the convergence rate. Our over-relaxation method, with relaxation parameter ω, $1 < \omega < 2$, is defined by

$$m_{ijk}^{\ell+1} = \frac{1}{6} \left(n_{i-1,jk}^{\ell+1} + n_{i,j-1,k}^{\ell+1} + n_{i,j,k-1}^{\ell+1} + n_{i+1,jk}^{\ell}\right.$$

$$\left. + n_{i,j+1,k}^{\ell} + n_{ij,k+1}^{\ell}\right), \quad i,j,k=1,\ldots, M,$$

$$n_{ijk}^{\ell} = n_{ijk}^{\ell+1} = u(x_{ijk}), \qquad x_{ijk} \in \partial\Omega, \tag{2.8}$$

$$\tilde{n}_{ijk}^{\ell+1} = \omega\, m_{ijk}^{\ell+1} + (1-\omega) n_{ijk}^{\ell}, \qquad i,j,k=1,\dots,M,$$

$$n_{ijk}^{\ell+1} = \frac{\tilde{n}_{ijk}^{\ell+1}}{|\tilde{n}_{ijk}^{\ell+1}|}, \qquad i,j,k=1,\dots,M.$$

Finally, the relaxation method can be extended for use with the general Oseen-Frank energy density. Our initial computational experience confirms the stability and efficiency of the relaxation method for the general Oseen-Frank energy density.

3. Fractional Step Method

To motivate our fractional step method, we define the set

$$S = \{ n \mid |n_{ijk}| = 1 \text{ for } i,j,k = 0, \dots, M+1\}.$$

Then we define and use formally the indicator function of S [4]

$$I(n) = \begin{cases} 0 & \text{if} \quad n \in S, \\ \infty & \text{if} \quad n \notin S. \end{cases} \tag{3.1}$$

Now n is a local minimum of $\tilde{W}(n)$ subject to the constraints (2.1) and (2.2) if and only if n is a local minimum of

$$\tilde{W}(n) + I(n) \tag{3.2}$$

subject to the boundary constraint (2.2).

Local minima of (3.2) can be found as steady states of the time-dependent problem for $n(t) = n_{ijk}(t)$ given by

$$\frac{\partial}{\partial t} n_{ijk}(t) = -\nabla \tilde{W}(n_{ijk}) - \nabla I(n_{ijk}),$$

$$i,j,k=1, \ldots, M;\ t > 0, \qquad (3.3)$$

$$n_{ijk}(t) = u(x_{ijk}), \qquad x_{ijk} \in \partial\Omega,$$

$$n_{ijk}(0) = \text{initial guess},$$

where $\nabla \tilde{W}$ is the first variation of $\tilde{W}$ with respect to n_{ijk} and ∇I is the (formal) first variation of I with respect n_{ijk}.

We use a fractional step scheme to compute a numerical solution to (3.3). Let $\tau_\ell > 0$ be the ℓth time step and set $t_{\ell+1} = t_\ell + \tau_{\ell+1}$ $(t_0 = 0)$. Then we can compute an approximation $n_{ijk}^{\ell+1}$ to $n_{ijk}(t_{\ell+1})$ by

$$\frac{\tilde{n}_{ijk}^{\ell+1} - n_{ijk}^{\ell}}{\tau_{\ell+1}} = -\nabla \tilde{W}\ (\tilde{n}_{ijk}^{\ell+1}),$$

$$i,j,k=1,\ldots,M; \quad \ell = 0,1,\ldots,$$

$$\frac{n_{ijk}^{\ell+1} - \tilde{n}_{ijk}^{\ell+1}}{\tau_{\ell+1}} = -\nabla I\ (n_{ijk}^{\ell+1}), \qquad (3.4)$$

$$n_{ijk}^{\ell+1} = \tilde{n}_{ijk}^{\ell+1} = u\ (x_{ijk}), \qquad x_{ijk} \in \partial\Omega.$$

We note that the second step of (3.4) is the (formal) Euler-Lagrange equation for the functional

$$J(m) = \frac{1}{\tau_{\ell+1}}\left(\frac{1}{2}\ |\ m\ |^2 + \tau_{\ell+1}\ I\ (m) - \langle\ m,\ \tilde{n}_{ijk}^{\ell+1}\ \rangle\right).$$

Thus, we see that

$$n_{ijk}^{\ell+1} = \frac{\tilde{n}_{ijk}^{\ell+1}}{|\tilde{n}_{ijk}^{\ell+1}|} \quad .$$

Note that the algorithm (3.4) is defined for the general Oseen-Frank free energy density. For the case of the results presented in section 4

($\kappa_1 = \kappa_2 = \kappa_3 = 1, \quad \kappa_4 = 0$), we have that

$$\nabla \mathbf{W}(\tilde{n}) = -\tilde{\Delta}\, \tilde{n} \quad .$$

Hence, in this case we obtain the algorithm

$$\frac{\tilde{n}_{ijk}^{\ell+1} - n_{ijk}^{\ell}}{\tau_{\ell+1}} = \tilde{\Delta}\, \tilde{n}_{ijk}^{\ell+1} \, ,$$

$$i,j,k=1,\ldots,M; \quad \ell=0,1,\ldots,$$

$$n_{ijk}^{\ell+1} = \frac{\tilde{n}_{ijk}^{\ell+1}}{|\tilde{n}_{ijk}^{\ell+1}|} \, , \tag{3.5}$$

$$n_{ijk}^{\ell+1} = \tilde{n}_{ijk}^{\ell+1} = u\,(x_{ijk}) \, , \qquad x_{ijk} \in \partial\Omega.$$

Although the results presented in section 4 use a constant time-step $\tau_\ell = \tau > 0$, it is clear that the steady-state (with respect to some error tolerance) can be obtained with the fewest number of time steps by a variable sequence of time steps.

4. Computational Results

In this section we present and discuss our computational experiments intended to test the stability of the critical points described in section 1.

We tested the stability of the critical points

$$n_r(x) = \pi^{-1} \circ r \circ \pi \frac{x-a}{|x-a|} \qquad (4.1)$$

where $a = (.5, .5, .5)$ and $r(z) = z, z^2, z^2, z^3$, and $z/2$. Thus, we have tested whether (4.1) are local minima of $\tilde{W}(n)$ subject to the constraint

$$|n_{ijk}| = 1, \qquad i,j,k=0,\ldots, M+1, \qquad (4.2)$$

and the boundary constraint

$$u(x_{ijk}) = n_r(x_{ijk}), \qquad x_{ijk} \in \partial\Omega . \qquad (4.3)$$

In each case and for each method, we set

$$n^o_{ijk} = n_r(x_{ijk}).$$

The steady states for each case were identical for each method. Thus in this section we will only present computational results for the fractional step method with $M = 20$ and $\tau = .005$. The iteration was terminated when

$$\tilde{W}(n^{\ell+1} - \tilde{n}^{\ell}) \leq \delta \qquad (4.4)$$

where $\delta = 10^{-7}$.

We recall that $\Omega \subset R^3$ and that $n: \Omega \to S^2 \subset R^3$. Our graphical output will represent the values of n on given planes through Ω. We have chosen symmetry planes so that $n(x)$ is tangent to the given plane for x in that plane. Our graphical output for other planes (which we do not present here) represents the values of n on given planes projected onto that plane. We also give the number, L, of iterations performed before the termination criterion was satisfied and the energy of the configuration, $\tilde{W}(n)$.

We first present in figure 1 the projection of n_z onto the plane $x_3 = 1/2$, $n_z(x_1, x_2, 1/2)$. For initial data

$$n^o_{ijk} = n_z(x_{ijk})$$

the fractional step algorithm converged (satisfied the termination criterion) for $L = 15$. We exhibit n^L_{ijk} in figure 2. Since $n^L_{ijk} \sim n^o_{ijk}$, we can conclude that n_z is a local minimum. Note that although n_z is a solution to the Euler-Lagrange equations for the continuous problem, $n^o_{ijk} = n_z(x_{ijk})$ is not (exactly) a solution to the Euler-Lagrange equations (2.4) for the discrete problem. However, n^L_{ijk} is the solution to the discrete Euler-Lagrange equation (2.4) which approximates n_z. We have done other numerical experiments with initial data

$$n^o_{ijk} = n_z(x_{ijk}) + p_{ijk}$$

where p_{ijk} is a perturbation satisfying

$$p_{ijk} = 0 , \qquad\qquad x_{ijk} \in \partial\Omega ,$$

and we have always found that n^ℓ converges to the solution given in Figure 2.

Next, we present in figure 3 the projection of n_{z^2} onto the plane $x_3 = 1/2$, $n_{z^2}(x_1, x_2, 1/2)$. Denote by n^L_r the solution to the fractional step method with initial data

$$n^o_{ijk} = n_r(x_{ijk})$$

after L iterations (time steps) where L is the iteration at which the method is terminated by (4.4). We present in figure 4, $n^L_{z^2}(x_1, x_2, 1/2)$ for $L = 152$. Our graphical output shows that n_{z^2} has evolved to the lower energy configuration, $n^L_{z^2}$, where $n^L_{z^2}$ has two singularities of "degree 1". There is a singularity which has the asymptotics of n_r for $r(z) = iz$ at a point $(\alpha_1, \alpha_2, 1/2)$ where $\alpha_1 > 1/2$, $\alpha_2 < 1/2$ and there is a singularity with the asymptotics of n_r for $r(z) = -iz$ at a point $(\alpha_1, \alpha_2, 1/2,)$ where $\alpha_1 < 1/2$, $\alpha_2 > 1/2$. Thus, n_{z^2} is <u>not</u> a local minimum.

The computed minimum configuration admits two singularities with the asymptotic behavior of n_{iz} and n_{-iz} which repel each other and which are confined by the boundary conditions. We note that

$$\tilde{W}(n^L_{z^2}) = 24.651 < 27.978 = \tilde{W}(n_{z^2}).$$

The results in figure 5 and figure 6 show that n_{z^3} is also not a local minimum and that the computed minimum configuration has three singularities with the asymptotic behavior of $e^{i\theta_j}z$ for $j=1,2,3$. The resolution of our numerical grid (M=20) is not sufficient for us to accurately estimate θ_j.

The results in figure 7 and figure 8 show that $n_{\bar{z}^2}$ is not a local minimum and that the computed minimum configuration with initial data $n^0 = n_{z^2}$ admits two singularities with asymptotic behavior given by $e^{-\frac{i\pi}{4}}\bar{z}$ and $e^{\frac{i\pi}{4}}\bar{z}$.

Finally, we see in figure 9 and figure 10 that $n_{z/2}$ is not a local minimum and that the singularity given by $r(z) = z/2$ moves to $(1/2, 1/2, x_3)$ for $x_3 > 1/2$ and becomes a singularity with the asymptotic behavior of $\underline{n}_z$. Figures 9 and 10 show $n_{z/2}(x_1, 1/2, x_3)$ and $n^L_{z/2}(x_1, 1/2, x_3)$, the projection of n on the longitudinal plane $x_2 = 1/2$.

Acknowledgements

We would like to thank Professor Jerrald Ericksen and Professor Roland Glowinski for their continuous interest in our research and their advice.

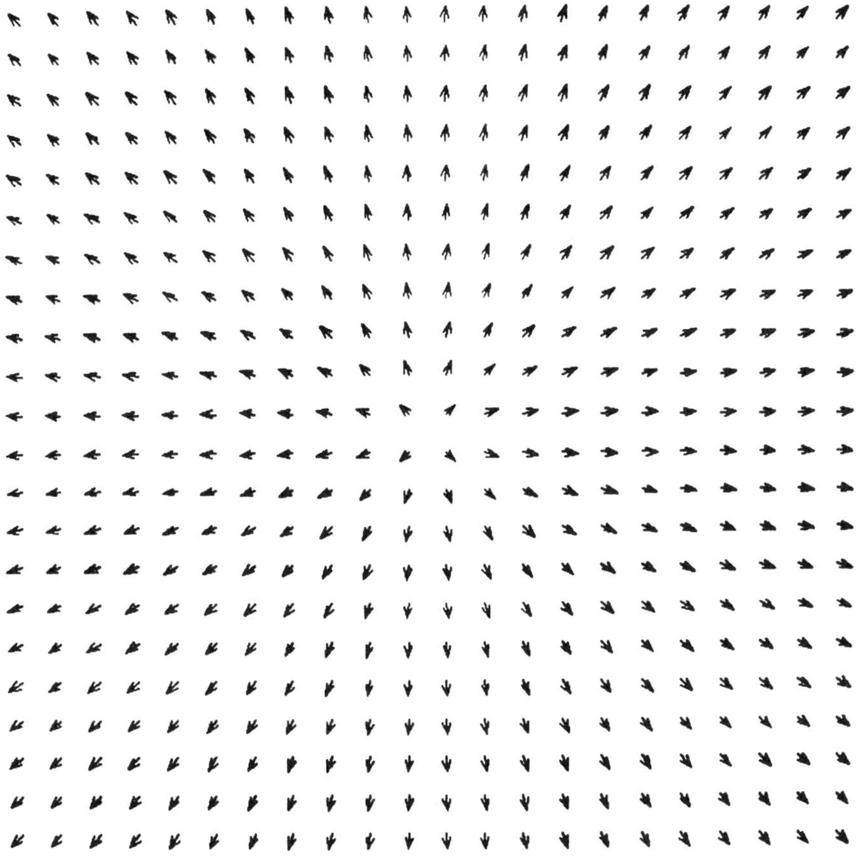

Figure 1. Vector field for $n_z\,(x_1, x_2, 1/2)$. $\tilde{W}(n_z) = 14.711$.

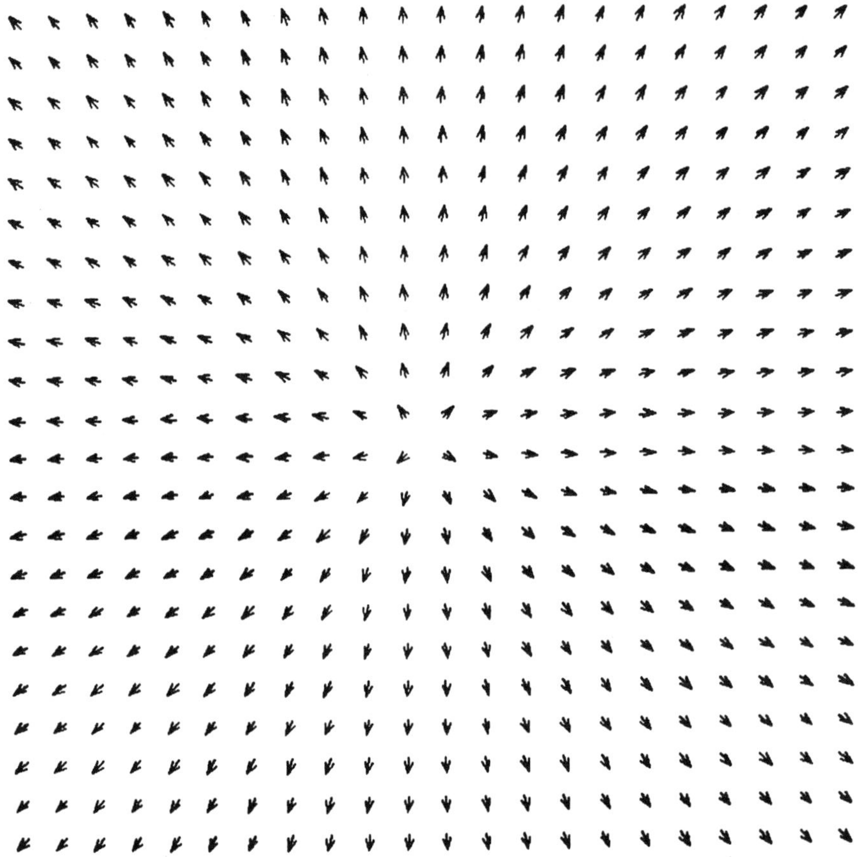

Figure 2. Vector field for $n^L_z\ (x_1, x_2, 1/2)$. $\tilde{W}(n^L_z) = 14.709$. L=15 .

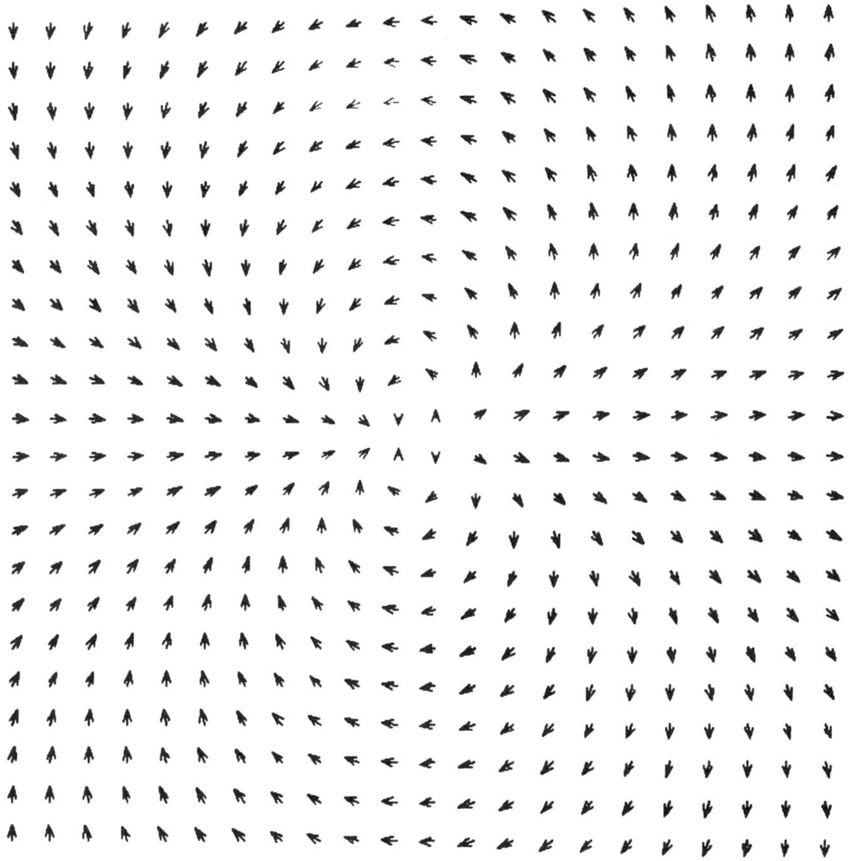

Figure 3. Vector field for $n_z{}^2\,(x_1,x_2,1/2)$. $\widetilde{W}(n_z{}^2) = 27.978$.

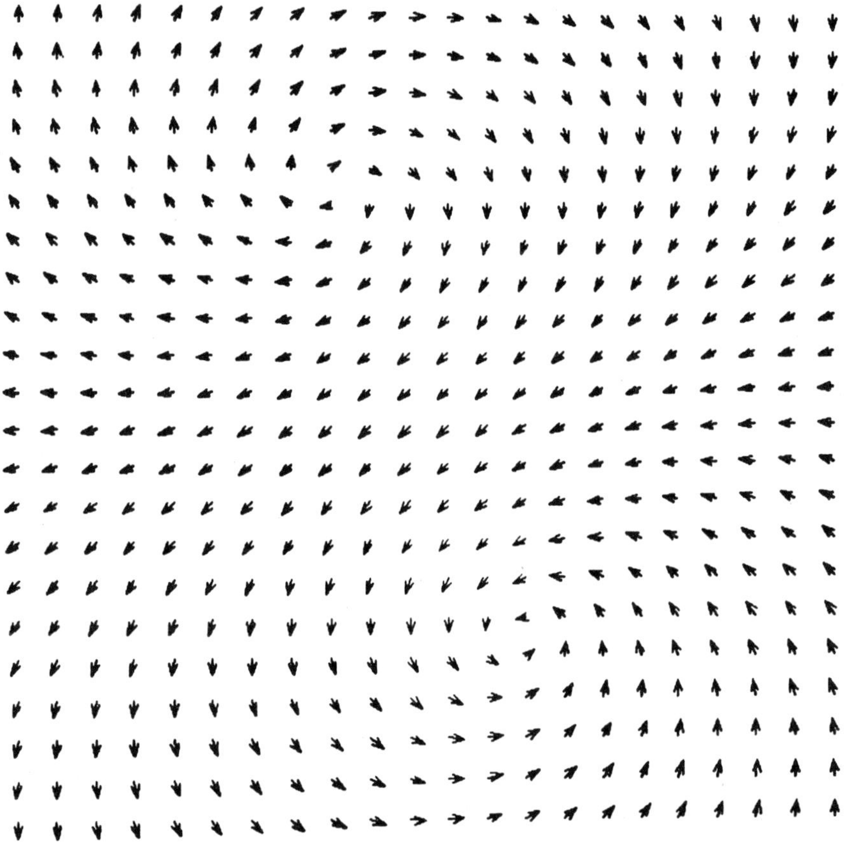

Figure 4. Vector field for $n^L_{z^2}(x_1,x_2,1/2)$. $\widetilde{W}(n^L_{z^2}) = 24.702$. L=152.

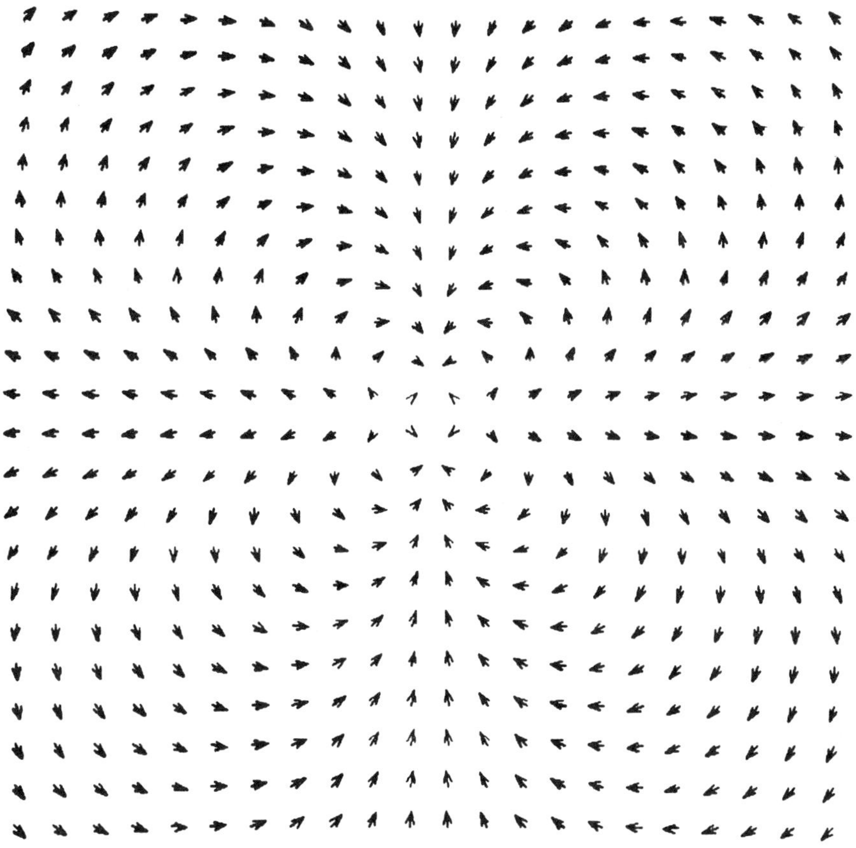

Figure 5. Vector field for $n_{z^3}(x_1, x_2, 1/2)$. $\widetilde{W}(n_{z^3}) = 39.081$.

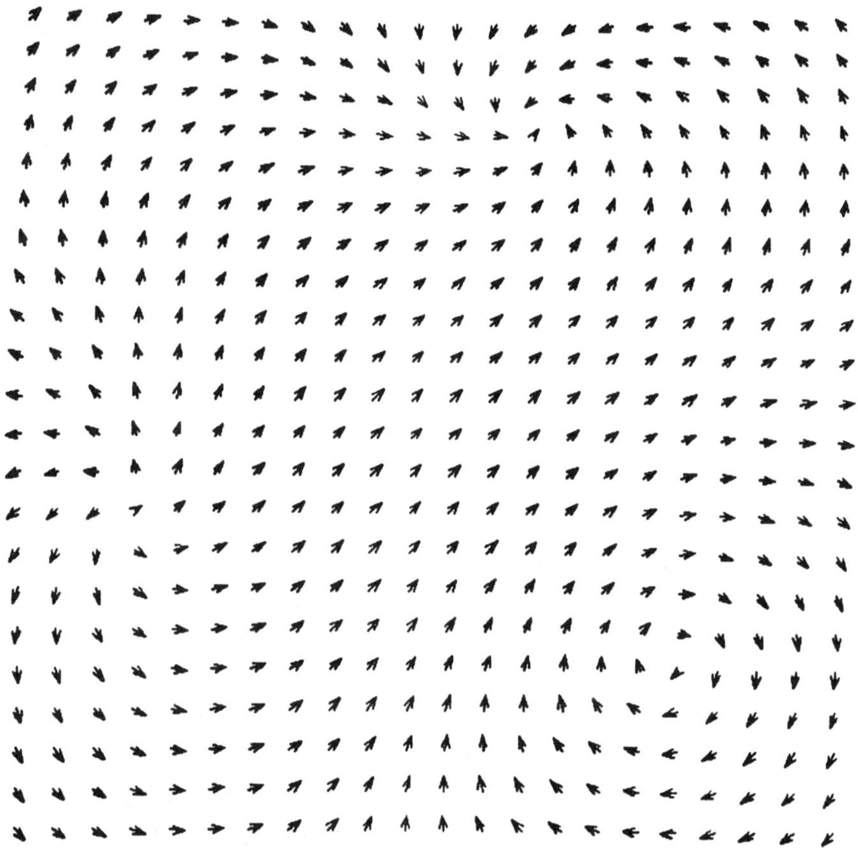

Figure 6. Vector field for $n^L_{z^3}(x_1,x_2,1/2)$. $\widetilde{W}(n^L_{z^3}) = 29.489$. L=114.

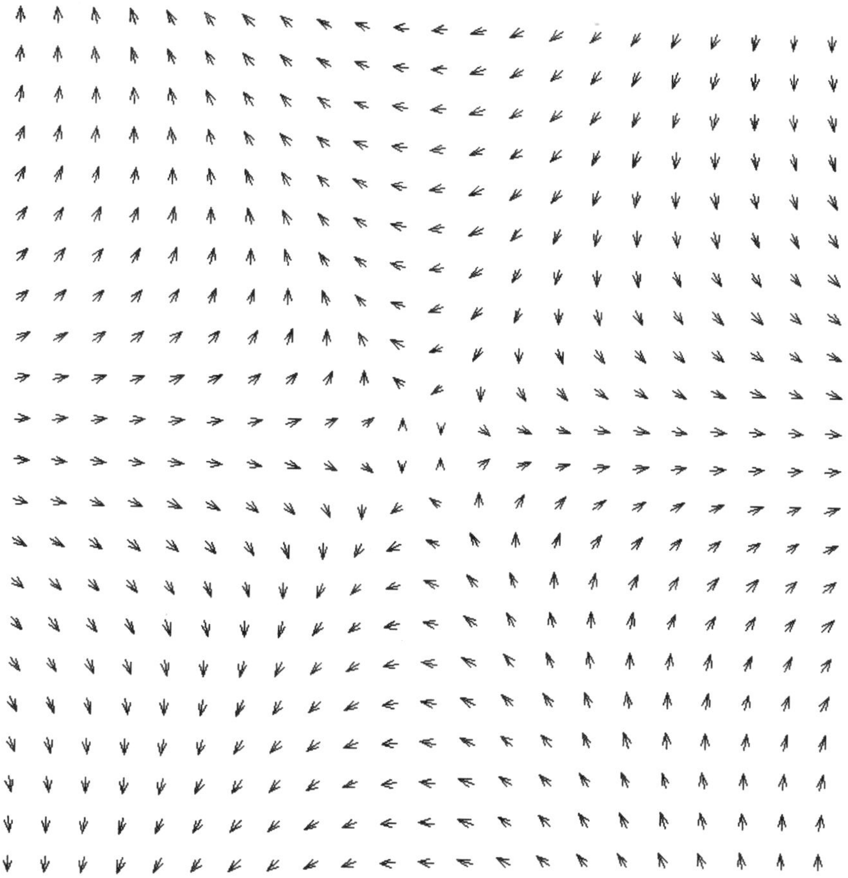

Figure 7. Vector field for $n_{\bar{z}^2}(x_1, x_2, 1/2)$. $\tilde{W}(n_{\bar{z}^2}) = 27.978$.

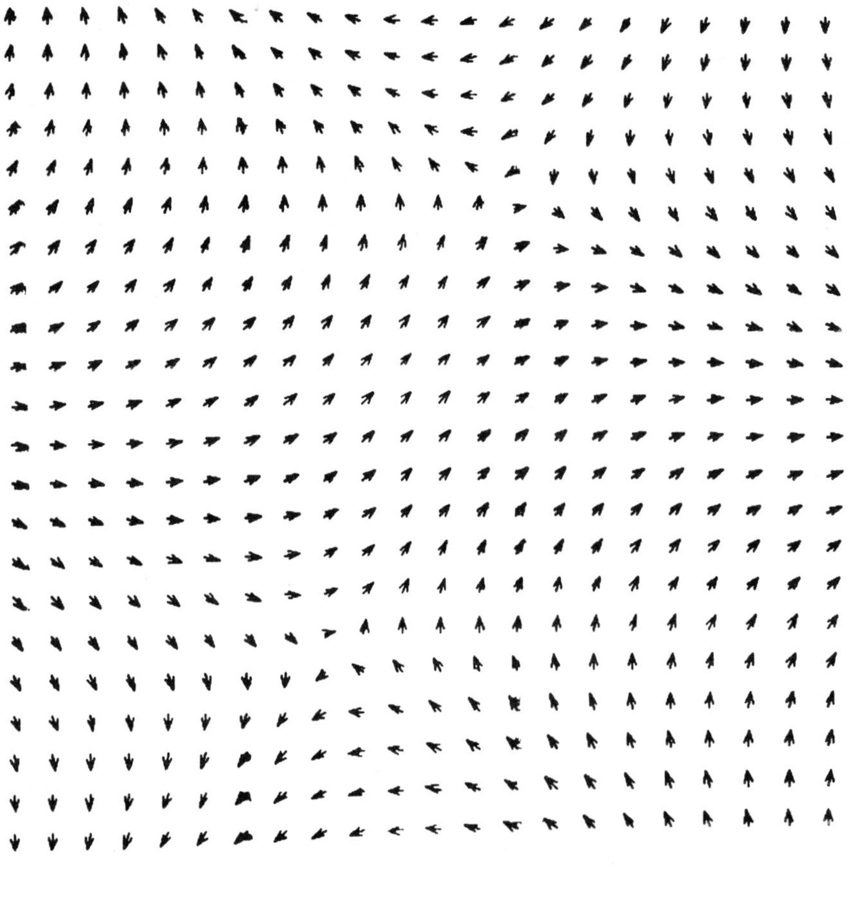

Figure 8. Vector field for $n^L_{\bar{z}^2}(x_1,x_2,1/2)$. $\tilde{W}(n^L_{\bar{z}^2}) = 24.704$. L=134.

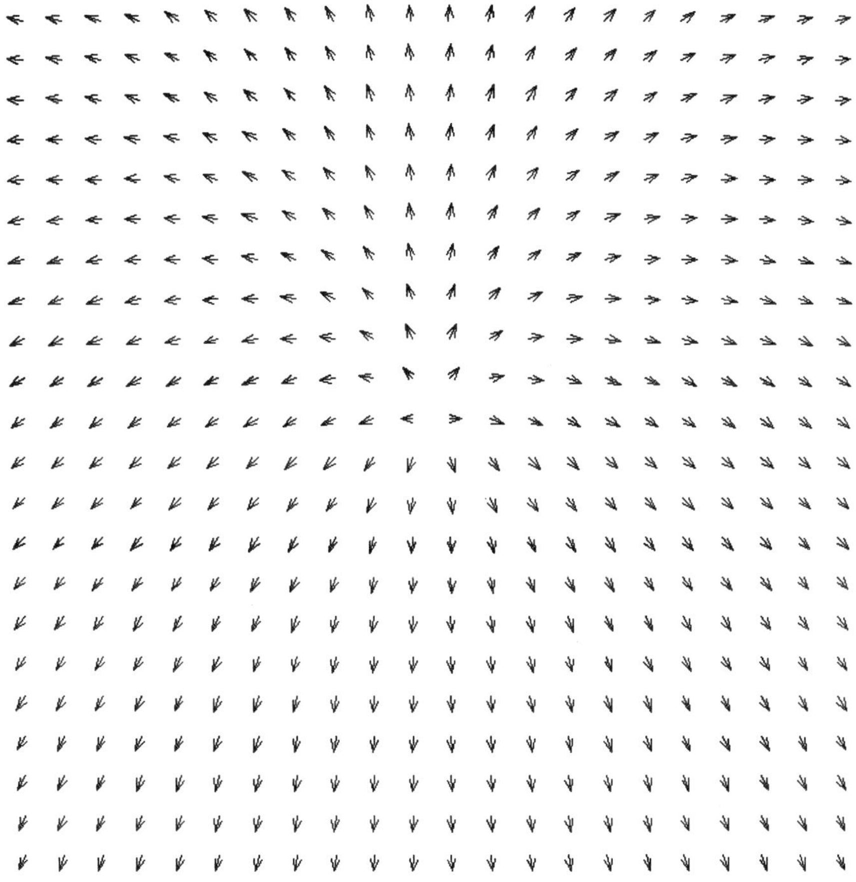

Figure 9. Vector field for $n_{z/2}(x_1, 1/2, x_3)$. $\tilde{W}(n_{z/2}) = 14.682$.

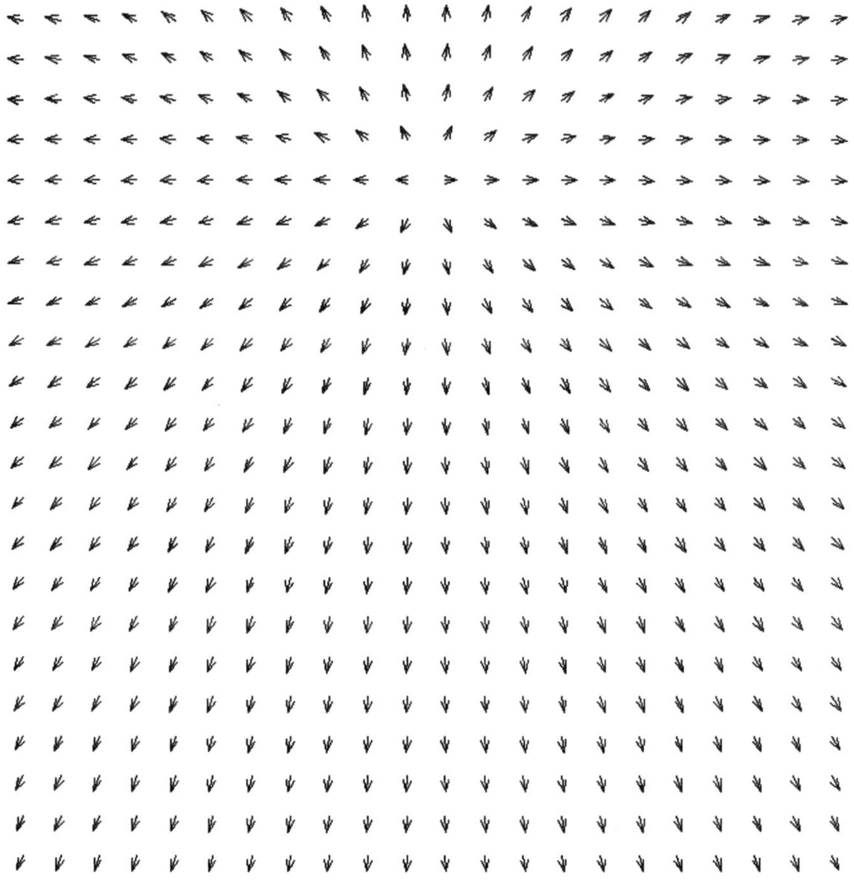

Figure 10. Vector field for $n^{L}_{z/2}(x_1,1/2,x_3)$. $\tilde{W}(n^{L}_{z/2}) = 12.982$. L=73.

References

1. H. Brezis, Liquid crystals and energy estimates for S^2-valued maps, this volume.

2. H. Brezis, J.-M. Coron, E. Lieb, Estimations d'energie pour des applications de R^3 a valeurs dans S^2, CRAS Paris (to appear)

3. J. L. Ericksen, Equilibrium theory of liquid crystals, in Adv. in Liq. Crystals 2, ed. G.H. Brown, Academic Press(1976)

4. R. Glowinski, Splitting methods for the numerical solution of the incompressible Navier-Stokes equations, MRC Report # 2741, 1984.

5. R. Hardt and D. Kinderlehrer, Mathematical questions of liquid crystal theory, this volume.

6. R. Hardt, D. Kinderlehrer, and F.-H.Lin, Existence and partial regularity of static liquid crystal configurations, Comm Math Physics 105 (1986), 547-570.

THE FLOW OF TWO IMMISCIBLE FLUIDS THROUGH A POROUS MEDIUM REGULARITY OF THE SATURATION

E. DiBenedetto

Department of Mathematics
Northwestern University
Evanston, Illinois 60201

1. Introduction

We give here a brief account of some recent results obtained jointly with H.W. Alt [2], concerning the regularity of the saturation in the flow of two immiscible fluids through a porous medium.

Let Ω, the porous body, be a bounded open set in R^N, $N > 2$ with smooth boundary $\partial\Omega$ and let $\Omega_T \equiv \Omega \times (0,T]$, $0 < T < \infty$. We suppose Ω is filled with two immiscible fluids of saturations v_i, hydrostatic pressures p_i, permeabilities $k_i(v_i)$ and gravity forces $\vec{e}_i(v_i)$, $i = 1,2$. Conservation of mass and Darcy's law yield (see [4,5,6]),

(1.1) $$\frac{\partial}{\partial t} v_i - \operatorname{div}\{k_i(v_i)[\nabla p_i + \vec{e}_i(v_i)]\} \quad \text{in} \quad \Omega_T,$$

with side condition

(1.2) $$v_1 + v_2 = 1.$$

We set $v = v_1$ and using (1.2) we redefine the functions $v_i \to k_i(v_i)$, $\vec{e}_i(v_i)$ as functions $v \to k_i(v)$, $\vec{e}_i(v)$. The difference $p_1 - p_2$ is the capillary pressure and it is a function $v \to \gamma(v)$ subject to the condition

(1.3) $$\gamma_{min} \le p_1 - p_2 = \gamma(v) \le \gamma_{max},$$

where γ_{min}, γ_{max} are given and $-\infty \le \gamma_{min} < 0 < \gamma_{max} \le +\infty$. The functions $v \to k_i(v)$, $\gamma(v)$ are given by experiments and their qualitative behaviour is shown in Fig. 1 (see [4,5,6]). Thus the system is degenerate since $k_i \to 0$ as $v_i \to 0$ $i = 1,2$. Because of the experimental and qualitative nature of these graphs, it is desirable to have information on the behaviour of solutions of (1.1)-(1.3) assuming the least possible of the functions $k_i(v)$, $\gamma(v)$. The results presented

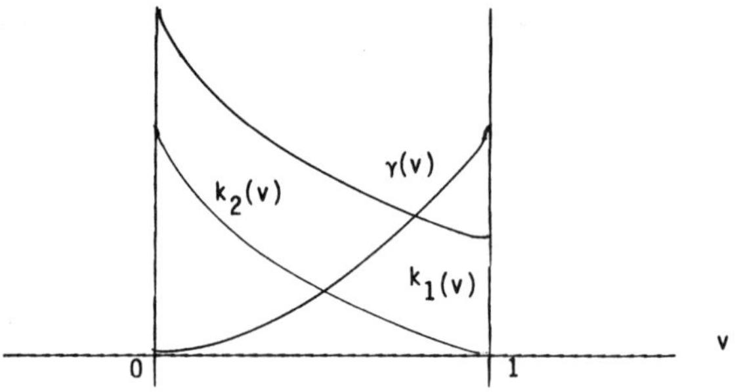

Figure 1

here are in this direction. We will prove that the saturations $(x,t) \to v_i(x,t)$ are continuous functions in Ω_T, without making any assumptions on the nature of the degeneracy of $v \to k_2(v)$ for v near one and assuming that $k_1(v) \sim v_1'(v)$, for v near zero.

When investigating regularity questions for the saturation v, it is more convenient to look at (1.1)-(1.3) in a transformed form.

Introduce a "mean pressure" u given by

$$u = p_2 + \int_0^{\gamma(v)} \frac{k_1(\gamma^{-1}(\xi))}{(k_1 + k_2)(\gamma^{-1}(\xi))} d\xi \tag{1.4}$$

and the quantities

$$A(v) = \frac{k_1(v) \cdot k_2(v)}{K(v)} \gamma'(v); \quad K(v) = k_1(v) + k_2(v)$$

$$\vec{B}(v) = \frac{k_1(v)k_2(v)}{K(v)} [\vec{e}_1(v) - \vec{e}_2(v)]$$

$$C(v) = \frac{k_2(v)}{K(v)} \quad \text{or} \quad -\frac{k_1(v)}{K(v)} \tag{!.5}$$

$$K(v)\vec{e}(v) = k_1(v)\vec{e}_1(v) + k_2(v)\vec{e}_2(v)$$

$$\vec{V} = k(v)[\nabla u + \vec{e}(v)]. \tag{1.6}$$

Then the system (1.1)-(1.3) is transformed into (see [1,7]),

$$\frac{\partial}{\partial t} v - \operatorname{div}[A(v)\nabla v + \vec{B}(v)] = \vec{V} \cdot \nabla C(v) \qquad \text{in } \Omega_T \tag{1.7}$$

$$\operatorname{div} \vec{V} = 0 \qquad \text{in } \Omega \quad \text{a.e. } t \in [0,T]. \tag{1.8}$$

Thus the system is separated into an elliptic equation for u and a parabolic equation for v. The quantity u can be regarded as a "mean pressure" since $p_1 < u < p_2$ if $p_1 < p_2$ and $p_2 < u < p_1$ if $p_1 > p_2$. Equation (1.8) can be interpreted as an equation of continuity for an "idealized" incompressible fluid with pressure u and velocity v, replacing the mixture of the two fluids.

The system (1.1)-(1.3) (or (1.7)-(1.8)) is complemented with initial data and boundary conditions on $S_T \equiv \partial\Omega \times (0,T]$ which might be of Dirichlet, Neumann mixed type, or even of Signorini (overflow) type. Here we will consider local weak solutions of (1.7)-(1.8) assuming that they do exist.

A local weak solution is defined as a pair (u,v) defined in Ω_T, satisfying the following

$$(x,t) \to v(x,t) \text{ is measurable; } v \in [0,1] \text{ and } v \in C(0,T;\, L^2(\Omega)). \tag{1.9}$$

Moreover defining

$$w = \int_0^v A(\xi)d\xi, \text{ we have} \tag{1.10}$$
$$w \in L^2(0,T;\, H^1_{loc}(\Omega)).$$

The function $u \in L^2(0,T;\, H^1_{loc}(\Omega)) \cap L^\infty_{loc}(\Omega_T)$ and if G is a subdomain of Ω, then u,v satisfy

$$\int_\Omega vf\Big|_{t_1}^{t_2} dx + \int_{t_1}^{t_2}\int_\Omega \{-vf_t + (\nabla w + \vec{B}(v) + \vec{V}c(v))\nabla f\}dxdt = 0 \tag{1.11}$$

for all $f \in L^2(0,T;\, H^1_0(G)) \cap H^1_{loc}(\Omega_T)$ and for all $0 < t_1 < t_2 < T$, and

$$\int_{t_1}^{t_2}\int_\Omega K(v)[\nabla u + \vec{e}(v)] \cdot \nabla g \; dxdt = 0 \tag{1.12}$$

for all $g \in L^2(0,T; H_0^1(G))$ and almost all $0 \leq t_1 < t_2 \leq T$. The various integrals in (1.11)-(1.12) converge, modulo basic assumptions listed below.

We assume throughout that $v \to K(v)$, $A(v)$, $\vec{B}(v)$, $\vec{e}(v)$ are continuous functions of $v \in [0,1]$ and there exist constants $0 < c_0 < C$ such that

$$[A_1] \qquad \begin{array}{ll} c_0 \leq K(v) \leq C, \quad c_0 \leq \gamma'(v), & \forall v \in (0,1) \\ A(v) + |\vec{B}(v)| + |C(v)| + |\vec{e}(v)| \leq C, & v \in [0,1]. \end{array}$$

On the degeneracy of $v \to A(v)$ for v near zero or v near one, we assume

$$[A_2] \qquad A(v) > 0, \quad v \in (0,1).$$

$\exists \delta_0 \in (0,1/2]$ such that

(a) For each $v, 0 \leq v \leq \delta_0$, $c_0\phi(v) \leq A(v) \leq C\phi(v)$,

where $v \to \phi(v)$ is a continuous increasing function of $v \in [0,\delta_0]$ $\phi(0) = 0$, and $\phi(v) \geq Cv\phi'(v)$.

(b) For each $v, 1 - \delta_0 \leq v \leq 1$, $c_0 \; \psi(1 - v) \leq A(v) \leq C\psi(1 - v)$, where $v \to \psi(1 - v)$ is a continuous decreasing function of $v \in [1 - \delta_0, 1]$, $\psi(0) = 0$.

We note that we make no assumptions on the decay of ψ. For example it might decay exponentially.

The main result is the following

<u>Continuity theorem</u> Let (u,v) be a local weak solution of (1.7)-(1.8) and let $[A_1]$-$[A_2]$ hold. Then $(x,t) \to v(x,t)$ is continuous in Ω_T, and for every compact set K Ω_T, there exist a continuous non-decreasing function $\omega_K(\cdot)$ depending upon c_0, C, ϕ, ψ such that $\omega_K(0) = 0$ and

$$|v(x_1,t_1) - v(x_2,t_2)| \leq \omega_K(|x_1 - x_2| + |t_1 - t_2|^{1/2})$$

for every pair of points $(x_i,t_i) \in K$, $i = 1,2$. The function $\omega_K(\cdot)$ depends also on the distance from K to the parabolic boundary of Ω_T.

The theorem has several consequences for which we refer to [2]. Here we briefly remark that if the behaviour of $\phi(v)$ and $\psi(1 - v)$ is power-like, then v is actually Hölder continuous and the convective velocity $\vec{V}$ in (1.8) is locally bounded.

The results have been stated under the assumption that a solution exists. The existence theory has been developed in [1] by working directly with the non-transformed system (1.1)-(1.3). It was also shown that the solutions constructed in [1] in $\mathring{\Omega}_T$, are local solutions of (1.7)-(1.8), so that local solutions of (1.7)-(1.8) do exist. On the other hand it is apparent that the continuity theorem can be used (as a compactness theorem) to prove existence of solutions to boundary value problems associated with (1.7)-(1.8), by employing a standard regularization technique.

An aspect of the theorem which is perhaps worth stressing is that it holds with no assumptions at all on the decay of $\psi(1 - v)$ for v near 1. To avoid assumptions on $\phi(\cdot)$ and $\psi(\cdot)$ is dictated by the rough experimental nature of the graphs in fig. 1. This fact however has a mathematical interest of its own in that it shows the strong smoothing effect of the diffusion terms in (1.7).

If the diffusion coefficient $A(v)$ vanishes only for one value of v, say for $v = 0$, and $A(1) \geq c_0 > 0$, then (1.7) is of "porous medium" type.

A good deal of complication, in proving regularity, is due to the fact that $A(v)$ degenerates at more than one value of v. Another source of difficulty is generated by the right hand side of (1.7), $\vec{v} \cdot \nabla C(v)$. This term is not in $L^1_{loc}(\Omega_T)$ and the result would not be possible (not even if $A(v) \equiv 1$) if it were not for the particular form of this term and the fact that $\operatorname{div} \vec{V} = 0$ (this allows us to write, at least formally $\vec{V} \cdot \nabla C(v) = \operatorname{div}[\vec{V} C(v)]$).

The new technique introduced seems to be suitable to study regularity of solutions of (1.7) in the case when $A(0)$ vanishes at an unspecified number of points. More generally for solutions of $\beta(u)_t - \Delta u \geq 0$ in $D'(\Omega_T)$ where β is any coercive maximal monotone graph on $R \times R$.

Below we outline the main arguments, without formal proofs, attempting to give heuristic justification of the various steps which imply continuity.

2. Some Heuristic Remarks on the Method

The continuity of the saturation at a point $z_0 \equiv (x_0,t_0) \in \mathcal{S}_T$ will follow if we can construct a sequence of nested cylinders Q_n around z_0, $n = 1,2,\ldots,$ such that the essential oscillation of v in Q_n decreases to zero as $n \to \infty$, in a way prescribed by the equation (1.7).

A basic fact in this context is the choice of the size of the cylinder Q_n. The size will reflect the degeneracy of $A(v)$. This idea of degeneracy-rescaled cylinders has been introduced in [3].

Let $z_0 \equiv (x_0,t_0) \in \mathcal{S}_T$, $R > 0$, $B_R \equiv \{|x - x_0| < R\}$, $\theta > 0$ and consider cylinders of the type

$$Q_R^\theta \equiv B_R \times \{t_0 - \theta R^2, t_0\}.$$

We will also consider coaxial cylinders contained in Q_R^θ and with same "vertex" z_0. Namely if $\sigma \in (0,1)$

$$Q_{\sigma R}^\theta \equiv B_{\sigma R} \times \{t_0 - \theta(\sigma R)^2, t_0\}.$$

If $\theta = 1$ then $Q_R^1 \overset{\text{def}}{\equiv} Q_R$ is the standard parabolic cylinder.

A general tool in the proof is the following fact which can be regarded as a weak version of the maximum principle. In view of its importance we will state if in detail.

Let $w \in V^{1,0}(Q_R) \equiv L^\infty(t_0 - R^2, t_0; L^2(B_R)) \cap L^2(t_0 - R^2, t_0; H^1(B_R))$.

The space $V^{1,0}(Q_R)$ is equipped with the norm

$$\|w\|^2_{\Gamma,Q_R} = \Gamma^{N+2/2} \operatorname*{ess\,sup}_{t_0-R^2<t<t_0} \|w\|^2_{2,B_R}(t) + \|\nabla w\|^2_{2,Q_R}, \tag{2.1}$$

where Γ is a number satisfying $\Gamma \geq 1$. If $\Gamma = 1$ we write

$$\|w\|^2_{1,Q_R} \overset{\text{def}}{\equiv} \|w\|^2_{V^{1,0}(Q_R)}$$

Set

$$M = \operatorname*{ess\,sup}_{Q_R} w; \quad N = \operatorname*{ess\,inf}_{Q_R} w; \quad L = \operatorname*{ess\,osc}_{Q_R} v = M - N.$$

For $s \geq 1$ define

$$(w - (M - \frac{L}{2^s}))^+ = \max\{w - (M - \frac{L}{2^s});0\}$$

$$(w - (N + \frac{L}{2^s}))^- = - \min\{(w - (N + \frac{L}{2^s}));0\}.$$

Suppose that w satisfies the inequality

$$\|(w - (M - \frac{L}{2^s}))^+\|^2_{\Gamma,Q_{\sigma\rho}} \leq \frac{C}{((1+\sigma)\rho)^2} \Gamma(\frac{L}{2^s})^2 |A^+_{M-\frac{L}{2^s},\rho}| \tag{2.2}$$

where

$$A^+_{s,p} \equiv \{(x,t) \in Q_p | w(x,t) > M - \frac{L}{2^s}\}$$

and $|\Sigma|$ denotes the Lebesque measure of the measurable set Σ. Suppose that (2.2) holds for $\Gamma > 1$ and

(a) for all $s \geq 1$

(b) for all $\sigma \in (0,1)$

(c) for all $\frac{R}{2} < \rho \leq R$.

Then there exist a (small) number $\lambda_0 \in (0,1)$ depending only upon C (and independent of Γ) such that if for some $s_0 \geq 1$

$$\begin{aligned}|A^+_{M-\frac{L}{2^{s_0}},R}| &= \text{meas}\{(x,t) \in Q_R | w(x,t) > M - \frac{L}{2^{s_0}}\} \\ &\leq \lambda_0 |Q_R|,\end{aligned} \tag{2.3}$$

then

$$w(x,t) \leq M - \frac{L}{2^{s_0+1}} \qquad \forall (x,t) \in Q_{R/2}. \tag{2.4}$$

The proof of this fact is an adaptation of standard iteration techniques. In the form given above it can be found in the appendix of [2].

It can be viewed as a weak form of the maximum principle since it says that if w is "mostly" small in Q_R (assumption (2.3)) then it can be large only in the vicinity of the parabolic boundary of Q_R and in fact in a small cylinder near z_0, w is "not too close" to the supremum (conclusion (2.4)).

A similar statement holds for $(w - (N + \frac{L}{2^s}))^-$. Precisely if for all $s > 1$, all $\sigma \in (0,1)$, all $\frac{R}{2} < \rho < R$

$$(2.5) \qquad \|(w - (N + \frac{L}{2^s}))^-\|^2_{\Gamma,Q_{\sigma\rho}} < \frac{C r}{(1-\sigma)^2\rho^2}\left(\frac{L}{2^s}\right)^2 |A^-_{N+\frac{L}{2^s},\rho}|,$$

then there exist $\lambda_0 \in (0,1)$, $\lambda_0 = \lambda_0(C)$ such that if

$$(2.6) \qquad |A^-_{N+\frac{L}{2^{s_0}},R}| < \lambda_0 |Q_R|, \text{ then}$$

$$(2.7) \qquad w(x,t) > N + \frac{L}{2^{s_0+1}}, \quad \forall (x,t) \in Q_{R/2}.$$

Here the definition of $A^-_{N+\frac{L}{2^{s_0}},R}$ is obvious.

If (2.4) holds, then we also have

$$\operatorname*{ess\,sup}_{Q_{R/2}} w < \operatorname*{ess\,sup}_{Q_R} w - \frac{1}{2^{s_0+1}} \operatorname*{ess\,osc}_{Q_R} w,$$

and subtracting $\operatorname*{ess\,inf}_{Q_{R/2}} w$ from both sides we find

$$(2.8) \qquad \operatorname*{ess\,osc}_{Q_{R/2}} w < \left(1 - \frac{1}{2^{s_0+1}}\right) \operatorname*{ess\,osc}_{Q_R} w.$$

An analogous fact holds if (2.7) is verified. If (2.8) holds $\forall R > 0$, then it is a well known fact that w is continuous at z_0 (in fact Hölder continuous).

As a rule in our proof we will try to derive inequalities similar to (2.2) or (2.5) and attempt to verify the smallness condition (2.3) or (2.6). This is impossible if $A(v)$ is near zero. Indeed, using (1.7) and standard energy estimates, the term involving ∇v in the definition (2.1) is of the form

$$\iint_{Q_R} A(v)|\nabla v|^2 dx dt.$$

To compensate for such a degeneracy we will derive inequalities like (2.2) in cylinders Q_R^θ, where θ is related to the degeneracy of $A(v)$. Then we will perform a change of variable in time to recover (2.2) or (2.5) in parabolic cylinders Q_R. Such a change of variable is analogous to the transformation $\tau \to at$ which transforms $u_t = a\Delta u$ into $u_\tau = \Delta u$ (a is a fixed positive constant). The difficulty consists in showing that for such "rescaled" cylinders a smallness condition of the type (2.3) or (2.6) can be proved.

3. The Main Points of the Proof

Let Q_R^θ be fixed (θ being a parameter to be chosen later), consider the cylinder Q_{2R}^ε defined by

$$Q_{2R}^\varepsilon \equiv B_{2R} \times \{t_0 - (2R)^{2-\varepsilon}, t_0\}$$

and set

$$\mu^+ = \operatorname*{ess\,sup}_{Q_{2R}^\varepsilon} v; \quad \mu^- = \operatorname*{ess\,inf}_{Q_{2R}^\varepsilon} v; \quad \omega = \operatorname*{ess\,osc}_{Q_{2R}^\varepsilon} v = \mu^+ - \mu^-.$$

The number θ is related to the oscillation of v in Q_{2R}^ε as follows

$$\theta^{-1} = \phi(\omega/2^m), \tag{3.1}$$

where m is a large number to be chosen later and $v \to \phi(v)$ "measures" the degeneracy of $A(v)$ for v near 0. We may assume that Q_R^θ Q_{2R}^ε. In fact if not $\phi(\omega/2^m) < CR^\varepsilon$ and ω is comparable to R and there would be nothing to prove.

Without loss of generality we may assume that $\omega \in [0, \delta_0]$, where δ_0 has been introduced in $[A_2]$. Thus if ω is near zero and m is large, then θ is large and Q_R^θ is quite long in the t-direction. Inside Q_R^θ we consider auxiliary cylinders $\overline{Q}_R$ whose size in the t direction is $\psi^{-1}(\omega/4)R^2$; namely

$$\overline{Q}_R \equiv B_R \times \{\overline{t} - \frac{R^2}{\psi(\omega/4)}, \overline{t}\}, \tag{3.2}$$

where $\overline{t}$ is any time level satisfying

$$t - \frac{R^2}{\psi(\omega/4)} > t_0 - \frac{R^2}{\phi(\omega/2^m)} .$$

The degeneracies of ψ and ϕ are unrelated; however by choosing m large enough in (3.1), we may insure that such cylinders can be constructed.

The worst situation occurs when in Q_R^θ we have $\mu^+ = 1$, $\mu^- = 0$, the other cases being obviously more favorable.

We first examine the behaviour of v near 1 in the subcylinders $\overline{Q}_R$ defined in (3.2). With C we denote a generic non-negative constant which can be determined only in terms of the data and independent of ω and R.

Lemma 3.1 There exist numbers $\alpha \in (0,1)$, $\eta_0 = \eta_0(\omega) \in (0,1)$ such that if

$$\text{(3.3)} \qquad \operatorname{meas}\{(x,t) \in \overline{Q}_R | v(x,t) > 1 - \tfrac{\omega}{2}\} < \eta_0 |\overline{Q}_\Omega|,$$

then either

$$\text{(3.4)} \qquad \min\{\psi(\tfrac{\omega}{4}); \tfrac{\omega}{2}\} \leq CR^\alpha, \quad \alpha \in (0,1), \quad \text{or}$$

$$\text{(3.5)} \qquad v(x,t) < 1 - \tfrac{\omega}{4}, \ \forall (x,t) \in \overline{Q}_{R/2}.$$

Sketch of the Proof. Set $v_\omega = \min\{v; 1 - \frac{\omega}{4}\}$ and let k be any number satisfying $1 - \omega/2 \leq k \leq 1 - \omega/4$. In the identity (1.11) we take test functions $(v_\omega - k)^+ \zeta^2$ where ζ is a cutoff function in $\overline{Q}_\rho$, $\frac{R}{2} \leq \rho \leq R$ which equals one on $Q_{\sigma\rho}$, $\sigma \in (0,1)$ and perform technical calculations for which we refer to [2]. Such calculations involve (1.7) and (1.8) simultaneously. If (3.4) is violated, we obtain

$$\text{(3.6)} \qquad \operatorname*{ess\,sup}_{\bar{t} - \frac{(\sigma\rho)^2}{\psi(\omega/4)} \leq t \leq \bar{t}} \|(v_\omega - k)^+\|^2_{2,B_{\sigma\rho}}(t) + \psi(\tfrac{\omega}{4}) \|\nabla(v_\omega - k)^+\|^2_{2,\overline{Q}_{\sigma\rho}}$$
$$\leq \frac{C}{[(1-\sigma)\rho]^2} \psi(\omega)\omega^2 |A_{k,\rho}|,$$

where $A_{k,\rho} \equiv \{(x,t) \in \overline{Q}_\rho : v(x,t) > k\}$.

In (3.6) we perform the change of variable

$$\tau = (t - \bar{t})\psi(\omega/4),$$

which transforms $\overline{Q}_\rho$, $\overline{Q}_{\sigma\rho}$ respectively into

$$Q_\rho \equiv B_\rho \times \{-\rho^2,0\};\ Q_{\sigma\rho} \equiv B_{\sigma\rho} \times \{-(\sigma\rho^2,0\}.$$

Therefore setting $w(x,\tau) = v_\omega(x,\overline{t} - \frac{\tau}{\psi(\omega/4)})$, inequality (3.9) can be written more concisely as

$$(3.7)\qquad \|(w-k)^+\|^2_{V^{1,0}(Q_{\sigma\rho})} < \frac{C}{((1-\sigma)\rho)^2}\left[\frac{\psi(\omega)}{\psi(\omega/4)}\right]\omega^2|\tilde{A}_{k,\rho}|$$

where $\tilde{A}_{k,\rho} \equiv \{(x,t) \in Q_\rho | w(x,t) > k\}$.

Starting from (3.7) and assumption (3.3), the lemma is a consequence of the "weak maximum principle", discussed in section 2.

Remark. Lemma 3.1 says that if in $\overline{Q}_R$, v is "mostly away from the degeneracy $v = 1$ (see assumption (3.3)), then in $\overline{Q}_{R/2}$ it is strictly away from 1.

Because of the parabolic nature of the problem, if $v < 1 - \omega/4$ in $\overline{Q}_{R/2}$ we expect that $v < 1$ for a certain length of time, around x_0. This is the meaning of the following lemma.

Lemma 3.2 Under the assumptions of lemma 3.1, we can find a cylinder $Q^*_R \equiv B_{R/8} \times \{t_0 - \frac{R^2}{8^2\psi(\omega/4)}, t_0\}$, around (x_0,t_0), and a large number s depending only upon ω, ϕ, ψ, such that either

$$\min\{\psi(\tfrac{\omega}{4});\ \tfrac{\omega}{2}\} < CR^\alpha, \text{ or}$$

$$v(x,t) < 1 - \frac{\omega}{2s},\quad \forall (x,t) \in Q^*_R.$$

The proof proceedes in two steps. In the first step, by means of a logarithmic estimate (lemma 3.2 of [2]), one shows that the set where v is near 1, relatively to $B_{R/2} \times [\overline{t},t_0]$, can be made arbitrarily small, provided one chooses s large enough. At this stage the fact that $v < 1$ in $\overline{Q}_{R/2}$ is essential. In the second step we work in the cylinder $B_{R/2} \times [\overline{t},t_0]$, derive inequalities analogus to (3.6) and then a rescaling argument as indicated above together with the "weak maximum principle" yield the lemma.

We summarize the analysis of the behaviour of v near 1, in the following

First alternative

Suppose that among all the cylinders $\overline{Q}_R$ we can find one for which (3.3) is satisfied. Then there is $s = s(\omega)$ such that either

$$\min\{\psi(\frac{\omega}{4}); \frac{\omega}{2}\} < CR^{\alpha}, \text{ or}$$

$$v(x,t) < 1 - \frac{\omega}{2^s} \quad \text{in } Q_R^* \equiv B_{R/8} \times \{t_0 - \frac{R^2}{8^2\psi(\omega/4)}, t_0\}.$$

This in turn implies

If such a $\overline{Q}_R$ can be found then either

$$\min\{\psi(\frac{\omega}{4}); \frac{\omega}{2}\} < CR^{\alpha}, \text{ or} \tag{3.8}$$

$$\operatorname*{ess\,osc}_{Q_R^*} v \leq (1 - \frac{1}{2^s}) \operatorname*{ess\,osc}_{Q_{2R}^{\varepsilon}} v.$$

Next we examine the case when for <u>no</u> cylinder $\overline{Q}_R$ the assumption (3.3) holds, by analyzing the behaviour of v near zero. If (3.3) is violated, then

$$\text{meas}\{(x,t) \in \overline{Q}_R | v(x,t) < \frac{\omega}{2}\} \leq (1 - \eta_0)|\overline{Q}_R| \tag{3.9}$$

for <u>all</u> cylinders $\overline{Q}_R$.

Fix a cylinder $\overline{Q}_R$. Since for $\overline{Q}_R$ (3.9) holds, for some time level t^*, we must have

$$\text{meas}\{x \in B_R | v(x,t^*) < \frac{\omega}{2}\} < (\frac{1 - \eta_0}{1 - \eta_0/2})|Q_R| \tag{3.10}$$

The time level t^* can be estimated by (see Lemma 4.1 of [2])

$$t^* \in [\overline{t} - \frac{R^2}{\psi(\omega/4)}, \overline{t} - \frac{\eta_0}{2}\frac{R^2}{\psi(\omega/4)}].$$

Estimate (3.10) says that for some time level t^* the set where v is near zero relatively to B_R, does not fill the full ball B_R. That is in B_R, at level t^* there is an appreciable set where v is away from the degeneracy.

We wish to exploit precisely the fact that v at t^* in B_R is "not completely degenerate".

First we show (see lemma 4.2 of [2]) that (3.10) roughly continues to hold for a short interval of time, starting from t^*. Precisely we have

$$(3.11)\quad \begin{cases} \text{there exist a number } m_1 = m_1(\omega) \text{ such that either} \\ \quad \min\{\phi(\omega/2^{m_1});\ \omega/2^{m_1}\} \le CR^{\alpha}, \text{ or} \\ \quad \operatorname{meas}\{x \in B_R \mid v(x,t) < \omega/2^{m_1}\} < [1 - (\frac{\eta_0}{2})^2](B_R) \\ \text{for all } t \in [\bar{t} - \frac{\eta_0}{2} R^2/\psi(\omega/4);\ \bar{t}]. \end{cases}$$

Now the statement (3.11) has been derived under the assumption (3.9) which holds for all cylinders $\bar{Q}_R$. Therefore in (3.11) the time $\bar{t}$ could be any time $t \in [t_0 - \theta R^2, t_0]$. This fact gives a "stability" information on v for all $t \in [t_0 - \theta R^2, t_0]$. It says that the set of "degeneracy" (i.e. v near zero), never fills the ball B_R for all $t \in [t_0 - \theta R^2, t_0]$.

The device of introducing the auxilliary subcylinders $\bar{Q}_R$ has the purpose of producing such a stability fact. Notice that such property, in general, is not expected since the length of Q_R^θ is large (see (3.1)). Even for solutions of the heat equation an estimate like (3.10) continues to hold only for a very small interval of time starting from t^*. In our case the "stability" for large intervals of time is a consequence of (3.9) valid for <u>all</u> $\bar{Q}_R \subset Q_R^\theta$.

We will exploit it in the following way. Set

$$w = \int_0^v A(\xi)d\xi; \quad v = \beta(w), \quad v \in [0, \delta_0],$$

where $\beta(\cdot)$ is a monotone increasing function of w. The new unknown w satisfies

$$(3.12)\qquad \frac{\partial}{\partial t}\beta(w) - \Delta w = \operatorname{div}[-\vec{B}(v) + \vec{V}C(v)] \quad \text{in } D'(\Omega_T).$$

Note that if v ranges over $[0, \omega/2^p]$, $p \ge m_1$, then

$$\beta'(w) > \frac{1}{\phi(\omega/2^p)}, \quad \text{on such a range.}$$

We set also

$$\bar{\omega} = \int_0^\omega A(\xi)d\xi.$$

The previous stability information can be translated in terms of w as follows

there exist a number $m_2 = m_2(\omega)$ such that

$$\text{meas}\{x \in B_R | w(x,t) < \frac{\bar{\omega}}{2^{m_2}}\} < [1 - (\frac{\eta_0}{2})^2] \, |B_R| \tag{3.13}$$

for all $t \in [t_0 - \theta R^2, t_0]$.

Next, having the "weak maximum principle" in mind, we wish to reduce the set where w is close to zero (or equivalently where v is close to zero). Precisely want to show

<u>Lemma 3.3</u> For every $\lambda_0 \in (0,1)$, we can find a (large) positive number $m_3 > m_2$ and a number m_0 such that if $m > m_0$, then either

$\frac{\bar{\omega}}{2^{m_3}} < CR^\alpha$ (α introduced in Lemma 3.1), or

$\text{meas}\{(x,t) \in Q_R^\theta | w(x,t) < \frac{\bar{\omega}}{2^{m_3}}\} < \lambda_0 |Q_R^\theta|.$

<u>Sketch of the proof.</u> We give a brief sketch of the proof since it explains why we did choose a large cylinder Q_R^θ to start with. Multiply formally (3.12) by $-(w - k)^- \zeta^2$, where $k = \bar{\omega}/2^n$, $n > m_2$ and ζ is a cutoff function in Q_{2R}^θ which equals one on Q_R^θ and such that

$$0 < \zeta_t < CR^2/\phi(\omega/2^m).$$

We integrate over Q_{2R}^θ and perform calculations described in detail in [2]. Such calculations, starting from (3.12) can be rigorously justified. We obtain

$$\|\nabla(w - k)^-\|^2_{2,Q_R^\theta} < \frac{C}{R^2} (\frac{\bar{\omega}}{2^n})^2 |Q_R^\theta| \{1 + (\frac{2^n}{\bar{\omega}}) \beta(\frac{\omega}{2^n}) \phi(\frac{\omega}{2^m})\}. \tag{3.14}$$

Notice that the coefficient $\phi(\omega/2^m)$ in the last factor of (3.14) comes from the estimate on ζ_t and depends crucially on the choice of long cylinders Q_R^θ (see (3.1)).

We recall a lemma of DeGiorgi. Let $w \in W^{1,1}(B_R)$ and $\ell > k$. Then

$$(3.15) \qquad (\ell - k)\operatorname{meas}\{x \in B_R | w(x) < k\} \le \frac{CR^{N+1}}{\operatorname{meas}\{x \in B_R | w(x) > \ell\}} \int_{[k<w<\ell]} |\nabla w| dx$$

We apply (3.15) to $x \to w(x,t)$ defined for a.e. $t \in [t_0 - R^2/\phi(\omega/2^m), t_0]$ and for the choice of levels

$$\ell = \frac{\bar\omega}{2^n}\ ; \quad k = \frac{\bar\omega}{2^{n+1}}\ , \quad n = m_2, m_2 + 1, \ldots .$$

By the stability information (3.13) we have

$$\operatorname{meas}\{x \in B_R | w(x,t) > \frac{\omega}{2^n}\} > \left(\frac{\eta_0}{2}\right)^2 |B_R|$$

for all $t \in [t_0 - R^2/\phi(\omega/2^m); t_0]$. Moreover setting, for notational simplicity

$$A_n(t) = \{x \in B_R | w(x,t) < \frac{\bar\omega}{2^n}\}; \quad A_n = \int_{t_0-(R^2/\phi(\omega/2^m))}^{t_0} |A_n(t)| dt,$$

from (3.15) with these choices we get

$$(3.16) \qquad \left(\frac{\bar\omega}{2^n}\right)|A_{n+1}(t)| \le \frac{C}{\eta_0^2} R \int_{A_n(t)\setminus A_{n+1}(t)} |\nabla w| dx.$$

Squaring both sides using Hölder inequality and integrating in t we have

$$(3.17) \qquad \left(\frac{\bar\omega}{2^n}\right)^2 |A_{n+1}|^2 \le C(\omega)R^2(A_n - A_{n+1}) \iint_{Q_R^\theta} |\nabla(w - \frac{\omega}{2^n})^-|^2 dx d\tau.$$

Divide by $(\bar\omega/2^n)^2$ and estimate the last integral by (3.14) to obtain

$$(3.18) \qquad (A_{n+1})^2 \le C(\omega)(A_n - A_{n+1})|Q_R^\theta|\{1 + \phi(\frac{\omega}{2^m})\beta(\frac{\omega}{2^n})\frac{2^n}{\bar\omega}\}.$$

We add these inequalities for $n = m_2, m_2 + 1, \ldots, m_3 - 3$. Then since

$\sum(A_n - A_{n+1}) < |Q_R^\theta|$, we have

$$|A_{m_3}| < \left(\frac{C(\omega)}{m_3}\right)^{1/2}\{1 + \phi\left(\frac{\omega}{2^m}\right)\beta\left(\frac{\omega}{2^{m_2}}\right)\frac{2^{m_3}}{\overline{\omega}}\}^{1/2}\,|Q^\theta|.$$

Choose m_3 so large that

$$\left[\frac{2C(\omega)}{m_3}\right]^{1/2} < \lambda_0,$$

and m_0 so large that $\phi\left(\frac{\omega}{2^{m_0}}\right)\beta\left(\frac{\omega}{2^{m_2}}\right)\frac{2^{m_3}}{\overline{\omega}} < 1$. This proves the lemma.

Remark. The choice of a long cylinder Q_R^θ has the technical purpose of controlling the right hand side of (3.18) as $n \to \infty$.

The lemma can be rephrased in terms of v by saying

> For every $\lambda_0 \in (0,1)$ there exist a number $m_4 > m_3$ and m_0, such that if $m > m_0$, then either

(3.19)
$$\min\{\frac{\omega}{2^{m_4}}\,;\ \phi\left(\frac{\omega}{2^{m_4}}\right)\} < CR^\alpha, \quad \text{or}$$

$$\text{meas}\{(x,t) \in Q_R^\theta | v(x,t) < \frac{\omega}{2^{m_4}}\} < \lambda_0\,|Q_R^\theta|.$$

We recall that the length of Q_R^θ is $R_\theta^2 = \dfrac{R^2}{\phi(\omega/2^m)}$. We now come to the choice of m. Such a choice will be affected by the lack of information on the nature of the degeneracy of $\psi(\cdot)$. Set

$$\phi_1 = \phi\left(\frac{\omega}{2^{m_4-1}}\right); \quad \phi_2 = \phi\left(\frac{\omega}{2^{m_4+2}}\right); \quad \nu = \phi_2\left(\frac{\phi_1}{\phi_2}\right)^{N+2/2}.$$

Then choose $m > m_0$ as the smallest real number such that $[\phi(\omega/2^m)]^{-1}$ is a multiple integer of ν^{-1}, i.e.

$$n\phi(\omega/2^m) = \nu, \quad \text{for some } n \in N.$$

Then split Q_R^θ in n subcylinders

$$Q_R^j \equiv B_R \times \{t_0 - \frac{jR^2}{\nu}\,;\ t_0 - \frac{(i-1)R^2}{\nu}\}; \quad j = 1,2,\ldots,n.$$

The cylinders Q_R^j are contiguous and their intersection has empty interior. Moreover $\sum_i^n |Q_R^j| = |Q_R^\theta|$.

Obviously for some j_0 we must have

(3.20) $$\text{meas}\{(x,t) \in Q_R^{j_0} \mid v(x,t) < \frac{\omega}{2^{m_4}}\} < \lambda_0 \, |Q_R^{j_0}|.$$

We want to show that a "maximum principle" statement holds in $Q_R^{j_0}$. For this we first derive the inequality for

$$X = \{ t_0-(j-1)\frac{R^2}{\nu} - \frac{\sigma\rho^2}{\nu} < t < t_0-(j-1)\frac{R^2}{\nu} \},$$

(3.21) $$\text{ess sup}_X \|(v_\omega - k)^-\|^2_{2,B_{\sigma\rho}}(t) + \phi(\frac{\omega}{2^{m_4+2}})\|\nabla(v_\omega - k)^-\|^2_{2,Q^{j_0}} \leq \frac{C}{[(1-\sigma)\rho]^2}\,\phi(\frac{\omega}{2^{m_4}})(\frac{\omega}{2^{m_4}})^2 \, |A^{j_0}_{k,\rho}|,$$

where $v_\omega = \max\{v; \frac{\omega}{2^{m_4+2}}\}$ and $\frac{\omega}{2^{m_4+1}} \leq k \leq \frac{\omega}{2^{m_4}}$.

Then we make the change of variable

$$\tau = [t - (t_0 - (j_0 - 1)\frac{R^2}{\nu})]\nu$$

which maps $Q_\rho^{j_0}$, $Q_{\sigma\rho}^{j_0}$ into Q_ρ and $Q_{\sigma\rho}$, so that (3.21) can be rewritten as

(3.22) $$(\frac{\phi_1}{\phi_2})^{N+2/2} \sup_{-\sigma\rho^2 \leq \tau \leq 0} \|(z-k)^-\|^2_{2,B_{\sigma\rho}}(\tau) + \|\nabla(z-k)^-\|^2_{2,Q_{\sigma\rho}} \leq$$

$$\leq \frac{C}{((1-\sigma)\rho)^2}(\frac{\phi_1}{\phi_2})(\frac{\omega}{2^{m_4}})^2|\tilde{A}_{k,\rho}|,$$

where $z(x,t) = v_\omega(x, \frac{\tau}{\nu} + t_0 - (j_0 - 1)R^2/\nu)$ and $\tilde{A}_{k,p} \equiv \{(x,t) \in Q_\rho \mid z(x,t) < k\}$.

Since (3.22) holds $\forall \sigma \in (0,1)$, all $\frac{R}{2} < \rho \leq R$ and all $\frac{\omega}{2^{m_4+1}} \leq k \leq \frac{\omega}{2^{m_4}}$, by the "weak maximum principle" we have

<u>Lemma 3.4</u> There exist $m_4 = m_4(\omega)$ such that either

$$\min\{\phi(\frac{\omega}{2^{m_4}}); \frac{\omega}{2^{m_4}}\} < CR^\alpha \quad \text{or}$$

$$v(x,t) > \frac{\omega}{2^{m_4+1}}, \quad \forall (x,t) \in Q^{j_0}_{R/2}.$$

The lemma establishes that there exist some time level in $Q^{\theta}_{R/2}$ where v is strictly away from zero, and hence (1.7) is not degenerate there. But then by a procedure similar to the one to establish the first alternative we conclude that

there exist $m_5 > m_4$ such that either

(3.23) $$\min\{\phi\left(\frac{\omega}{2^{m_5}}\right); \frac{\omega}{2^{m_5}}\} \leq CR^{\alpha}, \quad \text{or}$$

$$v(x,t) > \frac{m}{2^{m_5}}, \quad \forall(x,t) \in Q^{*}_{R} \equiv B_{R/8} \times \{t_0 - \frac{R^2}{8^2\psi(\omega/4)}; t_0\}.$$

We summarize.

Second alternative Suppose that (3.3) is violated for all cylinders of the type $\overline{Q}_R$ Q^{θ}_{R}. Then

(3.24) $$\operatorname*{ess\,osc}_{Q^{*}_{R}} v \leq \left(1 - \frac{1}{2^{m_5(\omega)}}\right) \operatorname*{ess\,osc}_{Q^{\varepsilon}_{2\Omega}} v.$$

4. Proof of the Theorem Concluded

Combining the two alternatives we have that in either case there exist continuous positive functions

$$\omega \to g(\omega),\ \eta(\omega),\ \sigma(\omega) \quad \text{such that}$$

$$g(\omega) \to \infty \quad \text{as} \quad \omega \to 0$$

$$\eta(\omega) = 2^{-m_5(\omega)} \to 0 \quad \text{as} \quad \omega \to 0$$

$$\sigma(\omega) = \min\{\psi\left(\frac{\omega}{4}\right); C\nu(\omega)\} \to 0 \quad \text{as} \quad \omega \to 0$$

such that either

$$\omega \leq g(\omega)R^{\alpha_0}, \quad \alpha_0 = \min\{\varepsilon, \alpha\}, \quad \text{or}$$

$$\operatorname*{ess\,osc}_{Q^{\sigma(\omega)}_{R}} v \leq (1 - \eta(\omega))\omega.$$

The process can now be iterated over a sequence of cylinders to yield the result. Precisely we might define sequences

$$R_0 = 2R; \quad R_1 = \frac{R_0}{16}; \ \ldots, \ R_{n+1} = \frac{R_n}{16}$$

$$\omega_0 = \operatorname*{ess\,osc}_{Q^{\varepsilon}_{2R}} v; \quad \omega_{n+1} = (1 - \eta(\omega_n))\omega_n,$$

and construct the sequence of cylinders

$$Q_0 \equiv Q^{\varepsilon}_{2R}; \ Q_{n+1} = B_{R_{n+1}} \times \{t_0 - \frac{1}{\sigma(\omega_n)} R^2 \ ; \ t_0\}.$$

For them we have that either

$$\omega_{n+1} \leq g(\omega_n) R_n^{\alpha_0}, \quad \text{or}$$

$$\operatorname*{ess\,osc}_{Q_{n+1}} v \leq (1 - \eta(\omega_n))\omega_n.$$

These inequalities imply that $\operatorname*{ess\,osc}_{Q_n} v \to 0$ as $n \to \infty$ thereby establishing the continuity of v, at z_0.

References

[1] H.W. Alt and E. DiBenedetto, 'Non steady flow of water and oil through inhomogeneous porous media', Ann. Sc. Norm. Sup. Pisa Serie IV Vol. XII no 3 (1985), 335-392.

[2] H.W. Alt and E. DiBenedetto, Regularity of the saturation in the flow of two immiscible fluids through a porous medium, Centre for Math. Analysis Australian Nat. Univ. Canberra, July 1985.

[3] E. DiBenedetto, Local behaviour of solutions of degenerate parabolic equations with measurable coefficients, Ann. Sc. Norm. Sup. Pisa (to appear).

[4] J. Bear, Dynamics of fluids in Porous media, American-Elsevier, New York 1972.

[5] R.E. Collins, Flow of fluids through Porous Media (3rd ed.) University of Toronto Press, Toronto Ontario, 1974.

[6] A.E. Scheidegger, The Physics of flow through Porous Media (3rd ed.) Univ. of Toronto Press, Toronto, Ontario, 1974.

[7] A. Spivak, H.S. Price, A. Settari, 'Solution of the equations for multidimensional, two-phase immiscible flow by variational methods, Soc. Pet. Eng. J. Feb. 1977, pp. 27-41.

MOLECULAR THEORY FOR THE NONLINEAR VISCOELASTICITY OF POLYMERIC LIQUIDS CRYSTALS

Masao Doi

Physics Department
Tokyo Metropolitan University
Setagaya, Tokyo
158 JAPAN

1. Introduction

The flow properties of nematic liquid crystals of small molecules have been successfully described by the phenomenological theory of Ericksen and Leslie [1-4]. However, for polymeric liquid crystals, such as concentrated solutions of rodlike polymers, the applicability of the theory is limited to low shear rate regimes. At higher shear rate, the polymeric liquid crystals usually show significant nonlinear effects, such as shear thinning [5], which is not explained by the Ericksen-Leslie theory.

A way of generalizing the Ericksen-Leslie theory to the nonlinear regimes was suggested by de Gennes [6]. He noticed that the order parameter in nematics is a tensor $Q_{\alpha\beta}$ rather than the director vector $\underset{\sim}{n}$, and set up a Landau type kinetic equation for $Q_{\alpha\beta}$. However, this phenomenological approach has not been pursued very far since the number of unknown parameters increases rapidly as more complexity is introduced.

In this paper, I describe a complementary approach, the molecular theory. Since this approach is based on a specific modeling of a system, it is less general, but it can give explicit results for rheological functions. I only give an outline of the theory. The detail is described in ref [7].

2. Kinetic Equation

Consider a solution of rigid rodlike polymers. Let $f(\underset{\sim}{u},t)$ be the probability that an arbitrarily chosen polymer is in the direction specified by a unit vector $\underset{\sim}{u}$ at time t. In a dilute solution, $f(\underset{\sim}{u},t)$ satisfies the following diffusion equation [7,8]

$$\frac{\partial f}{\partial t} = D_{r0}\underset{\sim}{\mathcal{R}} \cdot [\underset{\sim}{\mathcal{R}} f + \frac{f}{k_B T} \underset{\sim}{\mathcal{R}} V_e] - \mathcal{R} \cdot (\underset{\sim}{u} \times \underset{\approx}{\kappa} \cdot \underset{\sim}{u} f) \tag{1}$$

where

$$\mathcal{R} = \underset{\sim}{u} \times \frac{\partial}{\partial \underset{\sim}{u}} \tag{2}$$

is the operator of rotation, V_e, the potential of the external field, and D_{r0}, the rotational diffusion constant. In the nematic phase, the mean field approximation gives the following time evolution equation for $f(\underset{\sim}{u},t)$ [7,9].

$$\frac{\partial f}{\partial t} = \overline{D}_r \underset{\sim}{\mathcal{R}} \cdot [\underset{\sim}{\mathcal{R}} f + \frac{f}{k_B T} \underset{\sim}{\mathcal{R}} (V_e + V_{mf})] - \underset{\sim}{\mathcal{R}} \cdot (\underset{\sim}{u} \times \underset{\approx}{\kappa} \cdot \underset{\sim}{u} f) \tag{3}$$

where $\overline{D}_r$ is the effective rotational diffusion constant which accounts for the effect of entanglement [10], and V_{mf} is the mean field potential imposed on a polymer by the surrounding polymers. A commonly adopted form of V_{mf} is that given by Maier and Saupe [11]

$$V_{mf}(\underset{\sim}{u},[f]) = -\frac{3}{2} U k_B T \int d\underset{\sim}{u}'(\underset{\sim}{u} \cdot \underset{\sim}{u}')^2 f(\underset{\sim}{u}',t) \tag{4}$$

The parameter U characterizes the strength of the intermolecular potential which tends to orient the molecules in the same direction. Eqs. (3) and (4) give a nonlinear equation for $f(\underset{\sim}{u},t)$.

If the distribution function is known, the stress tensor is calculated by [9]

$$\underset{\approx}{\sigma} = 3ck_B T\langle \underset{\sim}{u}\underset{\sim}{u} - \frac{1}{3}\underset{\approx}{I}\rangle + c\langle \underset{\sim}{u}(\underset{\sim}{u} \times \underset{\sim}{\mathcal{R}}(V_{mf} + V_e))\rangle \tag{5}$$

where c is the number of polymers in unit volume.

3. Nonlinear Equation for the Order Parameter Tensor

From eq. (3) one can derive an approximate equation for the order parameter tensor $\mathcal{Q}_{\alpha\beta}$ which is defined by

$$\underset{\approx}{Q} = \langle \underset{\sim}{u}\underset{\sim}{u} - \frac{1}{3} \rangle \tag{6}$$

To get an equation for $\underset{\approx}{Q}$, one multiplies $\underset{\sim}{u}\underset{\sim}{u} - \underset{\approx}{I}/3$ on both sides of eq. (3) and integrates over u. Using the Hermitian property of $i\underset{\sim}{\mathcal{L}}$, i.e.,

$$\int d\underset{\sim}{u}\, A(\underset{\sim}{u})\underset{\sim}{\mathcal{L}}B(\underset{\sim}{u}) = - \int d\underset{\sim}{u}\, B(\underset{\sim}{u})\underset{\sim}{\mathcal{L}}A(\underset{\sim}{u}) \tag{7}$$

one gets

$$\dot{\underset{\approx}{Q}} = -6\overline{D}_r\underset{\approx}{Q} + \overline{D}_r U[\underset{\approx}{Q} \cdot \langle \underset{\sim}{u}\underset{\sim}{u} \rangle - \langle \underset{\sim}{u}\underset{\sim}{u}\underset{\sim}{u}\underset{\sim}{u} \rangle : \underset{\approx}{Q}]$$

$$+ \underset{\approx}{\kappa} \cdot \langle \underset{\sim}{u}\underset{\sim}{u} \rangle + \langle \underset{\sim}{u}\underset{\sim}{u} \rangle \cdot \underset{\approx}{\kappa}^{+} - 2\langle \underset{\sim}{u}\underset{\sim}{u}\underset{\sim}{u}\underset{\sim}{u} \rangle : \underset{\approx}{\kappa} \tag{8}$$

where the dot in $\dot{\underset{\sim}{Q}}$ denotes the time derivative, and V_e has been set to zero. By use of the following decoupling approximation

$$\langle \underset{\sim}{u}\underset{\sim}{u}\underset{\sim}{u}\underset{\sim}{u} \rangle : \underset{\approx}{Q} \cong \langle \underset{\sim}{u}\underset{\sim}{u} \rangle \langle \underset{\sim}{u}\underset{\sim}{u} \rangle : \underset{\approx}{Q}$$

$$\langle \underset{\sim}{u}\underset{\sim}{u}\underset{\sim}{u}\underset{\sim}{u} \rangle : \underset{\approx}{\kappa} \cong \langle \underset{\sim}{u}\underset{\sim}{u} \rangle \langle \underset{\sim}{u}\underset{\sim}{u} \rangle : \underset{\approx}{\kappa} \tag{9}$$

eq. (8) is rewritten as

$$\dot{\underset{\approx}{Q}} = \underset{\approx}{F} + \underset{\approx}{G} \tag{10}$$

where

$$\underset{\approx}{F} = -6\overline{D}_r[(1 - \frac{U}{3})\underset{\approx}{Q} - U(\underset{\approx}{Q} \cdot \underset{\approx}{Q} - \frac{1}{3}(\underset{\approx}{Q}:\underset{\approx}{Q})\underset{\approx}{I}) + U(\underset{\approx}{Q}:\underset{\approx}{Q})\underset{\approx}{Q}] \tag{11}$$

$$\underset{\approx}{G} = \frac{1}{3}(\underset{\approx}{\kappa} + \underset{\approx}{\kappa}^{+}) + \underset{\approx}{\kappa} \cdot \underset{\approx}{Q} + \underset{\approx}{Q} \cdot \underset{\approx}{\kappa}^{+} - \frac{1}{3}(\underset{\approx}{\kappa}:\underset{\approx}{Q})\underset{\approx}{I} - 2(\underset{\approx}{\kappa}:\underset{\approx}{Q})\underset{\approx}{Q} \tag{12}$$

Eq. (11) can be written as

$$F_{\alpha\beta} = -L \frac{\partial A}{\partial Q_{\alpha\beta}} + \lambda\delta_{\alpha\beta} \tag{13}$$

with

$$L = 6\bar{D}_r \tag{14}$$

and

$$A = \mathrm{Trace}\left[\frac{1}{2} \left(1 - \frac{U}{3}\right) \underset{\approx}{Q}\cdot\underset{\approx}{Q} - \frac{U}{3} \underset{\approx}{Q} \cdot \underset{\approx}{Q} \cdot \underset{\approx}{Q} + \frac{U}{3} \underset{\approx}{Q} \cdot \underset{\approx}{Q} \cdot \underset{\approx}{Q} \cdot \underset{\approx}{Q} \right] \tag{15}$$

The λ is chosen such that $F_{\alpha\alpha} = 0$.

If there is no flow field, eq. (10) becomes

$$\dot{Q}_{\alpha\beta} = -L \frac{\partial A}{\partial Q_{\alpha\beta}} + \lambda \delta_{\alpha\beta} \tag{16}$$

This is equivalent to the time evolution equation proposed by de Gennes. [4,6] Since $A(\underset{\approx}{Q})$ corresponds to the free energy, the equilibrium value of $\underset{\approx}{Q}$ is determined by the condition that A be minimized. Indeed if $\underset{\approx}{Q}(t)$ obeys eq. (16), $A(\underset{\approx}{Q}(t))$ changes as

$$\frac{d}{dt} A(\underset{\approx}{Q}(t)) = \frac{\partial A}{\partial Q_{\alpha\beta}} \dot{Q}_{\alpha\beta} = -L\left(\frac{\partial A}{\partial Q_{\alpha\beta}} \right)^2 < 0 \tag{17}$$

Thus if there is no field, $A(\underset{\approx}{Q})$ decreases monotonically and approaches the minimum, which is the equilibrium state.

It is easy to show that the equilibrium value of $\underset{\approx}{Q}$ is written as

$$\underset{\approx}{Q}_{eq} = S_{eq}(\underset{\sim}{n}\underset{\sim}{n} - \frac{1}{3} \underset{\approx}{I}) \tag{18}$$

where $\underset{\sim}{n}$ is an arbitrary vector and S_{eq} is given by

$$S_{eq} = \begin{cases} 0 & \text{for } U < 3 \\ \frac{1}{4} + \frac{3}{4}\left(1 - \frac{8}{3U}\right)^{1/2} & \text{for } U > \frac{8}{3} \end{cases} \tag{19}$$

This equation shows explicitly that the liquid crystalline phase appears for large value of U.

By the same approximation as that of eq. (9), eq. (5) is written as

$$\sigma_{\alpha\beta} = 3ck_BT \frac{\partial A}{\partial Q_{\alpha\beta}} \tag{20}$$

Eq. (10) and (20) give the constitutive equation: for given value of κ, Q is determined by eq. (10) and the stress by eq. (20).

It is important to note that the equations given above hold in both the isotropic and the nematic phases. [7] From now on, only the nematic phase is considered.

For a steady state of weak velocity gradient, the perturbation solution of eq. (10) gives [12]

$$[(1 - S_{eq})\kappa^{+} \cdot n + (1 + 2S_{eq})\kappa \cdot n] \times n = 0 \tag{21}$$

and

$$\sigma = - \frac{3ck_BT}{L} [\frac{1}{3} (\kappa + \kappa^{+}) + (\kappa \cdot nn + nn \cdot \kappa^{+})S_{eq} - 2S_{eq}^2 \, nn(\kappa:nn)] \tag{22}$$

This is a special case of the Ericksen-Leslie theory in the absence of the external field. The general case has been dealt with by Marrucci [13], who calculated the viscosity coefficients appearing in the Ericksen-Leslie theory.

5. Uniaxial Approximation

In order to see the results implied by eq. (10), it is convenient to use further approximation. Suppose that the solution of eq. (10) is written as

$$Q = S(nn - \frac{1}{3} I). \tag{23}$$

Substitution of eq. (23) into eq. (10) gives

$$\dot{S}(n_\alpha n_\beta - \frac{1}{3}\delta_{\alpha\beta}) + S(\dot{n}_\alpha n_\beta + n_\alpha \dot{n}_\beta) = -L \frac{\partial a}{\partial S} (n_\alpha n_\beta - \frac{1}{3}\delta_{\alpha\beta}) + G_{\alpha\beta} \tag{24}$$

with

$$a = \frac{1}{2} (1 - \frac{U}{3})S^2 - \frac{U}{9} S^3 + \frac{U}{6} S^4. \tag{25}$$

Multiplying $n_\alpha n_\beta$ to both sides of this equation and taking the sum over α and β, one gets

$$\hat{S} = -L\frac{\partial a}{\partial S} + \frac{2}{3}(1 - S)(1 + 2S)\underset{\approx}{\kappa}:\underset{\sim}{n}\underset{\sim}{n}. \tag{26}$$

From, eq. (24) and (25) it follows

$$\dot{\underset{\sim}{n}} = \frac{1 + 2S}{3S}\underset{\approx}{\kappa} \cdot \underset{\sim}{n} + \frac{1 - S}{3S}\underset{\approx}{\kappa}^{+} \cdot \underset{\sim}{n} - \xi\underset{\sim}{n} \tag{27}$$

where ξ is chosen so that $\dot{n} \cdot n = 0$.

Eq. (26) and (27) are a generalization of the Ericksen-Leslie theory. Instead of the equilibrium order parameter S_{eq}, S is now given by the solution of eq. (26).

In the case of steady shear flow, $v_x = \kappa y$, $v_y = v_z = 0$, $\underset{\approx}{n}$ is written as

$$n_x = \cos\chi \qquad n_y = \sin\chi \qquad n_z = 0 \tag{28}$$

The solution is given by

$$\frac{\kappa}{L} = \frac{(2 + S)S[(1 - \frac{U}{3}) - \frac{U}{3}\dot{S} + \frac{2}{3}US^2]}{(1 - S)^{3/2}(1 + 2S)^{3/2}} \tag{29}$$

and

$$\tan\chi = \left(\frac{1 - S}{1 + 2S}\right)^{1/2} \tag{30}$$

Fig. 1 shows the steady state viscosity $\eta = \sigma_{xy}/\kappa$ plotted against the reduced shear rate for various values of the initial order parameter S_{eq}. The shear thinning becomes less prominent as S_{eq} increases. This is qualitatively consistent with experimental results [5].

Acknowledgements:

The author thanks the Institute for Mathematics and its Applications and the Department of Chemical Engineering of the University of Minnesota for providing the opportunity to work at IMA. Thanks are also due to Professor M. Tirrell and Professor J.L. Ericksen for their stimulating discussions on the subject in this paper.

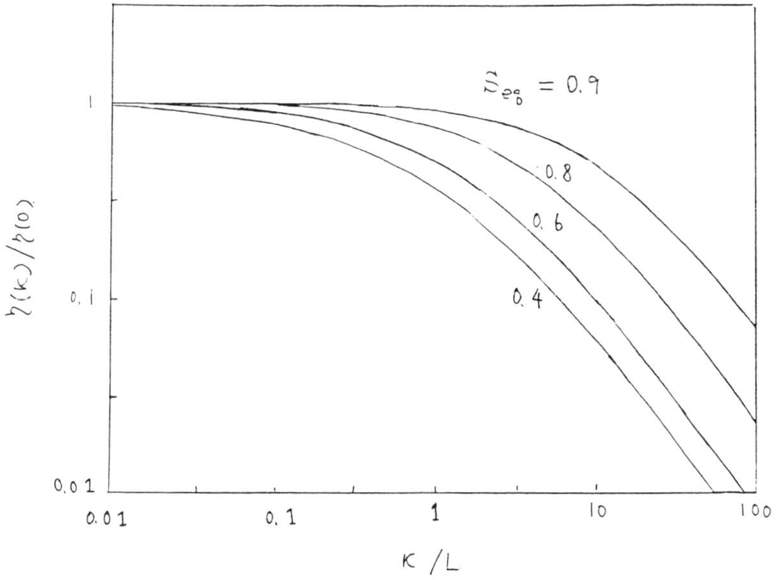

Figure 1

References

[1] J.L. Ericksen, Trans. Soc. Rheol. 5, 23 (1961).

[2] J.L. Ericksen, "Orienting Polymers" ed. by J.L. Ericksen, Springer, Berlin, p. 27.

[3] F.M. Leslie, Arch. Ration. Mech. Anal., 28, 265 (1968).

[4] P.G. de Gennes, "The Physics of Liquid Crystals", Clarendon Press, Oxford, 1975.

[5] K. Wissbrun, J. Rheol. 25, 619 (1981), see also "Orienting Polymers" ed. by J.L. Ericksen, Springer Berlin, p. 1.

[6] P.G. de Gennes, Mol. Cryst. Liq. Cryst. 12, 193 (1971).

[7] M. Doi and S.F. Edwards, "The Theory of Polymer Dynamics" Oxford University Press", to be published.

[8] J.G. Kirkwood and P.L. Auer, J. Chem. Phys. 19, 281 (1959), and J.G. Riseman and J.G. Kirkwood, J. Chem. Phys. 17, 442 (1949).

[9] M. Doi, J. Polymer Sci., 19, 229 (1981).

[10] M. Doi, and S.F. Edwards, J. Chem. Soc. Faraday Trans. II, 74, 568, 918 (1978).

[11] W. Maier and A.Z. Saupe, Z. Naturforsch., a14, 882 (1959).

[12] M. Doi, Faraday Symp. Chem. Soc. 18, 49 (1983).

[13] G. Marrucci, Mol. Cryst. and Liq. Cryst. 72, 153 (1982).

Mathematical Questions of Liquid Crystal Theory

Robert Hardt and David Kinderlehrer

School of Mathematics
University of Minnesota
Minneapolis, Minnesota 55455 USA

1 Introduction

A liquid crystal is a mesomorphic phase of a material which occurs between its liquid and solid phases. Frequently the material is composed of rod-like molecules which display orientational order, unlike a liquid, but lacking the lattice structure of a solid. It may flow easily and so may also be thought of as an anisotropic fluid. This anisotropy is evident in the way it transmits light; for example, a nematic liquid crystal is optically uniaxial. We take the opportunity of this note to discuss a few of the analytical issues which arise in the attempt to study static equilibrium configurations. One attractive feature of this subject is that it has a well developed continuum description in the Ericksen-Leslie [E_1],[L_1] theory. Some of the questions have significance in the context of harmonic mappings into spheres and we shall attempt to clarify these connections. We refer to the articles by F. Leslie [L_2],[L_3] in these proceedings both for other aspects of the static theory and for an introduction to flow problems.

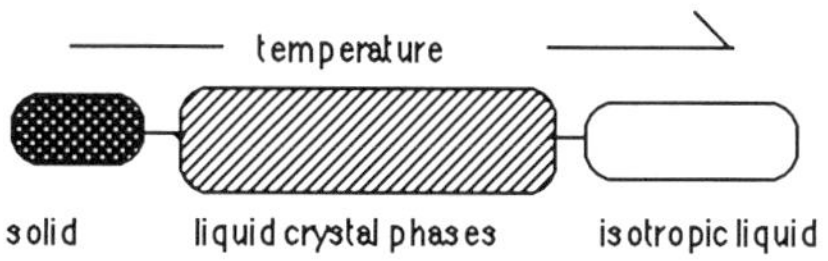

The kinematic variable in the nematic and cholesteric phases, which will be our major concern, may be taken to be the optic axis, a unit vector field n defined in the region Ω in $\mathbb{R}^3$ occupied by the material. Later on, when we discuss phase transitions and defects, we shall describe how this is connected to the Landau style order parameter associated to the molecular axis distribution. The bulk energy of the fluid at a fixed temperature is given in terms of the Oseen-Frank density, [O],[F],

$$2W(\nabla n,n) = \kappa_1(\operatorname{div} n)^2 + \kappa_2(n\cdot\operatorname{curl} n + q)^2 + \kappa_3|n\wedge\operatorname{curl} n|^2$$
$$+ (\kappa_2 + \kappa_4)(\operatorname{tr}(\nabla n)^2 - (\operatorname{div} n)^2) , \qquad (1.1)$$

where the constants κ_i, $i = 1,2,3,4$ are usually assumed to satisfy

$$\kappa_1 > 0 , \quad \kappa_2 > 0 , \quad \kappa_3 > 0 , \quad \text{and} \quad \kappa_2 \geq |\kappa_4| .$$

In this discussion we assume only that

$$\kappa_i > 0, \quad i = 1,2,3 . \qquad (1.2)$$

The liquid crystal is nematic if $q = 0$ and cholesteric, displaying some chiral character, otherwise. In the special case $\kappa_1 = \kappa_2 = \kappa_3 = \kappa$ and $\kappa_4 = q = 0$,

$$W(\nabla n, n) = \tfrac{1}{2}\kappa|\nabla n|^2 , \qquad (1.3)$$

which is the integrand for a harmonic mapping of Ω into S^2 .

Other contributions to the energy may be given by magnetic or electric fields or via boundary conditions. Or, for example, in the case of some blue phases, certain choices of the constants in (1.1) deprive the fluid of a smooth natural state, a condition called frustrated by physicists, cf. the papers by Cladis and Sethna [Cl], [Se]. We ought to keep in mind such additional possibilities since they are of practical and physical interest.

Let us set

$$\mathcal{W}(n) = \int_\Omega W(\nabla n, n)\, dx \qquad (1.4)$$

and denote by $\mathcal{U}$ the sum of additional energies, which may depend on other fields as well as n. In this generality, we may say that our concern is vector fields n and possibly other fields which render stationary the total energy,

$$\delta(\mathcal{W} + \mathcal{U}) = 0 , \qquad (1.5)$$

subject to suitable boundary conditions. It is apparent that we confront a nonlinear problem with a nonconvex constraint. There is a substantial literature about special solutions of (1.5), cf. [E_1],[E_2]. Primarily this concerns configurations arranged between two parallel plates and involves choosing particular forms of the solution which reduce the Euler-Lagrange equations of (1.5) to ordinary differential equations. These are especially important for applications in display devices. Bifurcation of solutions may occur and solutions need not be unique under these circumstances. Indeed, the exchange of stability is used to determine values of the constants κ_1, κ_2, and κ_3.

Parameters like n are frequently referred to as internal variables in mechanics. They are often microscopic quantities whose gross behavior influences that of the material. Void fractions in materials with voids, cf. Capriz-Giovine [CG], or lattice polarizations in multi-lattice crystals are other examples.

2 Existence theory for nematics and cholesterics

In this section we present an existence theory for some typical situations. For convenience, we arrange some notations. Suppose that $\Omega \subset \mathbb{R}^3$ is bounded with smooth boundary $\partial\Omega$ and let ([1])

$$H^1(\Omega;S^2) = \{ u \in H^1(\Omega;\mathbb{R}^3) : |u| = 1 \text{ in } \Omega \}$$

Given n_0 defined on $\partial\Omega$, consider the set of admissible variations, in this instance, virtual optic axes,

$$A(n_0) = \{ u \in H^1(\Omega;S^2) : u = n_0 \text{ on } \partial\Omega \}. \tag{2.1}$$

We first study

Problem 2.1   Find $n \in A(n_0)$ such that

$$W(n) = \inf_{A(n_0)} W(u) \tag{2.2}$$

So a solution of Problem 2.1 verifies (1.5) with $U = 0$. The Dirichlet boundary condition determined by $A(n_0)$ is called strong anchoring in the liquid crystal literature. The two immediate questions are

- When is $A(n_0)$ nonempty?

- When does a suitable minimizing sequence converge to a minimum value of W ?

One notices at the outset that $H^1(\Omega;S^2)$ as we have defined it is larger than the closure of smooth functions from Ω into S^2 in the H^1 topology. For example, if $0 \in \Omega$, $u(x) = x/|x| \in H^1(\Omega;S^2)$ but is not approximable by smooth functions ζ satisfying $|\zeta| = 1$. Moreover, liquid crystals display a variety of singularities and it is clear that in some cases they must be present simply for topological reasons. We shall discuss these issues in §4.

Lemma 2.1 *If $n_0 : \partial\Omega \to S^2$ is Lipschitz* (2), *then $A(n_0)$ is not empty.*

Proof: First note that in the special case where $\overline{\Omega}$ is homeomorphic to a closed ball by a Lipschitz mapping, a homogeneous extension of degree zero of n_0 has finite energy. The general case may be reduced to this one, cf. [HKL_2] Lemma 1.1. □

The second question amounts to determining the coerciveness of W. In most discussions of liquid crystals, the last term in W is set to zero, $\kappa_2 = -\kappa_4$. This is because

$$S(\nabla u) = \tfrac{1}{2}(\mathrm{tr}(\nabla u)^2 - (\mathrm{div}\,u)^2) = \tfrac{1}{2}\mathrm{div}((\nabla u)u - (\mathrm{div}\,u)u) \quad (2.3)$$

is a divergence, offering no contribution to the equilibrium equations. It is sometimes called a surface energy. As Oseen, and independently, Ericksen, have observed, it is a surface energy in the strictest sense, for

$$((\nabla u)u - (\mathrm{div}\,u)u)\nu$$

depends only on $u|_{\partial\Omega}$ and its tangential derivatives. It follows that if n_0 is Lipschitz on $\partial\Omega$, then the quantity

$$S(n_0) = \frac{1}{2}\int_\Omega (\mathrm{tr}(\nabla u)^2 - (\mathrm{div}\,u)^2)\,dx$$

is constant on $A(n_0)$.

On the other hand, this permits us to add multiples of (2.2) to W without changing the solution of Problem 2.1. Now for any $u \in H^1(\Omega;S^2)$,

$$|\nabla u|^2 = \mathrm{tr}(\nabla u)^2 + |\mathrm{curl}\,u|^2$$

$$= \quad tr(\nabla u)^2 \; + \; |u\cdot curl\, u|^2 \; + \; |u \wedge curl\, u|^2 \, .$$

Thus if $\alpha = \min\{ \kappa_1, \kappa_2, \kappa_3\}$ and

$$2W^*(\nabla u,u) \quad = \quad \kappa_1(div\, u)^2 + \kappa_2(u\cdot curl\, u + q)^2 + \kappa_3|u\wedge curl\, u|^2 + \alpha S(\nabla u) \, , \quad (2.4)$$

we have that

$$\begin{aligned} W^*(\nabla u,u) &\geq \tfrac{1}{2}\alpha|\nabla u|^2 - \kappa_2 q|u\cdot curl\, u| - \tfrac{1}{2}\kappa_2 q^2 \\ &\geq \tfrac{1}{4}\alpha|\nabla u|^2 - C\, , \qquad u \in H^1(\Omega;S^2)\, , \end{aligned}$$

for some $C > 0$. Consequently, with the obvious notation, the functional $\mathcal{W}^*$ is coercive on $H^1(\Omega;S^2)$. It is easy to check that it is lower semicontinuous on $\mathcal{A}(n_0)$, and so a solution of Problem 2.1 exists for $\mathcal{W}^*$ Finally,

$$\mathcal{W}(u) \quad = \quad \mathcal{W}^*(u) \; + \; (\kappa_2 + \kappa_4 - \alpha)\, \mathcal{S}(n_0),$$

so we arrive at

Theorem 2.2 *Let* n_0 *be Lipschitz on* $\partial\Omega$ *with* $|n_0| = 1$ *and assume that* W *satisfies* (1.2). *Then there exists an* $n \in \mathcal{A}(n_0)$ *such that*

$$\mathcal{W}(n) \quad = \quad \inf_{\mathcal{A}(n_0)} \mathcal{W}(u).$$

In order to have a basis for a wider discussion, let us immediately take into account the effects of magnetic and electric fields. The magnetization vector is given by

$$M \quad = \quad (\chi_0 \mathbf{1} + \chi_a n\otimes n)H \qquad (2.5)$$

where χ_0 and χ_a are real and $H = H(x)$, $x \in \Omega$, is a given magnetic field, usually constant. The energy contributed to the system is

$$W_M(u) \quad = \quad -\tfrac{1}{2} M\cdot H\, . \qquad (2.6)$$

Analogous to the previous theorem, we may state

Theorem 2.3 *Let* n_0 *be Lipschitz on* $\partial\Omega$ *with* $|n_0| = 1$ *and assume that* W *satisfies* (1.2). *Let* $H \in L^2(\Omega;\mathbb{R}^3)$. *Then there exists an* $n \in \mathcal{A}(n_0)$ *such that*

$$\mathcal{W}(n) - \int_\Omega W_M(n)\,dx = \inf_{\mathcal{A}(n_0)} \{\mathcal{W}(u) - \int_\Omega W_M(u)\,dx\}.$$

The electric displacement vector, however, should be considered as induced owing to the much larger size of the dielectric constants when compared to the magnetization ones, [dG], p.97. The displacement vector is given by

$$D = D(E,n) = (\alpha_0 \mathbf{1} + \alpha_a n \otimes n)E \tag{2.7}$$

for an electric field E, where $\alpha_0, \alpha_a \in \mathbb{R}$, and the energy density of the system is

$$W - \tfrac{1}{2} D \cdot E .$$

In the static case, which we shall assume, with no free charge, Maxwell's equations are

$$\operatorname{div} D = 0 \quad \text{and} \quad \operatorname{curl} E = 0. \tag{2.8}$$

We thus assume that $E = -\nabla\phi$, where ϕ is a new dependent variable and write $A(\nabla\phi, n) = \frac{1}{2} D \cdot E$. For the simplest sort of problem, let us prescribe $\phi_0 \in H^1(\Omega)$ and seek a couple (n,ϕ) so that

$$\delta\Big(\mathcal{W}(n) - \int_\Omega A(\nabla\phi, n)\,dx\Big) = 0 \quad \text{for } n \in \mathcal{A}(n_0) \text{ and}$$
$$\phi \in H^1(\Omega),\ \phi = \phi_0 \text{ on } \partial\Omega. \tag{2.9}$$

The functional determined by

$$W(\nabla u, u) - A(\nabla\phi, u)$$

is not bounded below since the electromagnetic term is highly competitive with the bulk energy. To resolve this difficulty, the first equation of (2.8) will be imposed as a constraint. Given a virtual optic axis u, we define

$$\zeta = \Phi(u)$$

by

$$\begin{aligned} \operatorname{div} D(\nabla\zeta, u) &= 0 \quad \text{in } \Omega \\ \zeta &= \phi_0 \quad \text{on } \partial\Omega. \end{aligned} \tag{2.10}$$

This equation is uniformly elliptic, independently of u, $|u| = 1$, provided

$$\alpha_0 > 0 \quad \text{and} \quad \alpha_0 + \alpha_a > 0, \tag{2.11}$$

which we assume henceforth. Now ζ is the solution of the minimization problem

$$\frac{1}{2}\int_\Omega D(\nabla\zeta, u)\cdot\nabla\zeta \, dx = \inf \frac{1}{2}\int_\Omega D(\nabla\eta, u)\cdot\nabla\eta \, dx,$$
$$\eta \in H^1(\Omega), \ \eta = \phi_0 \text{ on } \partial\Omega,$$

whence for some constants $C_0, C_1 > 0$,

$$\begin{aligned}\int_\Omega |\nabla\zeta|^2 \, dx &\le \text{const.} \int_\Omega D(\nabla\zeta, u)\cdot\nabla\zeta \, dx \\ &\le \text{const.} \int_\Omega D(\nabla\phi_0, u)\cdot\nabla\phi_0 \, dx \\ &\le C_0 \int_\Omega |\nabla\phi_0|^2 \, dx = C_1 .\end{aligned}$$

Thus the functional

$$W(u) - \int_\Omega A(\nabla\Phi(u), u) \, dx, \quad u \in A(n_0), \tag{2.12}$$

is bounded below. This brings us to

Theorem 2.4 *Let* n_0 *be Lipschitz on* $\partial\Omega$ *with* $|n_0| = 1$ *and assume that* W *satisfies* (1.2). *Let* (2.11) hold. *Then*

(i) (n, ϕ) *is a solution of* (2.9) *if and only if* $\phi = \Phi(n)$ *and*

$$\delta\left(W(n) - \int_\Omega A(\nabla\Phi(n), n) \, dx\right) = 0 \quad \text{on } A(n_0) ; \tag{2.13}$$

(ii) *there exists an* $n \in A(n_0)$ *such that*

$$W(n) - \int_\Omega A(\nabla\Phi(n), n) \, dx = \inf_{A(n_0)} \left\{W(u) - \int_\Omega A(\nabla\Phi(u), u) \, dx\right\}.$$

To show that the stationary points of the functional on the entire set $A(n_0) \times \{\phi \in H^1(\Omega) : \phi = \phi_0 \text{ on } \partial\Omega\}$ coincide with those on the submanifold determined by (2.10) is very easy, at least if one employs the equilibrium equations which we derive in the next section, cf. [HKL$_2$] §4 for details. The proof of (ii) follows from the preceding discussion. One must verify, of course, that $u_j \to n$ in $L^2(\Omega)$ implies that

$$\int_\Omega A(\nabla\Phi(u_j), u_j)\,dx \to \int_\Omega A(\nabla\Phi(n), n)\,dx.$$

The nature of boundary conditions is discussed by de Gennes and Ericksen [E$_1$] p. 241. Although strong anchoring is the most common, and has obvious mathematical advantages, interfacial energies of different sorts may also be pertinent. When a portion of the liquid corresponding to $\Gamma \subset \partial\Omega$ adjoins an isotropic fluid, for example, a surface energy of the form

$$\int_\Gamma f((n\cdot\nu)^2)\,dS \qquad (2.14)$$

is sometimes introduced. In such cases it seems reasonable to impose coerciveness on W. It would be of interest to have a good mathematical study of droplets immersed or floating in water, for instance. The shape of the droplet may be variable as well, leading to a free boundary problem.

Methods of functional analysis would suggest the natural existence condition in Problem 2.1 to be that n_0 have a finite energy extension to Ω, or in other words, that the Dirichlet Problem for ordinary harmonic functions have a solution in $H^1(\Omega;\mathbb{R}^3)$ with boundary values n_0. For example in the special case where Ω is smooth and contains the origin, the function

$$n_0(x) = (\cos\theta, \sin\theta, 0), \qquad x \in \Omega,$$

with θ the angle in the x_1, x_2 plane, is a solution of the nematic equilibrium equations away from the x_3-axis, but it does not have finite energy in Ω. However it does admit a finite energy extension to Ω, and Gulliver has shown that $\mathcal{A}(n_0)$ is not empty. His construction gives an interesting geometric picture of how these boundary values strive to attain finite energy. So Problem 2.1 has a solution, but it is not n_0.

More generally, one may show that for every boundary orientation $n_0 \in H^{1/2}(\partial\Omega; S^2)$, i.e., for any n_0 enjoying a finite energy extension to Ω, $\mathcal{A}(n_0)$ is not empty [HL$_1$, 6.2], cf. also [HKL$_2$] Lemma 2.3.

Line singularities, or disclinations, which are described by vectors like n_0 above are observed. Several ways to account for them energetically have been suggested. We shall return to this later.

Before closing this section, let us discuss uniqueness. Our results establish the existence of an energy minimizing configuration, but do not tell us how many there are, nor if there are equilibrium configurations which are not absolute minimizers. Almost nothing is known about this. Consider, for example, the nematic case where

$$n_0 = a \quad \text{on } \partial\Omega, \qquad a \in S^2, \text{ a constant.}$$

Clearly, $n(x) = a$ in Ω is the only minimizing solution. Consider the equal constant case, where the equilibrium equation is given by

$$\Delta u + |\nabla u|^2 u = 0 \quad \text{in } \Omega, \tag{2.15}$$

which we shall show in the next section. A smooth solution of (2.15) fulfills the identity

$$r\int_{\partial Br} |\nabla_{\tan} u|^2 \, dS = \int_{Br} |\nabla u|^2 \, dx + \int_{\partial Br} |\partial u/\partial r|^2 \, dS$$

where B_r is a ball of radius $r > 0$. Thus any smooth solution of (2.15) with constant boundary values is constant. If $|a|$ is sufficiently small, it is possible to deduce that all (weak) solutions of (2.15) are smooth, the details of which are not pertinent to us here.

On the other hand, some very simple boundary values lead to multiple solutions. We found this example with Mitchell Luskin. One may easily check, and it will be apparent from §3, that for any $q \in \mathbb{R}$,

$$n_q(x) = (\cos qx_3, \sin qx_3, 0), \qquad x \in \Omega, \tag{2.16}$$

is a solution of the nematic equilibrium equations with energy

$$\int_\Omega W(\nabla n_q, n_q) \, dx = \tfrac{1}{2}\kappa_2 |\Omega| q^2 . \tag{2.17}$$

Let us again restrict our attention to the equal constant case, with $\kappa_2 = 1$, and choose $\Omega = Q_h$, a cube of side length $2h$ centered at $x = 0$. Let $0 < \delta < 1$ and $\eta(x)$ be a Lipschitz function such that

$$\eta = 1 \quad \text{in } Q_{(1-\delta)h},$$

$$\eta = 0 \quad \text{on } \partial Q_h , \quad \text{and}$$
$$|\nabla\eta| \le 1/(\delta h).$$

Fix $a = (0,0,1)$ which is orthogonal to $n_q(x)$, $x \in Q_h$, and set

$$u = (\sin\eta) n_q + (\cos\eta) a .$$

Now $|u| = 1$ and

$$|\nabla u|^2 \le 2(\sin\eta)^2 |\nabla n_q|^2 + 2|\nabla\eta|^2 .$$

Hence

$$\int_{Q_h} |\nabla u|^2 \, dx \le \int_{Q_h - Q_{(1-\delta)h}} (|\nabla n_q|^2 + |\nabla\eta|^2) \, dx$$
$$\le 2|Q_h - Q_{(1-\delta)h}| \, (q^2 + (\delta h)^{-2})$$
$$= 16(3\delta - 3\delta^2 + \delta^3) h^3 (q^2 + (\delta h)^{-2}) .$$

In particular, we may assure that

$$\int_{Q_h} |\nabla u|^2 \, dx \le \int_{Q_h} |\nabla n_q|^2 \, dx ,$$

by choosing δ and q so that

$$16(3\delta - 3\delta^2 + \delta^3) h^3 (q^2 + (\delta h)^{-2}) < 8h^3 q^2$$

or

$$2(3\delta - 3\delta^2 + \delta^3)[\delta^2(2(1-\delta)^3 - 1]^{-1} < (qh)^2.$$

In this situation, n_q is not the minimum found by Theorem 2.3. So even for boundary values which are topologically trivial, that is, of degree zero as mappings of $\partial\Omega$ into S^2, there may be more than one solution of the equilibrium equations.

In [HKL_5] we generalize this example by considering, again for the equal constant case, an arbitrary planar solution, i.e. one in the form

$$n(x) = (n^1(x), n^2(x), 0) \quad \text{for } x \in \Omega .$$

One easily infers that

$$n = (\cos\theta, \sin\theta, 0) \quad \text{and} \quad |\nabla n|^2 = |\nabla\theta|^2$$

for some angle function $\theta \in H^1(\Omega;\mathbb{R})$ satisfying

$$-\Delta\theta = 0 \quad \text{in} \ \Omega .$$

In particular, all planar solutions are smooth. However, one finds in [HKL_5] a constant C so that n fails to minimize energy whenever

$$C \leq r^{-1}\int_{B_r} |\nabla\theta|^2 \, dx \quad \text{and} \quad B_{2r} \subset \Omega .$$

A consequence is that

a harmonic mapping of $\mathbb{R}^3$ into S^2 whose image lies in a plane and which minimizes energy with respect to its boundary values on arbitrarily large balls is constant.

Planar solutions, in the nonequal constant case have been studied by E. MacMillan [MM].

3 Equilibrium Equations

An elementary, if formal, way to calculate the equations satisfied by a stationary point of W, namely

$$\delta \int_\Omega W(\nabla n, n)\, dx = 0 \qquad \text{on} \quad \mathcal{A}(n_0) \tag{3.1}$$

is to introduce a Lagrange multiplier $-\frac{1}{2}\lambda(|n|^2 - 1)$. We then obtain the formal equations

$$\int_\Omega \{ W_p(\nabla n, n)\cdot\nabla\zeta + W_n(\nabla n, n)\cdot\zeta - \lambda n\cdot\zeta \}\, dx = 0 ,$$
$$\zeta \in H_0^1(\Omega;\mathbb{R}^3) \cap L^\infty(\Omega;\mathbb{R}^3), \tag{3.2}$$

or in weak form,

$$-\operatorname{div} W_p(\nabla n, n) + W_n(\nabla n, n) = \lambda n \qquad \text{in} \ \Omega \tag{3.3}$$

$$n = n_o \qquad \text{on } \partial\Omega ,$$

where λ is an unknown function. The expression

$$h = -\operatorname{div} W_p(\nabla n, n) + W_n(\nabla n, n)$$

is called the molecular field.

Another way to write (3.3) is to introduce a generalized Legendre transform

$$t = W\mathbb{1} - W_p \nabla n^T \tag{3.4}$$

or

$$t_{ij} = W\delta_{ij} - (\partial W/\partial p_j^k) n_{x_i}^k ,$$

whence

$$\operatorname{div} t = 0. \tag{3.5}$$

The geometrical content of (3.3) is clear, but it may be misleading to the analyst since the dependence of λ on ∇n and $\nabla^2 n$ is disguised. Choosing $v = \eta n$ in (3.2) where η has compact support, gives that

$$\int_\Omega \{\eta(W_p \cdot \nabla n + W_n \cdot n - \lambda) + W_p \cdot n \otimes \nabla \eta\} \, dx = 0.$$

With $\alpha = \min\{\kappa_1, \kappa_2, \kappa_3\}$, we now write

$$W(\nabla n, n) = \tfrac{1}{2}\alpha |\nabla n|^2 + V(\nabla n, n) \tag{3.6}$$

so

$$W_p(\nabla n, n) = \alpha \nabla n + V_p(\nabla n, n).$$

Now $n\nabla n = 0$, so $nW_p = nV_p$. This gives that

$$\lambda = -\operatorname{div} nV_p + W_p \cdot \nabla n + W_n \cdot n ,$$

and, after some manipulation,

$$-\operatorname{div}(W_p - n \otimes nV_p) + (\mathbb{1} - n \otimes n)W_n = \nabla n \cdot (nV_p) + (W_p \cdot \nabla n)n \quad \text{in } \Omega .$$

We abbreviate this by writing

$$-\operatorname{div}(W_p - n\otimes n V_p) + Y = 0 \quad \text{in } \Omega , \tag{3.7}$$

where, for constants C_0 (= 0 in the nematic case) and C_1 ,

$$|Y(\nabla n, n)| \le C_0 + C_1|\nabla n|^2 . \tag{3.8}$$

For example, in the equal constant case, which corresponds to a harmonic mapping, (1.3), the equilibrium equation is

$$-\Delta n + |\nabla n|^2 n = 0 \quad \text{in } \Omega . \tag{3.9}$$

All cases demonstrate quadratic growth of Y in ∇n, a known peril in the study of systems. Although (3.9) is elliptic, it is not clear that (3.7) is elliptic for every choice of constants $\kappa_1, \kappa_2, \kappa_3$ and n , $|n| = 1$. Moreover, we do not know how important it is that the equations be elliptic. Our existence theorem was based only on the coerciveness of W on the set $\mathcal{A}(n_0)$ and our regularity results, discussed in §5 et seq., make use only of the ellipticity of a certain limit blow up system. However it is clear that a numerical scheme based on iteratively solving the equilibrium system will be more robust if this system is elliptic. Let us attend to this question now in more detail. Writing again

$$W(\nabla n, n) = \tfrac{1}{2}\alpha|\nabla n|^2 + V(\nabla n, n) , \quad \alpha = \min\{\kappa_1, \kappa_2, \kappa_3\} ,$$

we express W and V as functions of a 3×3 matrix $p = (p^i_j)$, a vector $n = (n^i) \in \mathbb{R}^3$, and the totally antisymmetric matrix, which we denote by $\Lambda = (\lambda_{ijk})$. Recall that for a vector $a \in \mathbb{R}^3$, $\Lambda(a) + \Lambda(a)^T = 0$,

$$\Lambda(a)v = a \wedge v , \quad \text{for all } v \in \mathbb{R}^3 ,$$

and

$$\Lambda(\xi \otimes \eta) = \xi \wedge \eta .$$

Thus we may write

$$W(p,n) = \tfrac{1}{2}\alpha|p|^2 + V(p,n) ,$$

$$V(p,n) = \tfrac{1}{2}\beta_1(\operatorname{tr} p)^2 + \tfrac{1}{2}\beta_2(\Lambda(n)\cdot p)^2 + \tfrac{1}{2}\beta_3|\Lambda(n)\Lambda(p)|^2 + \tfrac{1}{2}\beta_4 S(p) , \tag{3.10}$$

$$S(p) = \operatorname{tr}(p^2) - (\operatorname{tr} p)^2,$$

where $\beta_i = \kappa_i - \alpha_i$, $i = 1,2,3$, and $\beta_4 = \kappa_2 + \kappa_4 - \alpha$. Note that $\beta_i > 0$ and that at least one of the $\beta_i = 0$, $i = 1,2,3$. We wish to evaluate the expression

$$U(\xi \otimes \eta, n) = \{W_p(\xi \otimes \eta, n) - n \otimes n\, V_p(\xi \otimes \eta, n)\} \cdot \xi \otimes \eta$$

$$= \alpha|\xi|^2|\eta|^2 + (\mathbf{1} - n \otimes n) V_p(\xi \otimes \eta, n) \cdot \xi \otimes \eta. \tag{3.11}$$

We consider each of the terms separately. Notice that the term involving S gives no contribution. Let $U_i(\xi \otimes \eta, n)$ denote the coefficient of β_i in (3.11). We have that

$$\begin{aligned} U_1(\xi \otimes \eta, n) &= (\mathbf{1} - n \otimes n)\, \mathrm{tr}\, \xi \otimes \eta\, \mathbf{1} \cdot \xi \otimes \eta \\ &= (\xi \cdot \eta)^2 - \xi \cdot n\, \eta \cdot n\, \xi \cdot \eta \ . \end{aligned} \tag{3.12}$$

For the second term,

$$\begin{aligned} U_2(\xi \otimes \eta, n) &= [(\mathbf{1} - n \otimes n)\, \Lambda(n) \cdot \xi \otimes \eta\, \Lambda(n)] \cdot \xi \otimes \eta \\ &= |\Lambda(n) \cdot \xi \otimes \eta|^2 - \Lambda(n) \cdot \xi \otimes \eta\, n \otimes n\, \Lambda(n) \cdot \xi \otimes \eta \ . \end{aligned}$$

Now $n \otimes n \Lambda(n) = n \otimes \Lambda(n)^T n = -n \otimes \Lambda(n) n = -n \otimes (n \wedge n) = 0$, so

$$U_2(\xi \otimes \eta, n) = |\Lambda(n) \cdot \xi \otimes \eta|^2 = (\det(\xi \eta n))^2 , \tag{3.13}$$

the square of the triple product of ξ, η, and n.

Finally, we consider U_3. For matrices p and q,

$$\begin{aligned} \tfrac{1}{2}\, {}^{d}\!/_{dt}\Big|_{t=0} |\Lambda(n)\Lambda(p + tq)|^2 &= \Lambda(n)\Lambda(p) \cdot \Lambda(n)\Lambda(q) \\ &= -\Lambda^T \Lambda(n)^2 \Lambda(p) \cdot q \ , \end{aligned}$$

which tells us that

$$U_3(\xi \otimes \eta, n) = -(\mathbf{1} - n \otimes n)\Lambda^T \Lambda(n)^2 \Lambda(\xi \otimes \eta) \cdot \xi \otimes \eta$$

$$= |\Lambda(n)\Lambda(\xi \otimes \eta)|^2 + n \otimes n \Lambda^T \Lambda(n)^2 \Lambda(\xi \otimes \eta) \cdot \xi \otimes \eta$$

$$= |\Lambda(n)\Lambda(\xi \otimes \eta)|^2 - \xi \cdot n\, \Lambda(n)\Lambda(\xi \otimes \eta) \cdot \Lambda(n)\Lambda(n \otimes \eta)$$

$$= |n \wedge (\xi \wedge \eta)|^2 - \xi \cdot n(\xi \wedge \eta \cdot n \wedge \eta - n \cdot (\xi \wedge \eta) n \cdot (n \wedge \eta))$$

$$= |n \wedge (\xi \wedge \eta)|^2 - \xi\cdot n(\xi\cdot n|\eta|^2 - \xi\cdot\eta\, n\cdot\eta)$$

$$= |n \wedge (\xi \wedge \eta)|^2 - (\xi\cdot n)^2|\eta|^2 + \xi\cdot\eta\, n\cdot\eta|\eta|^2.$$

Let us now calculate

$$\inf_{|\xi| = |\eta| = |n| = 1} U_3(\xi\otimes\eta, n)$$

for which it suffices to set $n = e_3$. In this case,

$$U_3(\xi\otimes\eta, e_3) = |e_3 \wedge (\xi\wedge\eta)|^2 - (\xi^3)^2 + \xi^3\eta^3\,\xi\cdot\eta$$

$$= (\eta^3)^2(1 - (\xi^3)^2) - \xi^3\eta^3(\xi^1\eta^1 + \xi^2\eta^2)\,.$$

This expression is most negative if (ξ^1, ξ^2) and (η^1, η^2) are parallel, so let

$$\begin{aligned} \xi &= \sqrt{1 - x^2}\; a + x e_3 \\ \eta &= \sqrt{1 - y^2}\; a + y e_3 \end{aligned} \qquad , \quad -1 \le x, y \le 1$$

where a is a unit vector in the x_1, x_2 plane and set

$$f(x,y) = (1 - x^2)y^2 - xy\sqrt{1 - x^2}\sqrt{1 - y^2}\ .$$

We found the minimum of this function, with apologies to the Minnesota Supercomputer Institute, by HP 15C to be

$$f(\sqrt{3/4}, 1/2) = -0.125,$$

in other words

$$\inf_{|\xi| = |\eta| = |n| = 1} U_3(\xi\otimes\eta, n) = -0.125.$$

The infimum of U_1 may be calculated in terms of $f(x, \operatorname{sgn} y\sqrt{1 - y^2})$, so in addition

$$\inf_{|\xi| = |\eta| = |n| = 1} U_1(\xi\otimes\eta, n) = -0.125.$$

Hence

$$U(\xi\otimes\eta,n) \geq \alpha + \beta_2(\det(\xi,\eta,n))^2 - 0.125(\beta_1+\beta_2)$$
$$\geq \alpha - 0.125(\beta_1+\beta_2)\ ,$$

so a sufficient condition for the ellipticity of the system is that

$$\alpha > {}^1/_8((\kappa_1-\alpha)^+ + (\kappa_3-\alpha)^+)\ , \quad \alpha = \min(\kappa_1,\kappa_2,\kappa_3).$$

The terms U_1 and U_3 actually compete somewhat, so a necessary condition may be derived by accounting for this.

4 Regularity theory

In this section we consider the question of the regularity of energy-minimizing configurations, such as those obtained in Theorems 2.2, 2.3, and 2.4.

First we again observe that one cannot expect, in general, complete smoothness, or even continuity, of a minimizer n on all of $\overline{\Omega}$. In fact, if the given boundary-value function $n_0 : \partial\Omega \to S^2$ has nonzero degree, then there is no continuous mapping $u : \overline{\Omega} \to S^2$ whose restriction to $\partial\Omega$ is n_0. The simplest case of this is where Ω equals the unit ball B and $n_0(x) = x$ for each $x \in \partial B = S^2$. The intriguing problem that the minimizer for this particular boundary data is simply the function $x/|x|$ has recently been settled by Brezis, Coron, and Lieb, cf. [B], [BCL$_1$], [BCL$_2$].

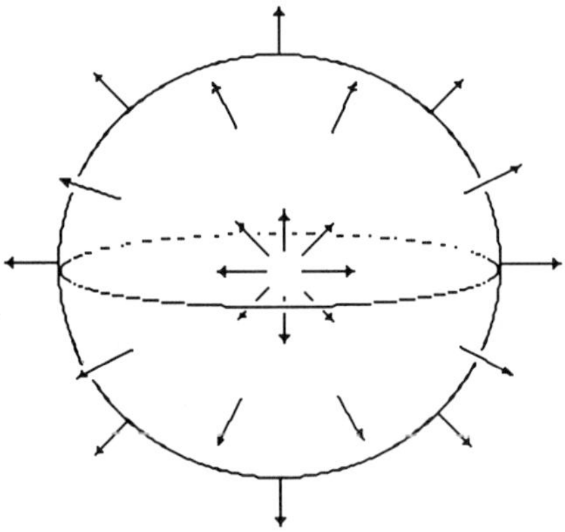

$x/|x|$

Topological restrictions may thus force the existence of singularities in minimizing solutions. However, singularities may also occur in the absence of such topological restrictions.

Theorem 4.1 There exists a smooth map $n_0 : \partial\Omega \to S^2$ which has degree 0 (and hence admits some continuous extension to $\overline{\Omega}$) but for which any energy minimizing extension must have a discontinuity.

Proof : Although Ω is simply the unit ball in the proof in [HL_1], it is easiest to describe the idea of the construction by considering Ω to be a dumbell-shaped region consisting of two unit balls, B_+ and B_-, joined by a long thin tube T . On $\partial B_+ \sim T$, let n_0 be the outward unit normal vectorfield, and, on $\partial B_- \sim T$, let n_0 be the inward unit normal vectorfield. On $\partial T \sim (B_+ \cup B_-)$, n_0 should be essentially the constant unit vectorfield parallel to the axis of T and pointing towards B_- . To make $\partial\Omega$ and n_0 smooth, some slight interpolation is needed near the two ends of T .

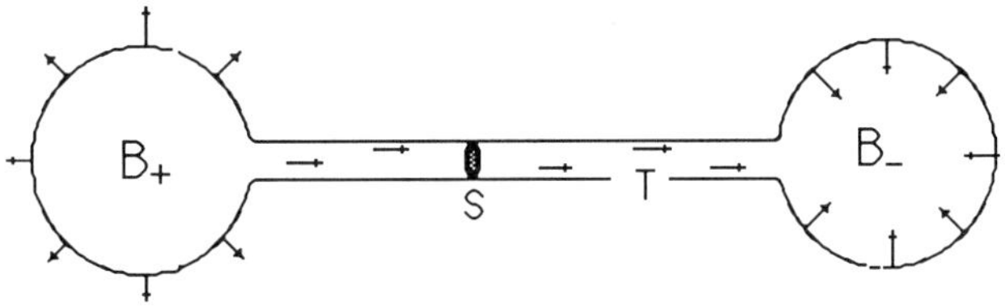

One may easily construct a specific H^1 extension of n_0 to $\overline{\Omega}$ that is approximately $x/|x|$ in B_+ , $-x/|x|$ in B_- , and constant in T , to deduce that the minimum W^* -energy is at most only slightly more than twice the W^* -energy of $x/|x|$ in the unit ball (e.g., 16π in the equal constant case). In particular, this minimum energy is bounded independent of the length of T.

On the other hand, n_0 , having degree 0 , admits some continuous extension to Ω . If an energy minimizer were itself continuous on $\overline{\Omega}$, then it would, by the regularity discussion below, be smooth on $\overline{\Omega}$. To complete the argument we now show that for any any smooth extension u of n_0 to $\overline{\Omega}$, $\int_\Omega |\nabla u|^2 dx$, and hence the W^* -energy of u , is bounded below by a constant times the length of T . Let S be any slice of the tube T by an orthogonal plane and let Ω_+ and Ω_- denote the two components of $\Omega \sim S$. Since $u|\overline{\Omega}_\pm$ is continuous, the degree of $u|\partial\Omega_\pm$ is zero. But this implies

$$u(S) \supset u(\partial\Omega_\pm \sim S) = n_0(\partial\Omega_\pm \sim S) \supset n_0(\partial B_\pm \sim T),$$

which contains more than half of the sphere. Then the energy of the slice $\int_S |\nabla u|^2 \, dy\, dz$, being at least twice the area of $u(S)$, must be greater than 4π. Integrating over all such slices gives the desired lower bound

$$\int_\Omega |\nabla u|^2 dx \geq \int_T |\nabla u|^2 dx > 4\pi \cdot \text{length}(T). \quad \square$$

Accepting now the necessary existence of singularities for some minimizers, we now turn to the question of the size of such singular sets. Here we will only treat interior singularities. For an analogous discussion of boundary singularities, where both n_0 and $\partial\Omega$ are smooth, see [HKL, §5]. We will also restrict our remarks to the nematic case with no magnetic or electric fields present. The generalizations to these other problems are handled in [HKL, §3-4]. To begin, we observe how the question of smoothness of a minimizer reduces to the question of continuity.

Lemma 4.2 *If $n \in H^1(\Omega; S^2)$ is a minimizer of W and if n is continuous near a point $a \in \Omega$, then n is analytic near a.*

Proof: Assuming that $n(a) = (0,0,1)$ and writing $n = (n_1, n_2, n_3)$, we may, in a small ball $B_r(a)$, make the substitutions

$$n_3 = \sqrt{1 - n_1^2 - n_2^2}, \quad \nabla n_3 = (-n_1 \nabla n_1 - n_2 \nabla n_2)/\sqrt{1 - n_1^2 - n_2^2}$$

in the weak equation (3.7) and use a smooth test function $\zeta = (\zeta_1, \zeta_2, 0)$ to infer that (n_1, n_2) is a critical point for $\int_{B_r(a)} J(\nabla u, u)\, dx$ where

$$J(p,u) = \tilde{W}[(p_1, p_2, (-u_1 p_1 - u_2 p_2)/\sqrt{1-u_1^2-u_2^2}), (u_1, u_2, \sqrt{1-u_1^2-u_2^2})].$$

Noting that $J(p,u)$ is quadratic in p for each u with $|u| < 1$ and

$$J(p,0) \geq \tfrac{1}{2}\alpha |p|^2,$$

we see that (n_1, n_2) satisfies, near a, a strongly elliptic system with analytic coefficients. By [M] §6.7, (n_1, n_2), and hence n, is analytic near a. $\square$

Note that this proof does not involve assuming that the original system of Euler-Lagrange equations is elliptic, which is, as we have seen in §3, false in general.

The singular set of a minimizer n may thus be characterized qualitatively as the closure of the set of discontinuities of n. However, to estimate the size of this set we need a quantitative characterization related to the energy minimality. For this we use the normalized Dirichlet energy on a ball

$$E_{r,a}(u) = r^{-1}\int_{B_r(a)} |\nabla u|^2\,dx$$

where $u \in H^1(\Omega)$ and $B_r(a) \subset \Omega$. In using this quantity in [HKL_2], as previously in [SU], one is motivated by the well known lemma of Morrey [M.3.52].

To study the behavior of a W-minimizer n on a ball $B_r(a) \subset \Omega$, we may use scaling to work on the unit ball B by noting that the expression $n_{r,a}(x) = n(rx + a)$ defines a W-minimizing function $n_{r,a} \in H^1(B;S^2)$ with

$$\int_B |\nabla n_{r,a}(x)|^2\,dx = E_{r,a}(n) .$$

The pivitol step in our regularity discussion is now the following:

<u>Energy Improvement Theorem</u> 4.3 <u>There are positive numbers</u> ϵ <u>and</u> $\theta < 1$ (<u>depending only on</u> κ_1, κ_2, <u>and</u> κ_3) <u>so that</u> $E_{\theta,0}(u) < \theta E_{1,0}(u)$ <u>for any</u> W <u>minimizer</u> $u \in H^1(B;S^2)$ <u>with</u> $E_{1,0}(u) < \epsilon^2$,

This leads to the

<u>Partial Regularity Theorem</u> 4.4 <u>Any</u> W <u>minimizer</u> $n \in H^1(\Omega;S^2)$ <u>is analytic on</u> $\Omega \sim Z$ <u>for some relatively closed subset</u> Z <u>of</u> Ω <u>which has one dimensional Hausdorff measure zero.</u>

To see that 4.3 implies 4.4 we first note that energy finiteness alone implies, by an elementary covering argument [SU, 2.7], that the set

$$Z = \{a \in \Omega : \limsup_{r \downarrow 0} E_{r,a}(n) > 0\}$$

has one dimensional Hausdorff measure zero. For $b \in \Omega \sim Z$ and s

sufficiently small one can apply 4.3 to the normalized function $n_{r,a}$ for $a \in B_{\frac{1}{2}s}(b)$ and $0 < r < \frac{1}{2}s$. By iteration and scaling, one then verifies the decay hypothesis of Morrey's lemma on $B_r(a)$. We conclude that $\Omega \sim Z$ is an open set and that $n \big| (\Omega \sim Z)$ is, by 4.2, analytic.

Sketch of the proof of 4.3. Were 4.3 false, there would be, for each θ with $0 < \theta < 1$, a sequence $u_1, u_2, \cdots$ of W-minimizers so that

$$\epsilon_i^2 = E_{1,0}(u_i) \to 0 \quad \text{as} \quad i \to \infty \quad \text{and} \quad E_{\theta,0}(u_i) > \theta \epsilon_i^2 \quad \text{for each} \quad i. \tag{4.1}$$

With $\overline{u_i}$ being the mean value $|B|^{-1}\int_B u_i \, dx$, the functions

$$v_i = (u_i - \overline{u_i})/\epsilon_i$$

have mean value zero and are, by Poincaré's inequality, bounded in H^1. A subsequence of v_i converges weakly in H^1, strongly in L^2, and pointwise almost everywhere to a function $v \in H^1(B;\mathbb{R}^3)$. Such a "blow-up limit function" v is better behaved than the original minimizers u_i in two ways:

(1) The image of v lies essentially in a plane through the origin.

(2) In coordinates on this plane the $\mathbb{R}^2$ valued function w determined by v satisfies a constant coefficient elliptic system.

The proof of (1) in [HKL$_2$] §2.1 is a calculation involving only Poincaré's inequality and the strong L^2 convergence of v_i to v. In the proof of (2) in [HKL$_2$] one simply substitutes u_i for n and $\epsilon_i \nabla v_i$ for ∇n in the weak version of the equilibrium equations (3.7). Dividing by ϵ_i, noting that $\epsilon_i^{-1}|Y(\epsilon_i \nabla v_i, u_i)| < C_1 \epsilon_i |\nabla v_i|^2$, and letting $i \to \infty$, we readily find that

$$\int_B [W_p(\nabla v, e) - e \otimes e \, V_p(\nabla v, e)] \cdot \zeta \, dx = 0 \quad \text{for} \quad \zeta \in H_0^1(B;\mathbb{R}^3) \cap L^\infty ,$$

where e is a unit vector orthogonal to the plane containing the image of v. Restricting to variations ζ with $\zeta \cdot e = 0$ then gives

$$\int_B W_p(\nabla v, e) \cdot \zeta \, dx = 0 .$$

Inasmuch as $W_p(\xi, e) \cdot \xi \geq \alpha |\xi|^2$, the corresponding constant coefficient system of 2 equations in three unknowns satisfied by w is thus elliptic.

In the equal constant case, where the u_i satisfy (3.9), one finds that any blow-up limit function v is simply a harmonic vector.

Returning to the proof of 4.3, our strategy now, as in all "blow-up" arguments, is to transfer estimates for the well-behaved blow-up limit function back to the original u_i and obtain, for i sufficiently large, a contradiction to (4.1). For example, it is natural to try to use the standard estimate

$$E_{\theta,0}(v) = \theta^{-1}\int_{B_\theta}|\nabla v|^2\,dx \le C\theta^2\int_B|\nabla v|^2\,dx = C\theta^2$$

where C depends only on κ_1, κ_2, and κ_3. If, on the sub-ball B_θ, the convergence of v_i to v were, not just weak, but <u>strong</u> in H^1, then, for i sufficiently large,

$$E_{\theta,0}(u_i) = \epsilon^2 E_{\theta,0}(v_i) \le 2C\theta^2\epsilon_i^2\,,$$

which would contradict (4.1) if θ were originally chosen smaller than $1/2C$.

The convergence here is actually strong in H^1 (see [HKL_3]), although this was not the tact taken in [HKL_2]. There, the strong convergence in L^2 was exploited, and an L^2 estimate

$$|B_\theta|^{-1}\int_{B_\theta}|v|^2\,dx \le C\theta^2\int_B|v|^2\,dx$$

was used. The crucial ingredient in either argument is the following important relation between the L^2 norms of a minimizer and its gradient.

<u>Lemma</u> 4.5 <u>For any</u> W <u>-minimizer</u> $u \in H^1(B;S^2)$ <u>and any vector</u> $v \in B$,

$$\|\nabla u\|^2_{L^2(B_{1/2})} \le C\,\|\nabla u\|_{L^2(B)}\cdot\|u-v\|_{L^2(B)}$$

<u>where the constant</u> C <u>depends only on</u> κ_1, κ_2, <u>and</u> κ_3.

The proof of 4.5 involves constructing a comparison $w \in H^1(B_r;S^2)$ with $w|\partial B_r = u|\partial B_r$ for some $r \in [\frac{1}{2},1]$. The function w is obtained by taking the the <u>harmonic function</u> $h : B_r \to \mathbb{R}^3$ having boundary values $u|\partial B_r$ and composing h with a projection onto S^2. The choice of a suitable projection is made using Fubini's Theorem. See [HKL_2] §2.3 for details.

Note that the argument in this construction gives another proof of Lemma 2.1. Moreover, this new proof has the advantage of generalizing to boundary-value functions n_0 belonging to the larger space $H^{\frac{1}{2}}(\partial\Omega;S^2)$, which is precisely the trace class of $H^1(\Omega;S^2)$ (See [HL_2]). One somewhat surprising corollary of 4.5 is

Corollary 4.6 The upper energy density $\limsup_{r\downarrow 0} E_{r,a}(n)$ is bounded absolutely in terms of the constants $\kappa_1, \kappa_2, \kappa_3$ and independent of the W-minimizer $n \in H^1(\Omega;S^2)$ and the point $a \in \Omega$.

To see this we assume that $\Omega = B$ and that $a = 0$ and abbreviate $E_r = E_{r,0}(n)$. Since $|n| \equiv 1$, 4.5 implies that

$$E_{\frac{1}{2}} \leq \tfrac{1}{2}C\,\|\nabla n\|_{L^2(B)} \cdot \|n\|_{L^2(B)} \leq cE_1^{1/2} .$$

By scaling and iteration

$$E_{2^{-k}} \leq c^{1-2^{-k}} E_1^{2^{-k}} .$$

For $0 < r \leq \frac{1}{2}$, we may choose $k \in \{1,2,\cdots\}$ so that $2^{-k-1} < r \leq 2^{-k}$ and note that trivially $E_r \leq 2E_{2^{-k}}$. Thus

$$\limsup_{r\downarrow 0} E_r \leq \limsup_{r\downarrow 0} 2c^{1-r} E_1^{r/2} = 2c .$$

Some other results obtained in [HKL_4] are

(1) For any compact $K \subset \Omega$ and any W-minimizer $n \in H^1(\Omega;S^2)$,

$$\int_K |\nabla n|^2\, dx \leq C_0 ,$$

where C_0 depends only on $\kappa_1, \kappa_2, \kappa_3$, Ω, and K.

(2) A W-minimizer $n \in H^1(\Omega;S^2)$ satisfies a reverse-Hölder inequality on any ball $B_r(a) \subset \Omega$: for $0 < \lambda < 1$,

$$\left(⨍_{B_{r/2}(a)} |\nabla n|^2 dx\right)^{1/2} \leq c_1\lambda \left(⨍_{B_r(a)} |\nabla n|^2 dx\right)^{1/2} + c_1\lambda^{-1}\left(⨍_{B_r(a)} |\nabla n|^p dx\right)^{1/p}$$

where $p = 2n/(n+1)$ and c_1 depends only on $\kappa_1, \kappa_2, \kappa_3$.

(3) $\nabla n \in L^q_{loc}(\Omega)$ <u>for some</u> $q > 2$.

(4) <u>The singular set</u> Z <u>of</u> n , <u>defined in 4.4, has Hausdorff dimension strictly less than one.</u>

For the regularity theory and singularity theory in the equal constant case, much more is known, thanks largely to the work of R. Schoen and K. Uhlenbeck [SU], L. Simon [S_1], and H. Brezis, J. Coron, and E. Lieb [BCL_2]. For the remainder of this section we will consider some of the special features of this case. Thus we are considering maps $n \in H^1(\Omega; S^2)$ which minimize the <u>Dirichlet integral</u>

$$\frac{1}{2}\int_\Omega |\nabla n|^2 \, dx \quad .$$

An important basic property is that now

<u>the normalized energy</u> $E_{r,a}(n)$ <u>is monotonically nondecreasing in</u> r <u>for</u> $0 < r < \operatorname{dist}(a, \partial\Omega)$.

From this and H^1 weak compactness, it follows [SU] that

<u>Any sequence of positive numbers approaching 0 contains a subsequence</u> $r_1, r_2, \cdots$ <u>so that the scaled functions</u> $n_{r_i,a} \in H^1(B; S^2)$ <u>converge weakly in</u> H^1 <u>to a homogeneous degree-0 function</u> $n_\infty \in H^1(B; S^2)$ <u>(called a tangent map of</u> n <u>at</u> a).

With the help of 4.5 and 4.2 it can be shown as in [SU], §4 that this convergence is strong in H^1 and that n_∞ is, like each of the $n_{r_i,a}$, energy minimizing on B . In particular, n_∞ is <u>stationary</u>, that is, a critical point for the Dirichlet integral under smooth variations. This implies that $n_\infty(x) = \omega(x/|x|)$ for some harmonic map $\omega : S^2 \to S^2$. If not a constant, ω must then be either conformal or anti-conformal [EL], §10.6. Since under stereographic projection

$$\Pi : S^2 \sim \{(0,0,1)\} \to \mathbb{R}^2 \simeq \mathbb{C} \ ,$$
$$\Pi(x_1, x_2, x_3) = 2[(1+x_3)/(1-x_3)]^{1/2}(x_1, x_2) \quad ,$$

conformal maps from S^2 to S^2 correspond to quotients of polynomials, we see that

Any stationary, homogeneous degree-0 map $h \in H^1(B;S^2)$ must be in the form $\omega_{p,q}(x/|x|)$ or $\omega_{p,q}((x_1,-x_2,x_3)/|x|)$ where, for some complex polynomials $p, q : \mathbb{C} \to \mathbb{C}$,

$$\omega_{p,q} = \Pi^{-1} \circ (p/q) \circ \Pi \ : S^2 \to S^2 \ .$$

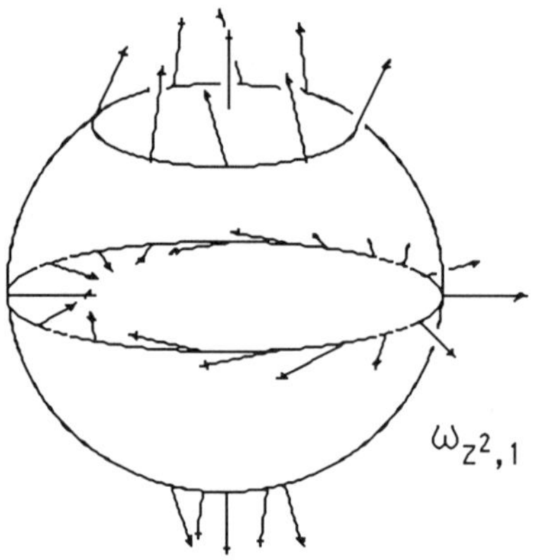

In case h is energy-minimizing (e.g., h is a tangent map of a minimizer), one may infer from Corollary 4.6 an upper bound for the degree of the corresponding $\omega_{p,q}$. Moreover, numerical investigations of Luskin et al. [CHKLL] indicated that even $\omega_{z^2,1}$ does not give an energy-minimizing map. Brezis, Coron, and Lieb [BCL_2] have now proven that any such homogeneous minimizer $\omega(x/|x|)$ must in fact reduce to

$$\omega(\xi) = Q\xi \qquad \text{for a rotation } Q \ .$$

Next, by the smoothness of the tangent map $n_\infty | (B \sim \{0\})$, the strong H^1 convergence of $n_{r_i,a}$ to n_∞, and the regularity theory, one may, as in [SU], Th. II, deduce the optimal result on the size of the singular set:

The singular set Z of any Dirichlet energy minimizing function $n \in H^1(\Omega;S^2)$ is an at most isolated subset of Ω.

Finally we make some remarks on the asymptotic behavior of n near a point a belonging to the singular set Z. From the general result of

L. Simon in $[S_1]$§8, it follows that any tangent map n_∞ of n at a is unique and that

$$\sup_{B_r(a)} \left| n(x) - n_\infty((x-a)/|x-a|) \right| \to 0 \text{ as } r \to 0 \ .$$

Moreover, the convergence here is in fact like r^α for some positive α. This rate is established in [GW] where the above $\omega_{p,q}$ classification is used to show that the argument of [S] is applicable.

It is a challenging problem to see how many of these regularity properties of Dirichlet minimizing functions carry over for minimizers of the general liquid crystal functional W.

Even in the equal constant case, little is known concerning the regularity of nonminimizing equilibrium solutions. Note that we have here considered the specific nonminimizing examples with the function n_q in (2.16) and the functions $\omega_{p,q}(x/|x|)$ of degree >1. For a small energy stationary harmonic map. G. Liao [L] has shown the removability of a singularity that is assumed a priori to be isolated. However, it is in general unknown if the singular set of a stationary harmonic map on a 3 dimensional domain even has 3 dimensional measure zero.

5 The order parameter and some applications

The energy of a liquid crystal configuration depends on its density ρ, the temperature θ, and an order parameter which is a matrix Q. The Oseen-Frank density (1.1) is derived from this general form by imposing constraints on the order parameter. Understanding this will assist us in studying phase transitions and more complicated defect behavior.

As mentioned in the introduction, the liquid crystals under consideration are thought of as rigid rod-like molecules. A configuration of the material may be described by a unit vectorfield which gives the direction of each molecule as a function of its center or mass $l(x)$. Now $l(x)$ is a molecular quantity and so it is certain statistical averages of it, or moments, which provide the kinematical variables of the macroscopic theory. Since one end of the molecule is indistinguishable from the other, l is equivalent to $-l$ and odd order moments must vanish. So

$$0 = \langle l \rangle = \langle l \otimes l \otimes l \rangle = \ldots \quad ,$$

where $\langle \ \rangle$ denotes the statistical average. The even order moment of lowest order is the most important contribution to the energy, hence,

$$L = \langle l \otimes l \rangle \tag{5.1}$$

plays the central role in the continuum theory. It has the properties

$$L = L^T , \quad \mathrm{tr}L = 1 , \quad L \geq 0 . \tag{5.2}$$

Modifying slightly, we define

$$Q = L - \tfrac{1}{3}\mathbf{1}, \qquad \mathrm{tr}\,Q = 0 . \tag{5.3}$$

The field Q, which depends on position, time, temperature, etc., is the customary order parameter of the theory. At the operational level, it is customary to identify it with the dielectric tensor, suitably normalized.

Consider a few simple cases.

$Q = 0$ (L has three equal eigenvalues)	random orientation: isotropic liquid
Q has two equal eigenvalues	characteristic of nematic and cholesteric phases: optically uniaxial with optic axis corresponding to the direction of the eignvector with the unequal eigenvalue.

In this second case, Q is written

$$Q = S(n \otimes n - \tfrac{1}{3}\mathbf{1}) , \qquad |n| = 1, \tag{5.4}$$

and the scalar S describes the degree of orientation. Thus

$$S = {}^3/_2 \langle l \cdot n \rangle^2 - \tfrac{1}{2} \tag{5.5}$$

and has the possible range $-\frac{1}{2} \leq S \leq 1$. Three special cases may be distinguished:

$S = -\frac{1}{2}$	l lies in a plane perpendicular to n
$S = 0$	random orientation
$S = 1$	perfect alignment.

In general it is found that $.3 \leq S \leq .7$ [E_1]. The bulk energy U is assumed to depend on Q, ρ, θ, and their gradients. We do not consider the most general case, however, and assume that ρ is constant and U is quadratic in ∇Q. Temperature is regarded as a control parameter. Hence,

$$U = U(\nabla Q, Q, \theta) . \tag{5.6}$$

It is subject to invariance principles and thermodynamical restrictions which need not concern us here, cf. Ericksen [E_1],[E_2].

For a nematic, one usually assumes that $S = S(\theta)$, so

$$\begin{aligned} U &= \tilde{U}(\nabla n, n, \theta) \\ &= \tilde{W}_0(n, \theta) + W(\nabla n, n, \theta) , \end{aligned} \tag{5.7}$$

where W is given by (1.1) with $q = 0$ and, from the invariance principles,

$$\tilde{W}_0(n, \theta) = W_0(\theta) . \tag{5.8}$$

In an isotropic liquid, $Q = 0$, so in our simple description here with density suppressed, $U = U_{isotropic}(\theta)$. The nematic/isotropic transition has been analyzed by Ericksen [E_3] from this point of view. Its primary mechanism may be seen as S changing from 0 to nonzero as the material is cooled below the critical temperature.

Casually speaking, smectic A is a lower temperature liquid crystal phase in which the molecules form identical coherent layers which glide easily over one another, similar to soaps from which the name is derived. The layers tend to be of a definite fixed thickness, the top the same as the bottom, infinitesimal by macroscopic standards. A more complete description may be found in Ericksen [E_1], Chandrasekhar [Ch], de Gennes [dG], or the article by C.C. Huang in this volume [H].

The order parameter Q in the smectic A phase is assumed to have two equal eigenvalues, like a nematic, but S varies slightly with $x \in \Omega$. Moreover, twist and bend deformations, which are those involving the curl in the energy density, are either absent or highly penalized. This is to reflect the kinematical properties mentioned above. When the curl terms are absent, we may write

$$Q = s_0(\sigma n \otimes \sigma n - {}^1/_3 \sigma^2 \mathbf{1}) , \tag{5.9}$$

where

$$s_0 = s_0(\theta)\ ,\quad \sigma = \sigma(x),\quad n = n(x),\ \text{and}\ \ \mathrm{curl}(\sigma n) \ = \ 0.$$

On setting $\nabla u = \sigma n$, Q assumes the form

$$Q \ = \ s_0(\nabla u \otimes \nabla u - {}^1/_3|\nabla u|^2\mathbb{1}) \tag{5.10}$$

and the energy U may be written

$$U(\nabla Q, Q, \theta) \ = \ U_0(|\nabla u|, \theta) + W(\nabla^2 u, \nabla u, \theta) \tag{5.11}$$

where

$$\begin{aligned} W(\nabla^2 u, \nabla u, \theta) \ &= \ \tfrac{1}{2}k_1(\Delta u)^2\ , \\ U_0(\sigma,\theta) \ &\geq \ 0,\ \ U_0(1,\theta) = 0. \end{aligned} \tag{5.12}$$

A variant of this is to permit a certain penetration of the twist and bend deformations within the smectic layers. In this case we write

$$\sigma n \ = \ \nabla u + \psi \tag{5.13}$$

and

$$W(\nabla^2 u, \nabla u, \nabla\psi, \theta) \ = \ \tfrac{1}{2}k_1(\Delta u)^2 \ + \ \tfrac{1}{2}k_2|\mathrm{curl}\,\psi|^2 \tag{5.14}$$

where k_2 is very large. The k_i are temperature dependent.

Pretransition effects and the ensuing phase transition from nematic to smectic A may be discussed in this context. A mathematical interpretation of this transition is that it arises by interchanging the nematic constraint $|n| = 1$ with the smectic constraint $\mathrm{curl}\,n = 0$ in the variational principle. It is possible to verify in a qualitative sense the experimental result that the effective k_2 diverges as the temperature decreases to the nematic/smectic A critical point.

Situations where constraints are relaxed or imposed or where minimum energy states suffer a change in symmetry are well known in the study of phase transitions, cf. [LL] or [A]. Here we encounter an interchange of constraints, which seems to be less studied. Perhaps a brief discussion of the ideas might be illuminating, which we limit to a one dimensional example in order to maintain a level of technical simplicity.

Consider a material whose high temperature phase is N and whose low temperature phase is A occupying the unit interval $\Omega = (0,1)$. Small amounts of A are observed in the material, whose fraction increases to 1 as temperature is decreased to the transition temperature $\theta = \theta_0$. We shall oversimplify our material by assuming that the A phase present above θ_0 may be regarded as an effective periodic structure embedded in the N phase.

We remark here that in the laboratory, cybotactic clusters of smectic A are observed in the nematic phase. Presumably they result from fluctuations in the order parameter. Indeed, fluctuations may be sufficiently large that there may be isotropic phase present as well, which is called a multicritical phenomenon [H]. A certain correlation length, rather than a simple proportion, becomes very large as transition temperature is approached and this property is used to predict the divergence of k_2, [dG], [Ch], [H].

Introduce $\phi, \psi \in H^1(\Omega)$ with values in $\mathbb{R}^3$ and appropriate energy densities:

- exact N-phase density

$$W_N(\phi', \psi') = \tfrac{1}{2}|\phi'|^2 + \tfrac{1}{2}|\psi'|^2 , \quad \text{with} \quad |\phi + \psi| = 1 ,$$

- relaxed N-phase density

$$U_N(\phi', \psi', \phi, \psi) = W_N(\phi', \psi') + \lambda g(|\phi + \psi|) ,$$

  where $\lambda > 0$ and $g(t)$ is a smooth, bounded function, constant for large values of t, satisfying $g(t) > g(1) = 0$ for $t \neq 0$,

- exact A-phase density

$$W_A(\phi', \phi) = \tfrac{1}{2}|\phi'|^2 + g(|\phi|), \quad \psi' = 0,$$

- relaxed A-phase density

$$U_A(\phi', \psi', \phi, \psi; \mu) = \tfrac{1}{2}|\phi'|^2 + \tfrac{1}{2}\mu|\psi'|^2 + g(|\phi|), \quad \mu > 0 ,$$

and finally a

- field energy

$$-\tfrac{1}{2}M\cdot H = -\tfrac{1}{2}\alpha_0|H|^2 - \tfrac{1}{2}\alpha_a(H\cdot\phi + H\cdot\psi)^2 ,$$

where $\alpha_0 > 0$, $\alpha_a > 0$, and H is a fixed vector.

Some observations about the relaxed functionals are appropriate. The density U_N is an ordinary penalization of W_N . If $(\phi_\lambda,\psi_\lambda)$ satisfies

$$I_\lambda = \int_\Omega U_N(\phi_\lambda',\psi_\lambda', \phi_\lambda,\psi_\lambda;\lambda)\, dx = \inf_A \int_\Omega U_N(\phi',\psi',\phi,\psi;\lambda)\, dx$$

for an admissible class A , then for a subsequence of λ,

$$(\phi_\lambda,\psi_\lambda) \to (\phi_0,\psi_0)$$

and

$$\lim I_\lambda = \int_\Omega W_N(\phi_0',\psi_0')\, dx ,$$

where (ϕ_0,ψ_0) is a minimizer of

$$\int_\Omega W_N(\phi',\psi')\, dx \qquad \text{in the class } A \cap \{|\phi + \psi| = 1\} .$$

However the analogous statement does not hold in the A-phase with respect to the term $\tfrac{1}{2}\mu|\psi'|^2$, in particular if the field energy is added or if boundary conditions are such that ψ = const. is not a minimum. Here the term $\tfrac{1}{2}\mu|\psi'|^2$ corresponds to a distortion of the A-phase.

For σ, $0 < \sigma < 1$, let $I_\sigma = [\tfrac{1}{2} - \tfrac{1}{2}\sigma, \tfrac{1}{2} + \tfrac{1}{2}\sigma] \subset \Omega$ be an interval governed by the relaxed A-phase density. Let χ_σ denote the characteristic function of I_σ , which we assume to be extended periodically to $\mathbb{R}$, and set $\chi_{\sigma,\epsilon}(x) = \chi_\sigma(x/\epsilon)$. For $\epsilon > 0$, we write a total energy

$$\begin{aligned} U_\epsilon(\phi',\psi',\phi,\psi;\sigma,\mu,\lambda) &= (1 - \chi_{\sigma,\epsilon})U_N(\phi',\psi',\phi,\psi;\lambda) + \chi_{\sigma,\epsilon}U_A(\phi',\psi', \phi, \psi;\mu) \\ &= \tfrac{1}{2}|\phi'|^2 + \tfrac{1}{2}(1 + (\mu - 1)\chi_{\sigma,\epsilon})|\psi'|^2 + \lambda(1 - \chi_{\sigma,\epsilon})g(|\phi+\psi|) + \chi_{\sigma,\epsilon}\, g(|\phi|) . \end{aligned} \tag{5.15}$$

The fraction σ will be regarded as the basic control variable.

Given $A \subset H^1(\Omega)$, a class of admissible functions satisfying fixed boundary conditions, say of Dirichlet type, let $(\phi_\epsilon,\psi_\epsilon)$ denote a minimizer of

$$\int_\Omega U_\epsilon \, dx$$

for fixed (σ,μ,λ). Applying a known standard reasoning from homogenization theory, [BLP], we infer that

$$(\phi_\epsilon,\psi_\epsilon) \rightarrow (\phi_0,\psi_0) \qquad \text{in } H^1(\Omega) \text{ weakly}$$

where (ϕ_0,ψ_0) realizes a minimum of the functional with the integrand

$$U_0(\phi',\psi',\phi,\psi;\ \sigma,\mu,\lambda) = \tfrac{1}{2}|\phi'|^2 + \tfrac{1}{2}(1 - \sigma + \sigma/\mu)^{-1}|\psi'|^2 + \lambda(1 - \sigma)\, g(|\phi+\psi|) + \sigma\, g(|\phi|) . \tag{5.16}$$

Moreover, if there is some pair $(\phi,\psi) \in A$ with $|\phi+\psi| = 1$, it is easy to establish that

$$E(\sigma,\mu,\lambda) = \int_\Omega U_0(\phi_0',\psi_0',\phi_0,\psi_0;\ \sigma,\mu,\lambda)\, dx \le C(1 - \sigma + \sigma/\mu)^{-1} . \tag{5.17}$$

Such a hypothesis is only the statement that the pure N-phase problem admits a solution.

There is considerable latitude in this formulation, so there is a choice of how to proceed from this point. In essence, E is-a function of three variables. For example, letting $\mu \rightarrow \infty$ in (5.16) illustrates that the effective coefficient of $\frac{1}{2}|\psi'|^2$,

$$k_{eff} \sim (1 - \sigma)^{-1} ,$$

while E remains finite. This expresses the desired qualitative behavior as $\sigma \rightarrow 1$. One may also attempt to follow the configuration through transition. We shall not pursue details of these arguments here.

The general multivariable problem based on interchanging the nematic and smectic constraints is susceptible to the same sort of treatment. Part of the analysis involves estimating the effective modulus associated to the functional

$$\tfrac{1}{2}\int_\Omega (1 + (\mu - 1)\chi_\sigma)|\mathrm{curl}\,\psi|^2 \, dx ,$$

where χ_σ is the characteristic function of a set of measure σ in a unit cube Ω. This may be achieved by convex duality, cf. eg. [KM].

(^{1}) Standard notation is employed for function spaces. $H^1(\Omega)$ denotes the functions which together with their first derivatives are square integrable in Ω.
(^{2}) This means that the boundary of the domain is locally representable as the graph of a Lipschitz function.

Acknowledgements The authors are pleased to acknowledge many stimulating and fruitful discussions with J. L. Ericksen. We also thank H. Brezis for many interesting conversations. Finally, we are indebted to our collaboraters F.-H. Lin and M. Luskin.

The preparation of this manuscript was partially supported by NSF grants DMS 85-11357 and MCS 83-01345.

References

[A] Antman, S., Ericksen, J. L., Kinderlehrer, D., and Muller, I., Metastability and incompletely posed problems, I.M.A. Vol. Math. Appl. 3, Springer, (1987)

[B] Brezis, H. Liquid crystals and energy estimates for S^2-valued maps, This volume

[BCL_1] Brezis, H. Coron, J. M., and Lieb, E. Estimations d'énergie pour des applications de R^3 a valeurs dans S^2, CRAS Paris, 303, (1986), 207-210

[BCL_2] ______________ Harmonic maps with defects, IMA preprint 253, to appear, Comm. Math. Physics

[BC] Brinkman, W. and Cladis, P., Defects in liquid crystals, Physics Today, May 1982, 48-54

[BLP] Bensoussan, A., Lions, J.-L., and Papanicolaou, Asymptotic analysis for periodic structures, North Holland, (1978)

[BZ] Bethuel, F. and Zheng, X Sur la densité des fonctions régulières entre deux variétés dans des espaces de Sobolev, CRAS Paris 303, 1986, 447-449

[CG] Capriz, G. and Giovine, P. On virtual inertia effects during diffusion of a dispersed medium in a suspension, This volume

[Ch] Chandrasekhar, S. Liquid Crystals, Cambridge, (1977)

[Cl] Cladis, P. A review of cholesteric blue phases, This volume

[CHKLL] Cohen, R., Hardt, R., Kinderlehrer D., Lin, S.-Y., and Luskin, M. Minimum energy configurations for liquid crystals: computational results, Theory and applications of liquid crystals, This volume

[dG] de Gennes, P. G. The physics of liquid crystals, Oxford, (1974)

[EL] Eels, J. and Lemaire, L. A report on harmonic maps, Bull. London Math. Soc. 10 (1978), 1-68

[E_1] Ericksen, J. L. Equilibrium theory of liquid crystals, Adv. in liquid crystals 2, (G. H.

Brown, ed) Academic Press (1976), 233-298

[E_2] ______________ Nilpotent energies in liquid crystal theory, Arch. Rat. Mech. Anal. 10 (1962), 189 -196

[E_3] ______________ A thermodynamic view of order parameters for liquid crystals, Orienting Polymers (Ericksen, J. L., ed) Springer Lecture Notes 1063, 1984, 27-36

[F] Frank, F. C., On the theory of liquid crystals, Discuss. Faraday Soc., 28, (1958), 19-28

[GW] Gulliver, R. and White, B., The rate of convergence of a harmonic map at a singular point, preprint

[H] Huang, C. C., The effect of the magnitude of the disordered phase temperature on the given phase transition, this volume

[HKL_1] Hardt, R. Kinderlehrer, D., and Lin F.-H., Existence et régularité des configurations statiques des cristaux liquides, CRAS Paris 301,(1985), 577-579

[HKL_2] ______________, Existence and partial regularity of static liquid crystal configurations, IMA preprint 175, Comm. Math. Physics 105 (1986), 547-570

[HKL_3] ______________ A remark about the stability of smooth equilibrium configurations of static liquid crystals, IMA preprint 231, Mol. Cryst Liq. Cryst. 136 (1986)

[HKL_4] ______________ Stable singularities of minimizing maps, to appear

[HKL_5] ______________ to appear

[HL_1] Hardt, R. and Lin, F.-H. A remark on H^1 mappings, Manus. math. 56, (1986), 1-10

[HL_2] Hardt, R. and Lin, F.-H. Mappings that minimize the p th power of the gradient, to appear in Comm. Pure and Appl. Math.

[KM] Kohn, R. and Milton, G. On bounding the effective conductivity of anisotropic composites, Homogenization and effective moduli of materials, I.M.A. Vol. Math. Appl. 1 (Ericksen, J. L., Kinderlehrer, D., Kohn, R., Lions, J.-L., eds) Springer, 1986, 97-125

[LL] Landau, L. D. and Lifshitz, E. M., Statistical Physics, 3rd Edition Part 1, Pergamon, 1980

[L_1] Leslie, F. Theory of flow phenomena in liquid crystals, Adv. Liq. Cryst. 4 (G. H. Brown, ed.) Academic Press (1979), 1-81

[L_2] ________ Theory of flow phenomena in nematic liquid crystals, Theory and application of liquid crystals, IMA Volumes in math and appl 5, (J.L. Ericksen and D. Kinderlehrer, eds)

[M] Morrey, C.B. Multiple integrals in the calculus of variations, Springer-Verlag, Heidelberg and New York, 1966

[MM] MacMillan, E. The statics of liquid crystals, Thesis, Master of Science, The Johns Hopkins University, 1982

[O] Oseen, C. W., The theory of liquid crystals, Trans. Faraday Soc., 29, (1933), 883-889

[SU] Schoen, R. and Uhlenbeck, K. A regularity theory for harmonic maps, J. Diff.

Geometry, 17 (1982), 307-335

[Si_1] Simon, L. Asymptotics for a class of nonlinear evolution equations, Ann. of Math, 118 (1983), 525-571

[Si_2] ________ Isolated singularities for extrema of geometric variational problems, Springer Lecture Notes 1161, 1985

[Se] Sethna, J. Theory of blue phases of chiral nematic liquid crystals, This volume

[WPC] Williams, C. Pieranski, P., and Cladis, P. E. Nonsingular $S = +1$ screw disclination lines in nematics, Phys. Rev. Letters, 29 (1972), 90-92

THE EFFECT OF THE MAGNITUDE OF THE DISORDERED PHASE TEMPERATURE RANGE ON THE GIVEN PHASE TRANSITION IN LIQUID CRYSTALS

C.C. Huang

School of Physics and Astronomy
University of Minnesota
Minneapolis, Minnesota 55455

I. Introduction

The difference between phases separated by a given phase transition can be characterized by one particular physical property called the order parameter. For example, the order parameter for distinguishing a ferromagnetic phase from a paramagnetic phase is the magnetization. Consequently, the order parameter (Ψ) is zero for the disordered phase $(T>T_c)$ and nonzero for the ordered phase $(T < T_c)$. Here T_c is the transition temperature. Depending on the way Ψ approaches zero at $T = T_c$, we have two types of phase transitions. The one with Ψ asymptotically approaching zero as T approach T_c^- is a continuous transition; otherwise, the transition is a first order one. For a continuous transition, the divergence of the order parameter-order parameter correlation length at $T = T_c$ results in anomalous behavior in many physical properties, e.g., heat capacity (C), susceptibility (χ), etc. For example, in the limit where the reduced temperature $t = ((T-T_c)/T_c)$ approaches zero, the critical exponent γ related to the diverging behavior of the susceptibility (χ) is defined as

$$\gamma \equiv - \lim_{t \to 0} \ln \chi / \ln t.$$

Consequently, in the temperature range sufficiently close to T_c, the susceptibility can be expressed as

$$\chi = Dt^{-\gamma}.$$

The similarity of this kind of diverging behavior in many physical properties near various and quite different continuous phase transitions has stimulated considerable experimental and theoretical effort to understand the fundamental physical principles behind the critical phenomena. In particular, the critical exponents characterizing anomalous behavior in the neighborhood of a second order phase transition are found to depend only on the spatial dimension of the physical system under study, the dimension of the order parameter and the range of interaction. Out of more than six static critical exponents, only two are independent, the rest can be related to these two by the scaling laws. Furthermore, the physical properties in the vicinity of a second order phase transition can be described by the scaling equations. All these experimental findings have been explained nicely by the renormalization group theory proposed by K. Wilson[2] and the subsequent expansion technique[3]. Although the renormalization group theory provides excellent explanation for many aspects of various phase transitions, with its highly anisotropic molecular structure and the existence of series of phase transitions in a relatively narrow temperature range, the liquid crystals have provided rich varieties and unconventional types of phase transitions and are unique systems for studying the effect of one phase transition on another.

Liquid crystals are organic molecules, anisotropic in shape, which have one or more mesophase between the crystalline state and the isotropic liquid (I).[4] N - (4-n-butyloxybenzylidene) - 4' - n - heptylaniline (40.7) is one of the liquid-crystal molecules and has the following molecular structure

$$C_4H_9\,O - (C_6H_4) - C\genfrac{}{}{0pt}{}{\diagup H}{\diagdown\!\!= N} - (C_6H_4) - C_7\,H_{15}\,.$$

Typical molecular length is about $20 \sim 30$ Å and cross section is about $6 \sim 8$ Å. For simplicity, in our schematic drawing of the molecular arrangements in a given mesophase, each molecule will be represented by a cigar-shaped object.

Liquid crystals are excellent systems for studying phase transitions, critical phenomena and melting processes. Unlike most other materials which have well separated phase transitions, liquid crystals can have a cascade of phase transitions in a relatively small temperature range. In other words, the disordered phase, assocaited with a given continuous phase transition at T_c, may undergo another transition at a higher temperature T_1 to an even more disordered phase. In most of the materials, the ratio T_c/T_1 is significantly smaller than one. In these situations, the order parameter associated with the phase transition at T_1 is almost saturated at T_c and will not dramatically affect the additional order parameter which sets in near the transition at T_c. However, in liquid-crystal compounds, the existence of many phase transitions in a relatively narrow temperature range leads to the ratio T_c/T_1 being very close to one. Thus fluctuations of the order parameter associated with the transition at T_1 may still be large enough to influence the nature of the transition occuring at T_c. Conversely, fluctuations associated with the transition at T_c may be already fairly important above the transition at T_1 and may affect the transition at T_1.

So far, more than twenty different liquid-crystal mesophases have been identified.[5] Five of them are relevant to our discussion here, namely, nematic (N), smectic-A (SmA), hexatic-B, smectic-C (SmC), and chiral-smectic-C (SmC*) phases. Schematic diagrams of some of these mesophases and isotropic phases (I) are shown in Fig. 1.

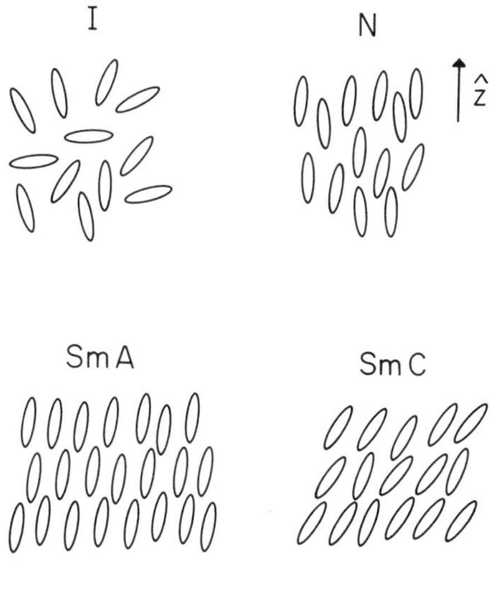

Figure 1. The schematic molecular arrangement in the isotropic (I), nematic (N), smectic-A (SmA), and smectic-C (SmC) phases.

In the isotropic phase, both the center of mass and long axis (molecular director) of each molecule are random. In the nematic phase, on the average, the molecular director is oriented along a preferred direction. The center of mass of each molecule remains random so that translational invariance is preserved. The SmA, hexatic-B, SmC, and SmC* phases have layered structures with weak coupling between the smectic layers. In addition, the molecules maintain translational invariance within the smectic planes. Both the SmA and hexatic-B phases have molecular directors, on average, parallel to the layer normal. However, the major difference between these two phases is that the hexatic-B phase is characterized by the long range bond orientational order.[6,7] The molecule bond is the one linking two adjacent molecules (see Fig. 2) and the bond angle (θ_b) is the angle between the molecule bond and some fixed laboratory axis (e.g. x-axis in Fig. 2).

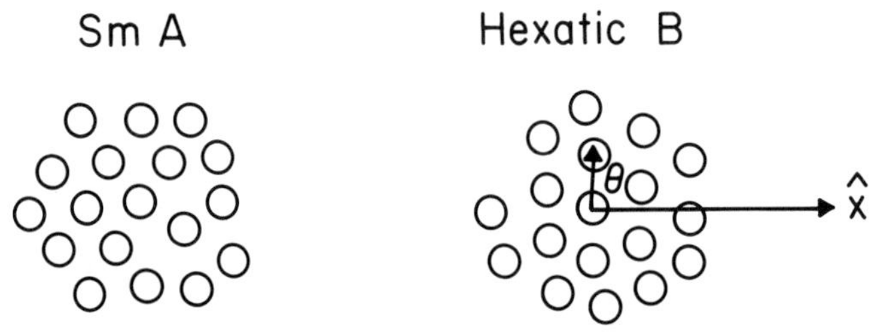

Figure 2. The schematic molecular arrangement within the smectic layer of the smectic-A and the hexatic-B phase

Then the long range bond orientational order means that the molecular bond angle remains fairly constant on a macroscopic scale. Both the SmC and SmC* phase have a finite tilt angle between the molecular director and the layer normal. If the constituent molecules are optically active, the SmC* phase will be observed instead of the SmC phase[8]. Because of the intrinsic molecular chirality, the molecular director will precess around the smectic layer normal with a typical pitch of about a few microns in the SmC* phase. In addition to the direct phase transition from the isotropic phase to these five mesophases, various phase transitions have been found between the mesophases. Among all these possible phase transitions, at least N - SmA, SmA - hexatic-B, SmA-SmC, and SmA-SmC* transitions have been demonstrated to be continuous in many liquid-crystal compounds. From our high-resolution heat-capacity, tilt-angel, and spontaneous polarization studies[9], we have demonstrated that the terms involving the helical pitch and the spontaneous polarization are only about 1% of the free energy related to the SmA-SmC* transition and are small corrections to the leading three terms involving the tilt-angle only. Consequently, we will use the SmA - SmC transition, in general, to represent both the SmA-SmC and SmA - SmC* transitions. In the following three sections, the effect of the disordered phase temperature range on the behavior of the N - SmA, SmA - SmC and SmA-hexatic-B transitions will be discussed separately. The conclusions will be drawn in Section V.

II. Nematic - Smectic - A Transition

In the nematic phase, the alignment of the molecules around the overall average molecular axis ($\hat{z}$ - axis) is described by the parameter

$$S_0(T) = \langle \frac{1}{2} (3\cos^2 \theta_n - 1)\rangle, \tag{1}$$

where θ_n is the angle between the $\hat{z}$ - axis and the molecular director. The formation of one-dimensional mass density wave along the $\hat{z}$-axis in the SmA phase can be represented by[10]

$$\rho(z) = \rho_0 \left[1 + \frac{1}{\sqrt{2}} |\psi| \cos (qz - \phi)\right]. \tag{2}$$

Here ρ_0 is the average density and $|\psi|$ measures the strength of the SmA order. $q = 2\pi/d$ is the wave vector of the density wave, d the interlayer spacing, and ϕ an arbitrary phase factor. Thus the SmA order parameter representing by the second term of Eq. (2) which is equivalent to $\rho_0|\psi|\, e^{i(qz - \phi)}$ and is n = 2 in the words of the n-vector model. In 1972, de Gennes[10] suggested a coupling Landau free energy expansion

$$\Delta G = \frac{\alpha}{2}\,|\psi|^2 + \frac{\beta_0}{2}\,|\psi|^4 + \frac{1}{2\chi_N}\,(\delta S)^2 - C(\delta S)\,|\psi|^2 \qquad (3)$$

to explain the existence of a tricritical point in the N-SmA transition. Earlier, employing molecular mean-field calculations, Kobayashi[11] and McMillan[12] showed that increasing coupling between the nematic order parameter and the SmA density wave amplitude drives the N-SmA transition from being second-order to being first order, which requires the existence of a tricritical point in between. Here in Eq. (3), $\alpha = \alpha_0(T - T_{NA})$ and α_0, β_0 and C are positive constants. $\delta S = S - S_0(T)$ is the change of nematic ordering induced by formation of smectic layering and χ_N the temperature dependent nematic susceptibility, which is larger near the I-N transition point (T_{IN}), and decreases as $T_{IN} - T$ increases.

Minimizing the free energy ΔG with respect to (δS) and eliminating (δS) from ΔG, we have

$$\Delta G = \frac{\alpha}{2}\,|\psi|^2 + \frac{\beta}{4}\,|\psi|^4.$$

Here $\beta = \beta_0 - 2\chi_N C^2$.

Thus, for a wide nematic range, $\chi_N(T_{NA})$ is small, β remains positive and the transition is second order; for a narrow nematic range $\chi_N(T_{NA})$ is large, β becomes negative and the transition is first order. The cross over from a second order N-SmA transition to a first order one as the nematic range decreases has been experimentally demonstrated and occurs in compounds or mixtures with fairly small nematic temperature range $(T_{IN} - T_{NA} \cong 1.7\text{K}$ or $1 - T_{NA}/T_{IA} \cong 0.006)$[13,14]. Thus the majority of the liquid-crystal compounds have second-order N-SmA transitions.

During the past fifteen years, numerous theoretical[15] and experimental[16,17] studies have been carried out to understand the nature of this intriguing phase transition. Among many unusual physical properties related to the N-SmA transition, the heat-capacity measurements have been studied most widely. Here we will discuss the critical exponent (α) and the amplitude ratio (A^+/A^-) characterizing the heat-capacity anomaly in the vicinity of the N-SmA transition. Table I is the list of the available data from single component compounds measured at one atmosphere pressure. The quantity r_{NA} ($=1 - T_{NA}/T_{IA}$), which measures the relative size of the disordered phase (i.e., nematic phase) temperature range, has been included in Table I. Actually the compounds are listed with decreasing r_{NA}.

TABLE I. Critical parameters from different liquid-crystal compounds for the heat-capacity anomaly near the N-SmA transition.

Compound	r_{NA}	α	A^+/A^-	Ref.
40.7	0.074	-0.03	1.04	17
$\overline{8}$S5	0.067	0±0.03	-	18
		-0.17±0.03	0.58	19
CBOOA	0.064	0.15±0.02	-	20
		-0.12±0.02	0.78	19
40.8	0.043	0.15±0.05	1.2	21
8OCB	0.038	0.23±0.04	1.21	22
		0±0.03	1.35	23
		0.03±0.04	1.39	19
		0.25±0.02	1.2	19
		0.16±0.03	-	24
		0.24±0.03	1.01	25
$\overline{9}$S5	0.036	0.22±0.03	-	26
8CB	0.023	0.30±0.05	1.08	27
		0.25±0.02	-	24
		0.29±0.03	0.97	19
		0.31±0.03	0.83	28
$\overline{10}$S5	0.017	0.45±0.05	-	26
9CB	0.006	0.50±0.05	~1	13

40.7: 4-n-butyloxybenzylidene-4'-n-heptylaniline

$\overline{8}$S5: 4-n-pentylphenylthiol-4'-n-octyloxybenzoate

CBOOA: n-p-cyanobenzylidene-p-n-octyloxyaniline

40.8: 4-n-butyloxybenzylidene-4'-n-octylaniline

8OCB: 4-cyano-4'-n-octyloxybiphenyl

$\overline{9}$S5: 4-n-pentylphenylthiol-4'-n-nonyloxybenzoate

8CB: 4-cyano-4'-n-octylbiphenyl

$\overline{10}$S5: 4-n-pentylphenylthiol-4'-n-decyloxybenzoate

9CB: 4-cyano-4'-n-nonylbiphenyl

Figure 3 shows the variation of α and A^+/A^- with r_{NA}. The amplitude ratio (A^+/A^-) data are very scattered and no systematic variation with r_{NA} can be identified. Even though the data are still relatively scattered, a systematic variation of the magnitude of α with r_{NA} seems to emerge. With a sufficiently small r_{NA}, i.e., a small nematic temperature range, the value of α approaches $1/2$ which characterizes a tricritical-like behavior. As r_{NA} increases, within relatively large experimental uncertainty, the experimental data seem to approach $\alpha = 0$ which is consistent with an XY-like exponent ($\alpha = -0.02$). In the case of 40.7, the amplitude ratio A^+/A^- is also consistent with the XY-model.[17] However, the compound 9CB has an amplitude ratio about equal to one which is in serious disparity with a theoretical calculation[29] and experimental results on He^3-He^4 mixture ($A^+/A^- \simeq 0.12$)[30] in the vicinity of the tricritical point.

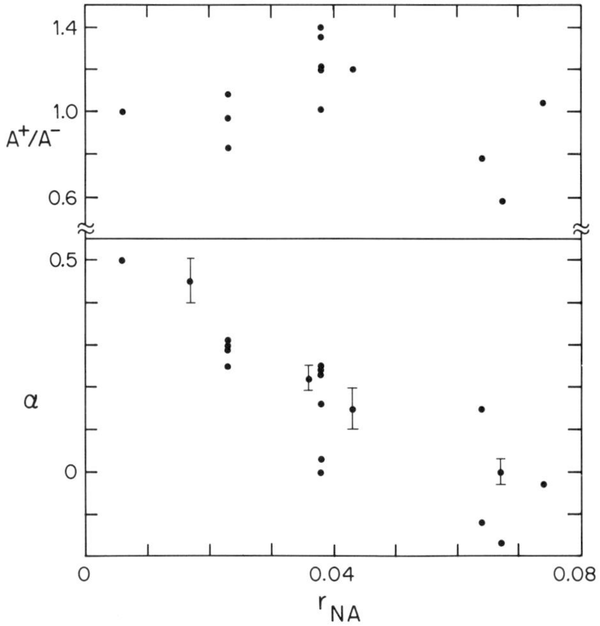

Figure 3. The amplitude ratio A^+/A^- and critical exponent α of the N-SmA transition vs $r_{NA}(= 1-T_{NA}/T_{IN})$. Typical reported error bars for α are shown.

Although the general trend of the critical exponent α strongly suggests the cross over from XY-like behavior to tricritical-like one and then to a first order phase transition as the nematic range decreases, the existence of a tricritical point, usually, requires that for a given compound, the critical exponent α should have a tricritical-like behavior ($\alpha \simeq 0.5$) in the temperature range far away from T_c and a critical-like behavior ($\alpha \simeq 0$) in the region near T_c. See Fig. 4 for a schematic drawing for these two regions.[31] That a single exponent with value in between 0 and 0.5 can fit the data very well over three decades in the reduced temperature scale[28] leads us to believe that either the cross over region with some effective critical exponent is very large and/or we simply don't understand the cross over behavior in this transition at all. The formation of one-dimensional density waves in the N-SmA transition indicates that the lower critical dimensionality of this transition is $d_L = 3$. But three dimensions is also the upper critical dimensionality (d_u) of a conventional tricritical point. Thus the cross over behavior we depict in Fig. 4 for a conventional tricritical point, e.g., $He_e^3 - He^4$ mixture, some antiferromagnetic systems, etc. may not be applicable in the N - SmA transition. The unresolved puzzle related to the effect of the nematic phase temperature range on the N-SmA transition which we summarize here is just the tip of the iceberg featuring the complicated and intriguing nature of the N-SmA transition[15-17].

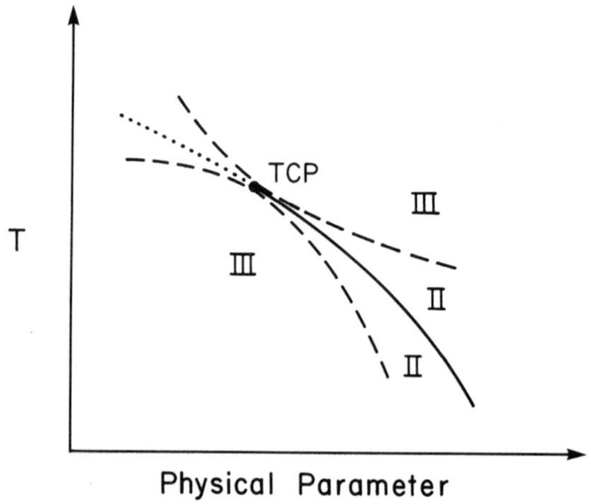

Figure 4. Schematic phase diagram near a conventional tricritical point (TCP) with tricritical-like (III) and critical-like region (II) being separated by the dashed lines. The solid line is the continuous transition line and the dotted one the first order transition which are separated by a tricritical point.

III. Smectic-A-Smectic-C Transition

Since de Gennes first suggested that the SmA-SmC transition may be continuous and may exhibit helium-like critical behavior[10], there has been considerable interest in the nature of this phase transition. As far as the critical behavior is concerned, conflicting results (i.e., helium-like, mean-field-like or cross over from mean-field-like to helium-like) were reported in various measured results.[32] Based on their high-resolution heat-capacity measurements near the SmA-SmC transition of one liquid-crystal compound, Huang and Viner[33] have proposed an extended mean-field model to give an excellent account of the nature of the SmA-SmC transition and explanation for the inconsistency among the reported critical exponents.[33,34] Subsequent high-resolution heat-capacity and tilt-angle measurements support the idea of the extended mean-field model in describing the SmA-SmC transition[35,36].

The appropriate order parameter for the SmA-SmC transition is $\Psi = \theta_C e^{i\phi}$ which is reminiscent of the XY-like transition. Here θ_C is the tilt-angle of the molecular director from the smectic layer normal and ϕ the azimuthal angle. The symmetry of Ψ requires that only even powers of Ψ should appear in the free-energy expansion. Thus the transition can be of second-order and this is the case for the majority of the SmA-SmC transitions which have been studied so far. In an ordinary second-order phase transition, in addition to the two leading terms in the free-energy expansion, i.e., $|\Psi|^2$ and $|\Psi|^4$, one adds $|\Delta\Psi|^2$ to account for the contribution of order-parameter fluctuation to the total free energy. Huang and Viner[33] have found that the SmA-SmC transition is mean-field like, i.e., the fluctuations contribution is small, but the $|\Psi|^6$ term has to be included in the free-energy expansion. Thus the extended mean-field model suggested by them has the following free energy expansion

$$G = G_0 + at\theta_C^2 + b\theta_C^4 + c\theta_C^6 . \tag{4}$$

Here G_0 is the background free energy in the absence of the SmA-SmC transition, the reduced temperature $t = (T-T_{AC})/T_{AC}$, T_{AC} is the SmA-SmC transition temperature, and the coefficients (a, b, and c) are positive constants for a continuous transition. The variable θ_c is the magnitude of the smectic-C order parameter. Now minimizing the free energy with respect to θ_c, we have $\theta_c = 0$ for $T > T_{AC}$ and

$$\theta_c = (R(1-3t/t_0)^{1/2} - 1)^{1/2} \tag{5}$$

for $T < T_{AC}$, where $R = b/3c$ and $t_0 = b^2/ac$. Substituting Eq. (5) into Eq. (4), the heat capacity can be calculated from $C = -T(\partial^2 G/\partial T^2)$ and

$$C = \begin{cases} C_0 & \text{for} \quad T > T_{AC}\ , \\ C_0 + AT(T_m - T)^{-1/2} & \text{for} \quad T < T_{AC}\ . \end{cases} \tag{6}$$

Here C_0 is the background heat capacity derived from the background free energy G_0 which should vary smoothly through the transition, $A = a^{3/2}/[2(3c)^{1/2}\ T_{AC}{}^{3/2}]$ and $T_m = T_{AC}(1+t_0/3)$ where T_{AC} is chosen to be the midpoint of the mean-field heat-capacity jump. In the reduced temperature scale, the parameter t_0 is equal to the full width at half height of the $(\Delta C/T)$ vs. T curve. Here $\Delta C = C - C_0$ is the anomalous part of the heat capacity.

From Eq. (5) an important property associated with the dimensionless parameter t_0 can be obtained. For $|t| \ll t_0$, $\theta_c \sim |t|^{1/2}$ and $|t| \gg t_0$, $\theta_c \sim |t|^{1/4}$, thus t_0 characterizes the cross over from an ordinary mean-field region $|t| \ll t_0$ to a tricritical-like region $|t| \gg t_0$. In all high-resolution studies near the SmA-SmC transition[36], one finds that $t_0 < 5 \times 10^{-3}$ which is about ten times smaller than the value for all the other mean-field transitions[33]. With transition temperatures being approximately equal to 350 K, the cross over from $\theta_c \sim |t|^{1/2}$ to $\theta_c \sim |t|^{1/4}$ occurs around $T_{AC} - T \simeq 1$ K. This explains why the reported results on the critical exponent β associated with the tilt-angle vary so much from one experiment to the other one. A well-characterized evolution of the effective critical exponent β in a log (θ_c) vs. log(t) plot is one of the prominent features of the extended mean-field model.[34]

Furthermore, within the mean-field model, one can derive a unique relationship between the amplitude of the order parameter (θ_c) and the anomalous part of the heat capacity ΔC $(=C-C_0)$,[37] i.e.,

$$\theta_c = \left(\frac{T_{AC}}{a} \int_T^{T_{AC}} \frac{\Delta C}{T} \, dT \right)^{1/2} \qquad (7)$$

Such a simple relation does not exist in the 3D XY model in which $\theta_c \sim |t|^{0.35}$ and $\Delta C \sim |t|^{-0.02}$. Thus Eq. (7) will provide a crucial test to see if the SmA-SmC transition should be described by the XY model or the extended mean-field model. Recently, we have carried out high-resolution heat-capacity and tilt-angle measurements near the SmA-SmC* transition of DOBAMBC (p-(n-decyloxybenzylidene) - p-amino-(2-methylbutyl) cinnamate). Our data are in excellent agreement with Eq. (7) (ref. 37). This result provides a strong support for describing the SmA-SmC transition using the extended mean-field theory. Moreover, Eq. (7) demonstrates that heat-capacity data have a stronger temperature dependence than the tilt-angle data and are more sensitive physical parameters to characterize the SmA-SmC transition. Experimentally, it is much easier to carry out the high-resolution heat-capacity measurement than the tilt-angle one. Thus, in search of the fluctuations dominant SmA-SmC transition, the heat-capacity measurement is a very powerful tool.

In fitting some of our high-resolution heat capacity data to the expression derived from the mean-field free energy, higher-order terms, i.e., $t^2\theta_c^2$, $t\theta_c^4$, $t\theta_c^6$ and θ_c^8 are added to the free-energy expansion (Eq. 4), individually, to test the relative importance of this individual term. From these fittings, we conclude that the extended mean-field free energy (Eq. 4) is sufficient to describe the SmA-SmC transition[36]. Moreover, the following three relations -- $t_0 = b^2/(ac)$, $\Delta C_J = a^2/(2bT_{AC})$, and $R = b/(3c)$ -- enable us to calculate all three free-energy expansion coefficients, i.e., a, b, and c from the tilt-angle and heat-capacity data. $a = 2T_{AC}t_0(\Delta C_J)/(3R)$, $b = at_0/(3R)$, and $c = b/(3R)$. Here ΔC_J is the mean-field heat-capacity jump at the transition temperature.

In order to make a systematic comparison between the mean-field coefficients of the SmA-SmC transitions of various liquid-crystal compounds, we have to know how to properly choose the normalization constant for the tilt-angle θ_c. In principle, the saturation tilt-angle at $T = 0°K$ is an ideal one. However, the SmC phase will transform to the other phase or become crystallized well above $T = 0°K$. But if we assume that the SmA - SmC could exist at $T = 0°K$ and could be well-described by the extended mean-field model, then for $|t|/t_0 \gg 1$, the asymptotic value of θ_c becomes $(b/(c(3t_0)^{1/2})|t|^{1/4}$. Now let us redefine θ_c, such that $\tilde{\theta} = |t|^{1/4}$ for $|t|/t_0 \gg 1$. With this normalized order parameter $\tilde{\theta}$, the extended mean-field free energy can be rewritten as[36]

$$G = G_0 + \tilde{a}t\tilde{\theta}^2 + \tilde{b}\tilde{\theta}^4 + \tilde{c}\tilde{\theta}^6. \quad (8)$$

Here, $\tilde{a} = a^{3/2}/(3c)^{1/2}$, $\tilde{b} = ab(3c)$, and $\tilde{c} = \tilde{a}/3$. Because the condition $b/(c(3t_0)^{1/2}) = 1$ is imposed, only two independent coefficients (i.e., $\tilde{a}$ and $\tilde{b}$) have to be determined in Eq. (8). The coefficient $\tilde{c}$ is proportional to $\tilde{a}$. Furthermore, $\tilde{a}$ and $\tilde{b}$ can be expressed in terms of the experimentally measurable quantities as follows:

$$\tilde{a} = 2\, T_{AC}\, (\Delta C_J)(t_0/3)^{1/2} \quad \text{and} \quad \tilde{b} = \tilde{a}(t_0/3)^{1/2}.$$

Consequently, only heat-capacity data are required to determine the coefficients $\tilde{a}$ and $\tilde{b}$. From our experience, it is much easier to obtain high-resolution heat-capacity data than the tilt-angle data. Thus the determination of the coeffients $\tilde{a}$ and $\tilde{b}$ will be more reliable than that of a, b, and c.

Based on all the available high-resolution heat-capacity measurements from our research group and others,[33,38-44] we have calculated $\tilde{a}$ and $\tilde{b}$. The results for those compounds with the I - N - SmA - SmC and the I - SmA - SmC transition sequence are shown in Table II and III, respectively. At the same time the relative magnitude (r_{AC}) of the disordered phase (SmA) temperature range is calculated for each compound. Here $r_{AC} = 1-(T_{AC}/T_{IA})$ and $1 - (T_{AC}/T_{NA})$ for the compounds with I - SmA - SmC and I-N-SmA-SmC transition sequence, respectively. The solid dots in Fig. 5 are the plots of the coefficient $\tilde{a}$ and $\tilde{b}$ versus r_{AC} for all the compounds listed in Table II. Except the compound 70.6, Fig. 5 clearly indicates that both $\tilde{a}$ and $\tilde{b}$ decrease almost linearly as $(1 - T_{AC}/T_{IA})$ and the compound MBRA 8 seems to be very close to a mean-field

TABLE II. The parameters obtained from heat-capacity measurements near the SmA-SmC(or SmC*) transitions of the compounds with the I-SmA-SmC (or SmC*) transition sequence.

Compound	T_{AC}(K)	T_{IA}(K)	$1-\frac{T_{AC}}{T_{IA}}$	$t_0 \times 10^3$	ΔC^a_J	$\tilde{a}^b$	$\tilde{b}^b$	Ref.
HOBACPC	353.240	409.2	0.137	5	$2.8X10^2$	8100	330	36
2M45OBC(R)	316.004	336.96	0.062	3.9	$1.24X10^2$	2830	102	37
2M45OBC(C)	316.350	336.96	0.061	3.5	$1.45X10^2$	3130	107	37
DOBAMBC	367.732	390.2	0.058	3.2	$1.25X10^2$	3000	98	38
70.6	343.301	353.5	0.029	1.6	$2.58X10^2$	4090	94	39
AMC-11	352.360	362.9	0.029	1.7	82	1380	33	39
MBRA8	322.918	329.2	0.019	0.9	71	790	14	40

[a] ΔC_J in J/mole K.

[b] $\tilde{a}$ and $\tilde{b}$ in J/mole.

HOBACPC: n-hexyloxybenzylidene-p'-amino-[2-chloro(n-propyl)]cinnamate.

2M45OBC: (2-methyl-butyl)-4'-(n-pentyloxybiphenyl)-4-carboxylate; R, racemic version; C, chiral version.

DOBAMBC: p-(n-decyloxybenzylidene)-p-amino-(2-methyl-butyl)cinnamate.

70.6: 4-n-heptyloxybenzylidene-4'-hexylaniline.

AMC-11: azoxy-4,4'-bi(n-undecyl-α-methylcinnamate).

MBRA8: S-4-0-(2-methyl-butyl)β-resorcylidene-4'-n-octylaniline.

TABLE III. The parameters obtained from heat-capacity measurements near the SmA-Sm C transitions of the pure compounds with the I-N-SmA-SmC transition sequence.

Compound	T_{AC}(K)	T_{NA}(K)	$1-\frac{T_{AC}}{T_{NA}}$	$t_0 X10^3$	ΔC^a_I	$\tilde{a}^b$	$\tilde{b}^b$	Ref.
2M4P9OBC	406.91	441	0.077	5.5	134	4670	200	32
70.7	344.64	356.6	0.034	2.8	123	2600	80	41
70.4	337.56	347.2	0.027	0.8	400	4400	70	39
$\overline{8}$S5	328.18	336.1	0.024	7.3	47	1500	74	18,42
40.7	323.11	329.8	0.022	1.3	65	870	18	39

[a] ΔC_J in J/mole K.

[b] $\tilde{a}$ and $\tilde{b}$ in J/mole.

2M4P9OBC: racemic 4-(2-methyl-butyl)phenyl-4-n-nonyloxybiphenyl-4-carboxylate.

70.7: 4-n-heptyloxybenzylidene-4'-n-heptylaniline.

70.4: 4-n-heptyloxybenzylidene-4'-n-butylaniline.

tricritical point ($\tilde{b} = 0$). The open circles in Fig. 5 show the variation of $\tilde{a}$ and $\tilde{b}$ with $(1 - T_{AC}/T_{NA})$ for the compounds with I - N - SmA - SmC transition sequence. The coefficients roughly follow the same trend as the compounds with the I-SmA-SmC transition sequence but with much larger scatter. Furthermore, the existence of the nematic phase seems to increase the value of $\tilde{a}$ and $\tilde{b}$, but no systematic relation with the size of the nematic range can be identified.

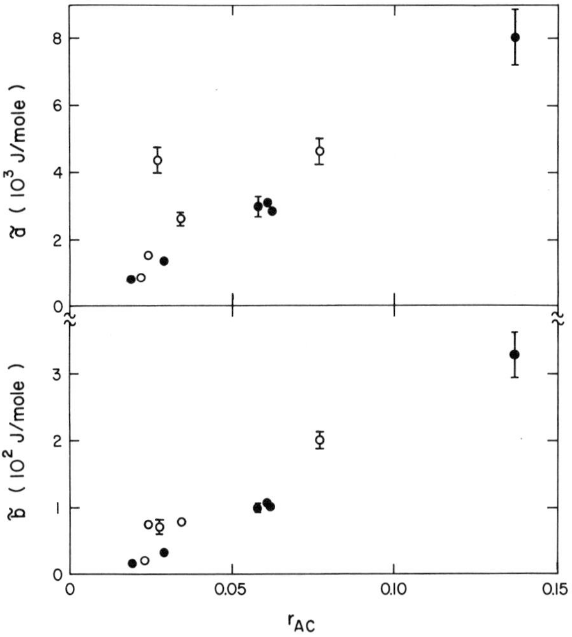

Figure 5. Plots of the mean-field coefficient a and b vs r_{AC} for twelve pure liquid-crystal compounds with the SmA-SmC transition. The solid dots are the compounds with I-SmA-SmC transition sequence and $r_{AC} = (1-T_{AC}/T_{IA})$. The open circles are the compounds with I-N-SmA-SmC transition sequence and $r_{AC} = (1-T_{AC}/T_{NA})$.

IV. Smectic-A-Hexatic-B Transition

The fundamental difference between the SmA and the hexatic-B phases is that the centers of mass of the molecules are randomly distributed like a two-dimensional liquid within the layer plane of the SmA phase and have long range bond orientational ordering in the hexatic-B phase. This difference can be characterized by X-ray diffraction studies. Pindak and coworkers[45] have first found the six-fold modulation scattering peaks featuring the long-range bond orientational ordering below the SmA-hexatic-B transition of n-hexyl-4'-n-pentyloxybiphenyl-4-carboxylate (65OBC). The order paramter representing the bond-orientational ordering can be written as $\Psi = |\Psi| e^{i6\phi}$. Again, this order parameter has XY-like symmetry. From its symmetry, the transition can be a continuous one. High-resolution heat-capacity studies by Huang et al.[46] have demonstrated that the SmA-hexatic-B transition of 65OBC is continuous. But instead of an XY-like critical exponent for the heat-capacity anomaly ($\alpha \simeq -0.02$), an exponent $\alpha \simeq 0.60 \pm 0.03$ is found.[47] From the viewpoint of the critical phenomena, it is strongly desirable to obtain two more critical exponents to test the scaling relationship. Although there exists theoretical arguments[48], none of the existing theory can satisfactorily explain the large value of α for the SmA-hexatic-B transition.

Unfortunately, similar to the superfluid transition in He^4, it is not trivial to measure the strength of the order parameter ($|\Psi|$) and there exists no obvious physical field which directly couples to the bond orientational order in the hexatic-B phase. Consequently, the heat-capacity exponent (α) has been the only critical exponent which can be obtained fairly reliably and with high resolution. Some attempts have been made to get the critical exponent β, associated with the temperature variation of the order parameter, from the birefringence measurements.[49] However, the data-interpretation may not be conclusive. Recently, we have successfully measured the divergence of thermal conductivity of 65OBC near its SmA-hexatic-B transition and acquired the critical exponent.[50] This will provide the theorists with further information toward our understanding of this intriguing phase transition.

The liquid-crystal compounds which show the existence of the hexatic-B phase are relatively rare. So far in the single component compounds, the hexatic-B phase has been identified only in two homologous series. The first one is 4-propionyl-4'-n-alkanoyloxyazobenzene which shows the first-order SmA-hexatic-B transition.[51,52] The more interesting one is the homologous series of n-alkyl-4'-n-alkoxybiphenyl-4-carboxylate[53] which, in most cases, has a continuous SmA-hexatic-B transition. In the second homologous series we have carried out heat-capacity measurements of the following eight compounds with various disordered phase (SmA) temperature ranges.[54] They are n-butyl (45OBC), n-hexyl (65OBC), and n-heptyl (75OBC) of the n-pentyloxy series, n-propyl (36OBC), and n-butyl (46OBC) of the n-hexyloxy series, n-propyl (37OBC) of the heptyloxy series, and n-ethyl [2(10)OBC] and n-propyl [3(10)OBC] of the decyloxy series. All of these compounds have an I-SmA-hexatic-B transition sequence as well as pronounced and approximately symmetric heat-capacity anomalies associated with the SmA-hexatic-B transition. Figure 6 shows one typical heat-capacity anomaly near the SmA-hexatic-B transition of 75OBC.

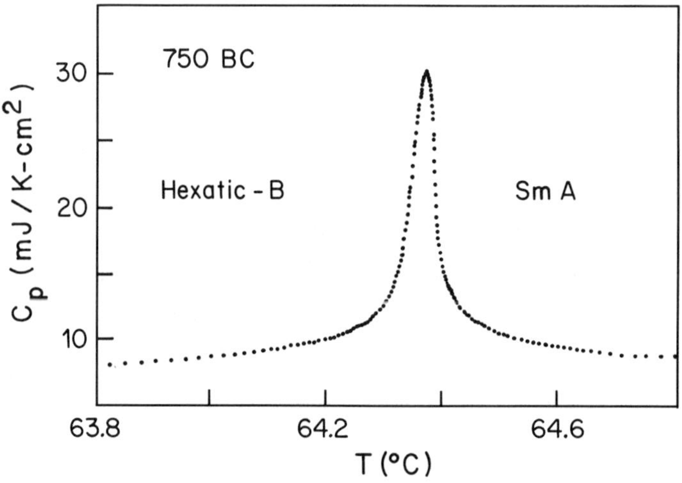

Fig. 6. Heat-capacity data near the SmA-hexatic-B transition of 75OBC. (Sample thickness = 50 μm)

We have performed nonlinear least-squares fittings of all our data to the following power-law expression with scaling correction terms:[55]

$$C_p = \begin{cases} A^+|t|^{-\alpha}\,(1 + E^+|t|^x) + B + Dt, & T > T_{AB} \\ A^-|t|^{-\alpha}\,(1 + E^-|t|^x) + B + Dt, & T < T_{AB} \end{cases}$$

Here $t = (T - T_{AB})/T_{AB}$ and $B + Dt$ is the background term. Here we have choose $x = 0.75$. The argument for choosing this value of x is given in Ref. 54. The fitting results are shown in Fig. 7 as the anomalous part of the heat capacity $\Delta C_p = C_p - B - Dt - A^{\pm}E^{\pm}|t|^{x-\alpha}$ vs $|T-T_c|$. Very good fitting is obtained for all our data. Table IV summarizes the parameters obtained from our heat-capacity

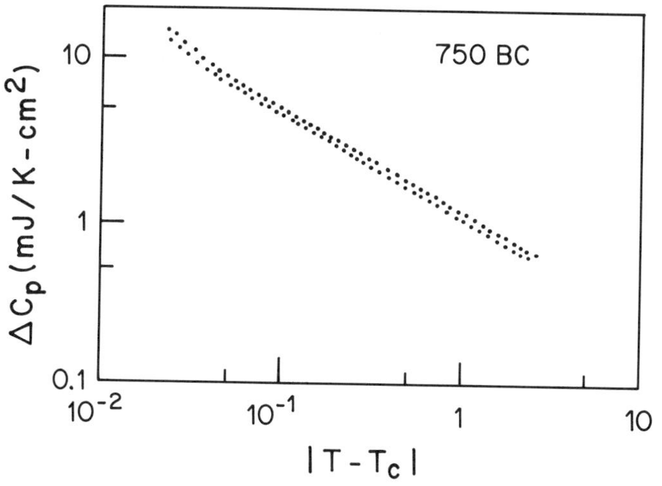

Figure 7. Anomalous part of the heat capacity $\Delta C_p = C_p - B - Dt - A^{\pm}E^{\pm}|t|^{x-\alpha}$ vs. $|T-T_c|$ for 750BC near the SmA-hexatic-B transition.

measurements near the SmA-hexatic-B transitions of various compounds. In Fig. 8, the critical parameters A^+/A^- and α are plotted as a function of $r_{AB}(= 1 - T_{AB}/T_{IA})$. Unlike the N-SmA and the SmA - SmC transitions, as the disordered phase (SmA) temperature range increases from 17 to 32 K by about a factor of two, the amplitude ratio A^+/A^- remains reasonably constant, within experimental error, and the critical exponent α also seems to be unrelated to this increase. We do not know why the compounds 46OBC and 45OBC have slightly smaller values of α.

TABLE IV. The parameters obtained from heat-capacity measurements near the SmA-hexatic-B transitions of various compounds.

Compound	T_{IA}(K)	T_{AB}(K)	$1 - \frac{T_{AB}}{T_{IA}}$	α	A^+/A^-
3(10)OBC	372.3	340.387	0.086	0.59±0.03	0.71±0.08
37OBC	375.4	344.141	0.083	0.56±0.03	0.78±0.08
36OBC	380.2	348.540	0.083	0.58±0.03	0.93±0.08
46OBC	365.0	340.408	0.067	0.49±0.03	0.77±0.08
45OBC	366.7	345.426	0.058	0.48±0.03	1.00±0.08
65OBC	358.2	340.441	0.050	0.60±0.03	0.85±0.08
75OBC	354.2	337.525	0.047	0.62±0.03	0.91±0.09
2(10)OBC	381.6	367.414	0.037	0.67±0.03	0.90±0.08

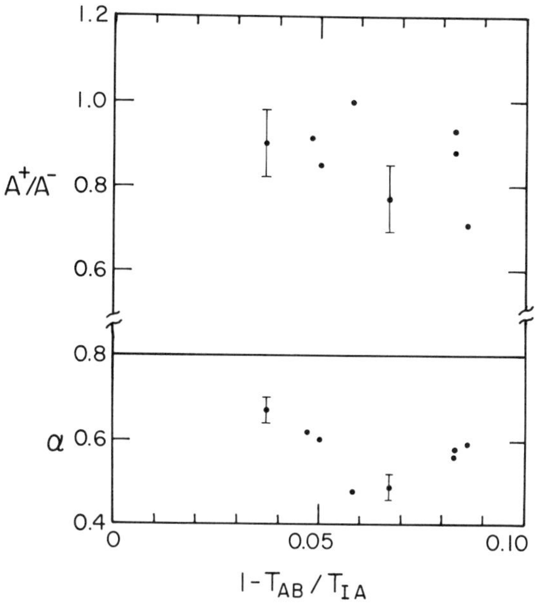

Figure 8. The amplitude ratio A^+/A^- and critical exponent α of the SmA-hexatic-B transition vs. r_{AB} (= $1 - T_{AB}/T_{IA}$). Typical reported error bars are shown here.

IV. Conclusions

Among various types of phase transitions which can occur between liquid-crystal mesophases, extensive studies, in particular, the high-resolution heat-capacity measurements, have been carried out near the N-SmA, SmA-SmC, and SmA-hexatic-B transitions. According to the symmetry of the order parameter, all these three phase transitions belong to the XY-like universality class, and should be characterized by the same set of critical exponents. To the contrary, we have found extremely rich varieties in the critical phenomena of these three phase transitions. Both the N-SmA and SmA-hexatic-B transition are found to be fluctuations dominated transitions and the SmA-SmC transition is mean-field like. Theoretically, we do not understand the nature of the N-SmA and SmA-hexatic-B transition as well as the mean-field behavior of the SmA-SmC transition with a large $|\Psi|^6$ term. Furthermore, although a phenomenological approach can be given to argue the evolution of the critical exponent associated with the N-SmA heat-capacity anomaly, there exists no theoretical prediction indicating how the critical exponent should evolve with the nematic temperature range. In the case of the SmA-SmC transition, a similar phenomenological argument was given by the author[56], but it is linked to the fluctuations related to the I-N transition, but not to the I-SmA or N-SmA transition. Also no explanation can be given to the variation of the mean-field coefficients $\tilde{a}$ and $\tilde{b}$ with the size of the SmA temperature range. Finally, why is the critical exponent α for the N-SmA transition and the mean-field coefficients for the SmA-SmC transition approximately linear functions of the disordered phase temperature ranges but not the critical exponent α for the SmA-hexatic-B transition? All of those pertinent experimental findings have opened up entirely new frontiers for theoretical work.

Acknowledgments

This work was partially supported by the National Science Foundation, Solid State Chemistry Program, Grant No. DMR 85-03419.

Reference

1. For general reference of critical phenomena, see "Introduction to Phase Transition and Critical Phenomena", H.E. Stanley (Oxford University Press) 1971.
2. K.G. Wilson, Phys. Rev. B 4, 3174, 3184 (1971).
3. K.G. Wilson and M.E. Fisher, Phys. Rev. Lett. 28, 240 (1972); K.G. Wilson and J. Kogut, Phys. Reports 12, 75 (1974).
4. "The Physics of Liquid Crystals", P.G. de Gennes (Clarendon Press, Oxford) 1974; "Liquid Crystals", S. Chandrasekhar (Cambridge University Press) 1977.
5. "Smectic Liquid Crystals", G.W. Gray and J.W. Goodby, (Leonard Hill) 1984.
6. B.I. Halperin and D.R. Nelson, Phys. Rev. Lett. 41, 121 (1978).
7. R.J. Birgeneau and J.D. Litster, J. Phys. Lett. (Paris) 39, 339 (1978).

8. R.B. Meyer, L. Liebert, L. Strzelecki, and P. Keller, J. Phys. Lett. (Paris) 36, 69 (1975).

9. S. Dumrongrattana and C.C. Huang, Phys. Rev. Lett. 56, 464 (1986); C.C. Huang and S. Dumrongrattana, Phys. Rev. A. 34, 5020 (1986).

10. P.G. de Gennes, Mol. Cryst. Liq. Cryst. 21, 49 (1973).

11. K. Kobayashi, Mol. Cryst. Liq. Cryst. 13, 137 (1971); Phys. Lett. 31A, 125 (1970); J. Phys. Soc. Jpn. 29, 101 (1970).

12. W.L. McMillan, Phys. Rev. A4, 1238 (1971).

13. J. Thoen, H. Marynissen, and W. Van Dael, Phys. Rev. Lett. 52, 204 (1984).

14. B.M. Ocko, R.J. Birgeneau, J.D. Litster, and M.E. Neubert, Phys. Rev. Lett. 52, 208 (1984).

15. T.C. Lubensky, J.de Chimie Phys. 80, 31 (1983) and references found therein.

16. D.L. Johnson, J. de Chimie Phys. 80, 45 (1983) and references found therein.

17. C.W. Garland, M. Meichle, B.M. Ocko, A.R. Kortan, C.R. Safinya, L.J. Yu, J.D. Litster and R.J. Birgeneau, Phys. Rev. A. 27, 3234 (1983) and references found therein.

18. C.A. Schantz and D.L. Johnson, Phys. Rev. A 17, 1054 (1978).

19. J.D. LeGrange and J.M. Mochel, Phys. Rev. Lett. 45, 35 (1980); Phys. Rev. A23, 3215 (1981).

20. D. Djurek, J. Baturic-Rubcic, and K. Franulovic, Phys. Rev. Lett. 33, 1126 (1974).

21. R.J. Birgeneau, C.W. Garland, G.B. Kasting, and B.M. Ocko, Phys. Rev. A24, 2624 (1981).

22. C.W. Garland, G.B. Kasting, and K.J. Lushington, Phys. Rev. Lett. 43, 1420, (1979)

23. D.L. Johnson, C.F. Hayes, R.J. DeHoff, and C.A. Shantz, Phys. Rev. B18, 4902 (1978).

24. I. Hatta and T. Nokayama, Mol. Cryst. Liq. Cryst. 66, 417 (1981).

25. J.M. Viner and C.C. Huang, Solid State Commun. 39, 789 (1981).

26. D. Brisbin, R. DeHoff, T.E. Lockhart, and D.L. Johnson, Phys. Rev. Lett. 43, 1171 (1979).

27. G.B. Kasting, C.W. Garland, and K.J. Lushington, J. Phys (Paris) 41, 879 (1980).

28. J. Thoen, H. Marynissen, and W. Van Dael, Phys. Rev. A26, 2886 (1982).

29. E.E. Gorodetskii and V.M. Zaprudskii, Zh. Eksp. Teor. Fiz. 72, 2299 (1977)(Sov. Phys. JETP 45, 1209 (1977)).

30. S.T. Islander and W. Zimmermann, Jr., Phys. Rev. A 7, 188 (1973).

31. The cross over from a tricritical-like behavior to an ordinary mean-field-like behaviour can be seen clearly in the temperature dependence of the tilt-angle below the SmA - SmC* transition of DOBAMBC. See ref. 40.

32. Summary of former results on the SmA-SmC transition can be found in Ref. 34.

33. C.C. Huang and J.M. Viner, Phys. Rev. A 25, 3385 (1982).

34. C.C. Huang and J.M. Viner, "Liquid Crystals and Ordered Fluid" Vol. 4 Ed. by A.C. Griffin and J.F. Johnson (Plenum) 1984.

35. R.J. Birgeneau, C.W. Garland, A.R. Kortan, J.D. Litster, M. Meichle, B.M. Ocko, C. Rosenblatt, L.J. Yu, and J. Goodby, Phys. Rev. A 27, 1251 (1983).

36. C.C. Huang and S.C. Lien, Phys. Rev. A 31, 2621 (1985) and references found therein.

37. S. Dumrongrattana, G. Nounesis, and C.C. Huang, Phys. Rev. A 33, 2181 (1986); C.C. Huang, Mol. Cryst. Liq. Cryst. (in press).

38. S.C. Lien, J.M. Viner, C.C. Huang, and N.A. Clark, Mol. Cryst. Liq. Cryst. 100, 145 (1983).

39. S.C. Lien, C.C. Huang, and J.W. Goodby, Phys. Rev. A 29, 1371 (1984).

40. S. Dumrongrattana, C.C. Huang, G. Nounesis, S.C. Lien, and J.M. Viner, Phys. Rev. A. 34, 5010 (1986).

41. M. Meichle and C.W. Garland, Phys. Rev. A 27, 2624 (1983).

42. S.C. Lien, C.C. Huang, T. Carlsson, I. Dahl, and S.T. Lagerwall, Mol. Cryst. Liq. Cryst. 108, 148 (1984).

43. J. Theon and G. Seynhaeve, Mol. Cryst. Liq. Cryst. 127, 229 (1985).

44. C.C. Huang and S.C. Lien, in "Multicritical Phenomena", Vol. 106, of the Proceedings of the North Atlantic Treaty Organization Advanced Studies Institute, ed. by R. Pynn and A. Skjeltorp (Plenum, N.Y. 1984). Ser. B, p. 73.

45. R. Pindak, D.E. Moncton, S.C. Davey, and J.W. Goodby, Phys. Rev. Lett. 46, 1135 (1981).

46. C.C. Huang, J.M. Viner, R. Pindak, and J.W. Goodby, Phys. Rev. Lett. 46, 1289, (1981).

47. J.M. Viner, D. Lamey, C.C. Huang, R. Pindak, and J.W. Goodby, Phys. Rev. A 28, 2433 (1983).

48. R. Bruinsma and G. Aeppli, Phys. Rev. Lett. 48, 1625 (1982).

49. C. Rosenblatt and J.T. Ho, Phys. Rev. A 26, 2293 (1982).

50. G. Nounesis, C.C. Huang, and J.W. Goodby, Phys. Rev. Lett. 56, 1712 (1986)

51. G. Poeti, E. Fanelli, and D. Guillon, Mol. Cryst. Liq. Cryst. Lett. 82, 107 (1982).

52. C.C. Huang, G. Nounesis and D. Guillon, Phys. Rev. A. 33, 2602 (1986).

53. J.W. Goodby and G.W. Gray, J. Phys. (Paris) Coll. 37, C3-17 (1976).

54. T. Pitchford, G. Nounesis, S. Dumrongrattana, J.M. Viner, C.C. Huang, and J.W. Goodby, Phys. Rev. A. 32, 1938 (1985).

55. F.J. Wegner, Phys. Rev. B 5, 4529 (1972).

56. C.C. Huang, Solid State Commun. 43, 883 (1982).

SOME TOPICS IN EQUILIBRIUM THEORY OF LIQUID CRYSTALS

Frank M. Leslie

Department of Mathematics
University of Strathclyde
Livingstone Tower
26 Richmond Street
Glasgow GI 1XH, Scotland

1. Introduction

Our aim here is firstly to describe the continuum equations that arise in equilibrium theory for nematic and cholesteric liquid crystals, and secondly to give an indication of the variety of problems that occur in this area of research. The theory itself is not without interest being a somewhat rare example of a theory of liquids with microstructure involving body moments and couple stresses that actually models properties of certain materials rather well. Also the problems that one meets even in static situations are by no means trivial, and can give rise to some unexpected results.

Our starting point is naturally a careful presentation of the partial differential equations that describe equilibrium configurations in these transversely isotropic liquids. However, it is helpful to indicate at least some of the assumptions behind the theory to give a better understanding of it, and also to some extent to motivate certain boundary conditions and stability criteria. Before turning to what is perhaps the main theme of static theory, effects involving external magnetic and electric fields, it is useful and not without interest to discuss equilibrium configurations likely to occur when one imposes certain surface alignments through prior treatment of the solid surfaces confining the liquid crystal. Even in relatively simple situations non-trivial problems can arise that lead to results that surprise. This is particularly true when the imposed conditions tend to produce line singularities in the alignment.

Certain changes in alignment due to the application of magnetic or electric fields to liquid crystals have been known and understood for some fifty years, but the more recent use of these Freedericksz transitions in electro-optic display devices has led to a marked increase in interest in this topic. Essentially these transitions require a magnetic or electric field applied perpendicular to an initial alignment, which begins to rotate towards the field direction once the field strength exceeds a particular threshold value. Initially it is helpful to confine the discussion to problems with magnetic fields, since in this case the theory is more straightforward, not involving complications due to interaction between the liquid crystal and the field on account of the relevant material parameters being small. Having described the basic effects for this simpler case, one then indicates the more complex calculations for electric fields where the aforesaid complications are present, the latter of course being more relevant for applications.

One naturally assumes some familiarity with certain physical properties of liquid crystals, accounts of these being readily available in the books by de Gennes [1] and Chandrasekhar [2]. These authors also discuss continuum theory, as do Stephen and Straley [3] in their lengthy review. However, an extensive review article dealing exclusively with continuum theory as it applies to equilibrium configurations is that by Ericksen [4], while Deuling [5] gives a somewhat briefer account of some of the topics discussed below.

2. Equations of Static Theory

Continuum theory of liquid crystals dates from the work of Oseen [6] and Zocher [7] some sixty years ago, the former essentially deriving the equilibrium equations for nematics and cholesterics in current usage. Much later Frank [8] gave a more direct derivation of this theory, and subsequently Ericksen [9] set it within the framework of classical mechanics by a formulation based on virtual work, which we present briefly below.

To describe the orientation of the anisotropic axis in these transversely isotropic liquids, all of the above employ a unit vector field $\underline{n}$, frequently referred to as a director, and thus for static theory denoting position by $\underline{x}$

$$\underline{n} = \underline{n}(\underline{x}), \qquad \underline{n} \cdot \underline{n} = 1. \tag{2.1}$$

Also they introduce an energy density associated with distortions of the alignment of the anisotropic axis of the form

$$W = W(\underline{n}, \nabla\underline{n}). \tag{2.2}$$

Since liquid crystals lack polarity, the theory does not distinguish between $\underline{n}$ and $-\underline{n}$, and thus the above functional dependence is restricted by the requirement that

$$W(\underline{n}, \nabla\underline{n}) = W(-\underline{n}, -\nabla\underline{n}), \qquad (2.3)$$

and also independence of arbitrary superposed rigid rotations further requires that

$$W(\underline{n}, \nabla\underline{n}) = W(Q\underline{n}, Q\nabla\underline{n}Q^T), \qquad (2.4)$$

where Q denotes any proper orthogonal second order tensor, and Q^T its transpose. Oseen and Frank both consider a quadratic dependence upon gradients and thereby find that for cholesterics

$$2W = K_1(\text{div } \underline{n})^2 + K_2(\tau + \underline{n} \cdot \text{curl } \underline{n})^2 + K_3(\underline{n} \times \text{curl } \underline{n}) \cdot (\underline{n} \times \text{curl } \underline{n})$$
$$- (K_2 + K_4)\text{div}[(\text{div } \underline{n})\underline{n} - (\underline{n} \cdot \text{grad})\underline{n}], \qquad (2.5)$$

the K's and τ material parameters.

For nematics, however, the constituent molecules lack chirality, and for this class of liquid crystal, therefore, material symmetry further restricts the energy density requiring that the condition (2.4) hold for the full orthogonal group. Hence for this type of liquid crystal the coefficient τ is necessarily zero, and thus the counterpart of the expression (2.5) is

$$2W = (K_1 - K_2 - K_4)(n_{i,i})^2 + K_2 n_{i,j} n_{i,j} + K_4 n_{i,j} n_{j,i}$$
$$+ (K_3 - K_2) n_i n_j n_{k,i} n_{k,j}, \qquad (2.6)$$

where Cartesian tensor notation is employed. Since nematics tend to align uniformly, Ericksen [10] argues that this must represent a state of minimum energy and thus for these materials therefore assumes that

$$W(\underline{n}, \nabla\underline{n}) > W(\underline{n}, 0) = 0, \quad \nabla\underline{n} \neq 0, \qquad (2.7)$$

which leads to

$$K_1 > 0,\ K_2 > 0,\ K_3 > 0,\ K_2 > |K_4|,\ 2K_1 > K_2 + K_4 \tag{2.8}$$

for the expression (2.6). Jenkins [11] discusses corresponding restrictions upon the coefficients in the form (2.5).

For his derivation of static theory, Ericksen [9] begins by assuming a principle of virtual work of the form

$$\delta \int_V W\,dV = \int_V(\underline{F} \cdot \delta\underline{x} + \underline{G} \cdot \Delta\underline{n})dV + \int_{\partial V}(\underline{t} \cdot \delta\underline{x} + \underline{s} \cdot \Delta\underline{n})dS, \tag{2.9}$$

where

$$\Delta\underline{n} = \delta\underline{n} + (\delta\underline{x} \cdot \text{grad})\underline{n}, \tag{2.10}$$

V denotes a volume of liquid crystal with surface ∂V, $\underline{F}$ body force per unit volume, $\underline{t}$ surface traction per unit area, and $\underline{G}$ and $\underline{s}$ generalized body and surface forces, respectively. Given the constraint (2.1b) and the usual assumption of incompressibility, the variations in the director $\underline{n}$ and position $\underline{x}$ are not arbitrary but are subject to

$$\underline{n} \cdot \delta\underline{n} = \underline{n} \cdot \Delta\underline{n} = 0, \quad \text{div}\ \delta\underline{x} = 0. \tag{2.11}$$

Consideration of an arbitrary, infinitesimal, rigid translation in which $\Delta\underline{n}$ is zero leads to the rather obvious balance of forces

$$\int_V \underline{F}\,dV + \int_{\partial V} \underline{t}\,dS = 0. \tag{2.12}$$

Similarly, consideration of an arbitrary, infinitesimal, rigid rotation $\underline{\omega}$ in which

$$\delta\underline{x} = \underline{\omega} \times \underline{x}, \quad \Delta\underline{n} = \underline{\omega} \times \underline{n} \tag{2.13}$$

leads to a balance of moments

$$\int_V(\underline{x} \times \underline{F} + \underline{n} \times \underline{G})dV + \int_{\partial V}(\underline{x} \times \underline{t} + \underline{n} \times \underline{s})dS = 0. \tag{2.14}$$

From the latter one sees that the generalized body and surface forces are related to the body couple $\underline{K}$ and the couple stress vector $\underline{1}$, respectively, by

$$\underline{K} = \underline{n} \times \underline{G}, \quad \underline{l} = \underline{n} \times \underline{s}, \tag{2.15}$$

relationships required below.

By appropriate manipulation of the left hand side of the variational principle (2.9), Ericksen [9] deduces that the stress vector $\underline{t}$ is given by

$$t_i = t_{ij}\,\nu_j, \quad t_{ij} = -p\,\delta_{ij} - \frac{\partial W}{\partial n_{k,j}}\,n_{k,i}, \tag{2.16}$$

$\underline{\nu}$ denoting the unit normal to the surface ∂V, and the pressure p arising from the incompressibility assumption. Also, after due allowance for the constraint (2.11b), the generalized stress vector $\underline{s}$ takes the form

$$s_i = \beta n_i + \hat{s}_i, \quad \hat{s}_i = s_{ij}\,\nu_j, \quad s_{ij} = \frac{\partial W}{\partial n_{i,j}}, \tag{2.17}$$

β an arbitrary scalar. In addition, the variational principle yields two balance laws

$$\begin{aligned} F_i - p_{,i} - \left(\frac{\partial W}{\partial n_{k,j}}\,n_{k,i}\right)_{,j} &= 0, \\ G_i + \gamma n_i - \frac{\partial W}{\partial n_i} + \left(\frac{\partial W}{\partial n_{i,j}}\right)_{,j} &= 0, \end{aligned} \tag{2.18}$$

the term with the scalar γ also a consequence of the constraint upon the director. The former is clearly the point form of the balance of forces (2.12), and less obviously the latter is the point form of the balance of moments (2.14), as one can show with the aid of the following identity

$$e_{ijk}\left(n_j \frac{\partial W}{\partial n_k} + n_{j,p}\frac{\partial W}{\partial n_{k,p}} + n_{p,j}\frac{\partial W}{\partial n_{p,k}}\right) = 0, \tag{2.19}$$

which is a consequence of the invariance restriction (2.4) as Ericksen [12] shows.

Commonly, magnetic or electric fields are used to align the anisotropic axis in samples of liquid crystal, and one can readily explain such effects from the fact that a magnetic field $\underline{H}$ induces a magnetisation $\underline{M}$ and an electric field $\underline{E}$ creates a polarisation $\underline{P}$, both leading to a body couple. For example, the simplest assumption for the magnetisation in the present context leads to

$$\underline{M} = \chi_\perp \underline{H} + \chi_a(\underline{n}\cdot\underline{H})\underline{n}, \quad \chi_a = \chi_\parallel - \chi_\perp, \tag{2.20}$$

the constants $\chi_{\parallel}$ and $\chi_{\perp}$ denoting the magnetic susceptibilities with the field parallel and perpendicular to the director, respectively, with the associated magnetic energy ψ_m given by

$$\psi_m = \frac{1}{2}\underline{M} \cdot \underline{H} = \frac{1}{2}(\chi_{\perp}\underline{H} \cdot \underline{H} + \chi_a(\underline{n} \cdot \underline{H})^2). \tag{2.21}$$

By an elementary argument it follows that the resultant body couple takes the form

$$\underline{K} = \underline{M} \times \underline{H} = \chi_a(\underline{n} \cdot \underline{H})\underline{n} \times \underline{H}, \tag{2.22}$$

and hence from the expression (2.15) that

$$\underline{G} = \chi_a(\underline{n} \cdot \underline{H})\underline{H}, \tag{2.23}$$

any contribution parallel to the director being inconsequential. For most nematics, the diamagnetic anisotropy χ_a is generally positive. Also similar reasoning leads to a body force given by

$$F_i = M_j H_{i,j} = M_j H_{j,i} \tag{2.24}$$

since the magnetic field is irrotational. It is of interest to note that

$$G_i = \frac{\partial \psi_m}{\partial n_i}, \quad F_i = \frac{\partial \psi_m}{\partial x_i}, \tag{2.25}$$

regarding ψ_m as a function of the director and position. The counterparts of the above in the case of an electric field follow quickly by writing

$$\underline{D} = \varepsilon_{\perp}\underline{E} + \varepsilon_a(\underline{n} \cdot \underline{E})\underline{n}, \quad \varepsilon_a = \varepsilon_{\parallel} - \varepsilon_{\perp}, \tag{2.26}$$

$\varepsilon_{\parallel}$ and $\varepsilon_{\perp}$ denoting the dielectric permittivities with the field parallel and perpendicular to the director, respectively, and $\underline{D}$ the electric displacement. The dielectric anisotropy can be either positive or negative depending on the nematic considered.

From the equations (2.18) it quickly follows that

$$F_i + G_k n_{k,i} - p_{,i} - \frac{\partial W}{\partial n_k} n_{k,i} - \frac{\partial W}{\partial n_{k,j}} n_{k,ji} = 0, \tag{2.27}$$

and hence when relations of the form (2.25) hold for the body force and generalized body force one obtains

$$(\psi - p - W)_{,i} = 0, \tag{2.28}$$

ψ denoting the magnetic or electric energy. Hence, in cases of interest, the balance of forces (2.18a) integrates to yield

$$p + W - \psi = \text{const.}, \tag{2.29}$$

and the balance of moments (2.18b) may be written as

$$\left(\frac{\partial W}{\partial n_{i,j}} \right)_{,j} - \frac{\partial W}{\partial n_i} + \frac{\partial \psi}{\partial n_i} + \gamma n_i = 0. \tag{2.30}$$

The latter are simply the Euler-Lagrange equations associated with the energy integral

$$E = \int_V (W - \psi) dV, \tag{2.31}$$

this being the approach adopted by Oseen [6]. However, Ericksen's formulation gives a mechanical interpretation to this equation.

The boundary condition for the director in liquid crystal theory can be of one of two types. Most analyses employ a strong anchoring condition, described for example by de Gennes [1], which assumes that the alignment of the director takes some prescribed, fixed orientation at the interface with a solid boundary, and that this given surface alignment is insensitive to the presence of an electromagnetic field. Some, however, employ a weak anchoring condition of the type discussed by Jenkins and Barratt [13], which assigns an interfacial energy to the boundary, and leads to conditions on both the stress and couple stress in general. Analyses employing the former condition appear to describe many experimental observations rather well, although the latter may prove more appropriate at a free surface or at an interface with another liquid. Given more than one solution for the boundary conditions selected, it is customary to discriminate on stability grounds in favour of that which minimises the energy integral (2.31), this containing an interfacial contribution when appropriate.

It is straightforward to verify that the term containing the coefficient K_4

in the energy (2.5) or (2.6) makes no contribution to the equation (2.30), and for this reason this term is often omitted from the theory. However, while this coefficient does not appear in the equations of static theory, it can play a role in two ways. If one employs a weak anchoring boundary condition, it can occur in the boundary conditions, and also it can contribute to the energy in a discussion of stability.

Given the constraint (2.1b) on the director, it is frequently convenient to describe it in terms of two angles, so that

$$\underline{n} = \underline{f}(\theta^\alpha), \quad \alpha = 1,2, \tag{2.32}$$

where

$$\underline{n} \cdot \frac{\partial \underline{f}}{\partial \theta^\alpha} = 0, \qquad \frac{\partial \underline{f}}{\partial \theta^1} \times \frac{\partial \underline{f}}{\partial \theta^2} \neq 0. \tag{2.33}$$

In this event, with the notation

$$W = \overline{W}(\theta^\alpha, \theta^\alpha_{,i}), \quad \psi = \overline{\psi}(\theta^\alpha), \tag{2.34}$$

it is relatively straightforward to show that the equation (2.30) reduces to

$$\left(\frac{\partial \overline{W}}{\partial \theta^\alpha_{,i}}\right)_{,i} - \frac{\partial \overline{W}}{\partial \theta^\alpha} + \frac{\partial \overline{\psi}}{\partial \theta^\alpha} = 0, \quad \alpha = 1,2. \tag{2.35}$$

More generally, if one employs curvilinear coordinates so that

$$\underline{x} = \underline{x}(\underline{\xi}), \quad \underline{n} = \underline{F}(\theta^\alpha, \underline{\xi}), \quad \alpha = 1,2, \tag{2.36}$$

and introduces the notation

$$JW = \tilde{W}(\theta^\alpha, \theta^\alpha_{,i}, \xi_i), \quad J\psi = \tilde{\psi}(\theta^\alpha, \xi_i), \tag{2.37}$$

where J denotes the jacobian

$$J = \frac{\partial(\underline{x})}{\partial(\underline{\xi})}, \tag{2.38}$$

the equations similarly become

$$\left(\frac{\partial \tilde{W}}{\partial \theta^{\alpha}_{,i}} \right)_{,i} - \frac{\partial \tilde{W}}{\partial \theta^{\alpha}} + \frac{\partial \tilde{\psi}}{\partial \theta^{\alpha}} = 0, \quad \alpha = 1,2, \tag{2.39}$$

as Ericksen [14] discusses. Clearly the forms (2.35) and (2.39) provide a more efficient means of deriving the relevant differential equations in specific problems.

3. Some Equilibrium Configurations

Much of the interest in static theory of liquid crystals concerns the competition that can occur between prescribed surface alignments and the use of an electromagnetic field to orientate the sample in a desired manner. However, before turning to such problems in the following section, it is of interest first to look at some questions that can arise due solely to the particular types of surface alignment selected, where even in some relatively simple situations the configuration to be adopted by the sample is not entirely clear. These problems essentially divide into two types, ones not involving any singularities in the alignment and other with such defects. We consider the former initially, and throughout this section confine our attention to nematics.

Consider a thin layer of nematic liquid crystal confined between two parallel, stationary glass plates, each having uniform surface alignment. Neglecting end effects, it is therefore natural to seek solutions of the equations of the previous section in which the components of the director relative to Cartesian axes take the forms

$$n_x = \cos\theta(z)\cos\phi(z), \quad n_y = \cos\theta(z)\sin\phi(z), \quad n_z = \sin\theta(z), \tag{3.1}$$

the angles θ and ϕ depending solely upon the z coordinate which is measured normal to the plates. Clearly the angle θ represents the tilt of the director out of the plane of the plates, and the angle ϕ the twist about the normal. One finds straightforwardly that the energy (2.6) here reduces to

$$2W = F(\theta)\theta'^2 + G(\theta)\phi'^2 \tag{3.2}$$

where the functions $F(\theta)$ and $G(\theta)$ are defined by

$$F(\theta) = K_1\cos^2\theta + K_3\sin^2\theta, \quad G(\theta) = (K_2\cos^2\theta + K_3\sin^2\theta)\cos^2\theta, \tag{3.3}$$

and the prime denotes differentiation with respect to z. Employing the equations (2.35) one quickly obtains

$$2[F(\theta)\theta']' - \frac{dF(\theta)}{d\theta}\theta'^2 - \frac{dG(\theta)}{d\theta}\phi'^2 = 0,$$
$$[G(\theta)\phi']' = 0, \tag{3.4}$$

and these readily yield the integrals

$$G(\theta)\phi' = a, \quad F(\theta)\theta'^2 + G(\theta)\phi'^2 = b, \tag{3.5}$$

a and b being constants of integration.

Consider first for completeness the situation where the prescribed alignment is the same at both plates and there is no net tilt or twist in the sample. In this event, with an appropriate choice of axes having the origin mid-way between the plates, the boundary conditions are

$$\theta(\pm h) = \theta_0, \quad \phi(\pm h) = 0, \tag{3.6}$$

$2h$ denoting the distance between the plates and θ_0 a given acute angle. Since the functions $F(\theta)$ and $G(\theta)$ are both positive, it quickly follows from the forms (3.5) and the above boundary conditions that

$$\theta = \theta_0, \quad \phi = 0, \tag{3.7}$$

the alignment necessarily uniform across the gap as one might anticipate.

Consider next the situation just described but with the alignment parallel at both plates so that θ_0 is zero, and imagine that one plate is rotated through some angle about one of its normals. Again with an appropriate choice of axes, this leads to boundary conditions of the form

$$\theta(\pm h) = 0, \quad \phi(h) = -\phi(-h) = \phi_0, \tag{3.8}$$

where ϕ_0 is given positive constant. One solution of the equations (3.4) satisfying these conditions is quickly found, namely

$$\theta = 0, \quad \phi = \phi_0 z/h, \tag{3.9}$$

of the type first discussed by Ericksen [15]. The energy per unit area of plate associated with this twisted planar structure is

$$E_p = K_2\phi_0^2/h, \tag{3.10}$$

and thus, as the twist parameter ϕ_0 increases, the energy in the sample clearly increases. This latter consideration led Leslie [16] to investigate nonplanar twisted solutions satisfying the conditions (3.8) in which

$$\theta(z) = \theta(-z), \quad \phi(z) = -\phi(-z), \tag{3.11}$$

and hence

$$\theta'(0) = 0, \quad \theta(0) = \theta_m, \quad \phi(0) = 0, \tag{3.12}$$

there being no loss of generality in regarding the parameter θ_m as positive given the symmetry of the problem. Using these latter conditions, the equations (3.5) quickly yield

$$F(\theta)\theta'^2 = a^2(1/G(\theta_m) - 1/G(\theta)), \tag{3.13}$$

and for solutions the right hand side must of course always be positive. This can lead to some restriction upon possible solutions depending upon the relative magnitudes of the two Frank constants appearing in the function $G(\theta)$. One quickly finds that when K_3 is greater than $2K_2$ one must restrict θ_m by

$$\theta_c < \theta_m < \pi/2, \tag{3.14}$$

where the acute angle θ_c is defined by

$$\sin^2\theta_c = (K_3 - 2K_2)/(K_3 - K_2). \tag{3.15}$$

By integration of (3.13) one finds that

$$ah/G(\theta_m)^{1/2} = \int_0^{\theta_m} \left[\frac{F(\theta)G(\theta)}{G(\theta) - G(\theta_m)} \right]^{1/2} d\theta, \tag{3.16}$$

and combining (3.5a) and (3.13) that

$$\phi_0 = G(\theta_m)^{1/2} \int_0^{\theta_m} \left[\frac{F(\theta)}{G(\theta)(G(\theta) - G(\theta_m))} \right]^{1/2} d\theta. \tag{3.17}$$

The energy per unit area of plate associated with these non-planar twisted solutions is

$$E_{n-p} = a^2 h/G(\theta_m), \tag{3.18}$$

and hence one determines the relative stability of the two solutions by comparing the two integrals just given. When the coefficients K_2 and K_3 are equal, Leslie shows that a complete analysis of the problem is possible, but otherwise he simply considers small departures from the planar configuration. In this way he finds that distortion of the twisted planar structure can occur at an overall twist less than π, but that in general this requires that the coefficient K_2 be larger than the other Frank constants, a condition not commonly realized in nematic materials.

Porte and Jadot [17] consider an interesting variant of the above problem in which the alignment at the plates is tilted rather than parallel. In their problem, therefore, the boundary conditions are

$$\theta(\pm h) = \theta_0, \quad \phi(h) = -\phi(-h) = \phi_0, \tag{3.19}$$

where θ_0 is a given acute angle and ϕ_0 a positive parameter. They also examine solutions with the properties (3.11) and (3.12) to obtain as above the equation (3.13) and ultimately

$$\phi_0 = G(\theta_m)^{1/2} \int_{\theta_0}^{\theta_m} \left[\frac{F(\theta)}{G(\theta)(G(\theta) - G(\theta_m))} \right]^{1/2} d\theta, \tag{3.20}$$

which essentially relates the maximum tilt θ_m to the parameters θ_0 and ϕ_0. For the special case in which the Frank constants satisfy

$$K_1 = K_3 = 2K_2, \tag{3.21}$$

they express the above integral in terms of elliptic integrals and thereby determine the tilt θ_m for given values of θ_0 and ϕ_0. When θ_0 and ϕ_0 are both acute angles so that the overall twist is less than π, they find that θ_m always lies between θ_0 and $\pi/2$. However, when ϕ_0 is a right angle so that the overall twist is π, there is a critial value of θ_0 at which θ_m attains the value $\pi/2$. With this overall twist, therefore, the tilt θ_m remains less than $\pi/2$ for angles of tilt at the surfaces smaller than this critical value, but, when the surface tilt exceeds this value, θ_m is equal to $\pi/2$. From this Porte and Jadot conclude that when the net twist is π, a sufficiently large tilt at the surfaces leads to their twisted nonplanar structure reverting to a simpler planar configuration in a plane normal to the plates. They also discuss the general case when the restrictions (3.21) no longer apply and reach similar conclusions, as does Thurston [18] by a different approach.

Fischer [19] discusses a somewhat similar problem to those above, but requiring continuous variation of the tilt at the boundaries, which is rather difficult to achieve in practice, although his motivation lies in flow considerations. Other problems not dissimilar to the above occur when weak anchoring boundary conditions are used, and Jenkins and Barratt [13] give examples. Little appears to have been attempted in this direction with cholesterics, largely due to the more complex configurations that occur in these materials.

To conclude this section we turn to a rather different type of problem that can arise when, for example, one confines a nematic in a circular capillary. As above let us ignore end effects and to fix ideas assume that the alignment at the capillary wall is normal to the surface. Choosing cylindrical polar coordinates (r,ϕ,z) with the z-axis coincident with the axis of the capillary, one solution of the equilibrium equations meeting the relevant boundary condition is

$$n_r = 1, \quad n_\phi = n_z = 0, \tag{3.22}$$

this being evident from the analysis due to Ericksen [20]. Rather clearly this solution is singular on the z-axis, and is in fact a member of a family of such solutions described by Frank [8]. However, these solutions have infinite energy associated with them, and so are physically unacceptable unless modified in some way, Ericksen [21] discussing one possibility. Another is that proposed independently by Cladis and Kleman [22] and Meyer [23], who consider solutions of the form

$$n_r = \cos\theta(r), \quad n_\phi = 0, \quad n_z = \sin\theta(r), \tag{3.23}$$

but, for ease of illustration, we confine our attention to the special case in which the Frank constants are all equal. In this event, the associated energy (2.6) reduces to

$$W = K/2r^2[(r\frac{d\theta}{dr})^2 - \sin 2\theta\, r\frac{d\theta}{dr} + \cos^2\theta], \tag{3.24}$$

K the Frank constant, and the resultant form of equation (2.39) is

$$r^2\frac{d^2\theta}{dr^2} + r\frac{d\theta}{dr} + \sin\theta\cos\theta = 0, \tag{3.25}$$

which integrates once to give

$$(r\frac{d\theta}{dr})^2 = c - \sin^2\theta, \tag{3.26}$$

with c a constant. One obtains a finite energy by choosing c equal to unity, and a further integration yields

$$\theta = 2\tan^{-1}R/r - \pi/2, \tag{3.27}$$

where R denotes the radius of the capillary. Since θ is equal to $\pi/2$ on the axis, this solution has no singularity and appears to model observed configurations rather well. Cladis and Kleman [22] and Meyer [23] also discuss the general case when the Frank constants are not equal, and consider in addition solutions when the surface alignment is parallel and azimuthal. Barratt [24,25] describes somewhat similar solutions that incorporate the modification proposed by Ericksen [21].

4. Influence of External Fields upon Equilibrium Configurations

In this section we turn to perhaps the most interesting aspect of static theory, the competition between electromagnetic fields and surface alignments to dictate particular orientations to a sample of liquid crystal. Certain effects arising from such competition were first discovered more than fifty years ago, and their satisfactory explanation using continuum theory was in some respects the first important test of Oseen's theory. More recently, the successful exploitation of such phenomena in display devices has greatly stimulated interest in this topic. Below we consider such problems in some detail, largely confining attention to magnetic fields for reasons that become apparent. Also, once again the greater part of our discussion concerns nematics.

Consider a layer of nematic liquid crystal confined between two parallel plates with its initial orientation uniform and parallel to the plates due to appropriate surface alignments, and suppose that a uniform magnetic field is applied normal to the plates. Given the symmetry of the problem, an obvious choice of Cartesian axes leads one to examine solutions in which the director components are simply

$$n_x = \cos\theta(z), \quad n_y = 0, \quad n_z = \sin\theta(z), \tag{4.1}$$

and

$$H_x = H_y = 0, \; H_z = H, \tag{4.2}$$

z being measured normal to the plates. Since the diamagnetic susceptibilities of liquid crystals are in general very small, it appears reasonable to assume that the presence of the liquid crystal does not significantly interact with the magnetic field, and therefore that the field strength H remains constant. Also, the diamagnetic anisotropy of nematics is invariably positive, and hence our discussion is confined to this case. Straightforwardly from the expression (2.6) the energy related to the above distortion is given by

$$2W = F(\theta)\theta'^2, \quad F(\theta) = K_1\cos^2\theta + K_3\sin^2\theta, \tag{4.3}$$

the prime again denoting differentiation with respect to z, and from the expression (2.21) the magnetic energy is readily found to be

$$2\psi_m = \chi(\theta)H^2, \quad \chi(\theta) = \chi_\perp + \chi_a\sin^2\theta. \tag{4.4}$$

In this case, therefore, the resultant form of equation (2.35) is

$$2[F(\theta)\theta']' - \frac{dF(\theta)}{d\theta}\theta'^2 + \frac{d\chi(\theta)}{d\theta}H^2 = 0, \tag{4.5}$$

and the assumption of strong anchoring at the plates leads to the boundary conditions

$$\theta(\pm h) = 0, \tag{4.6}$$

the origin mid-way between the plates, which are a distance $2h$ apart.

One solution of equation (4.5) meeting the boundary conditions is of course

$$\theta = 0, \tag{4.7}$$

so that the initial alignment may remain unperturbed. However, one can also have distorted solutions, the simplest being symmetric so that

$$\theta(z) = \theta(-z), \quad \theta'(0) = 0, \quad \theta(0) = \theta_m, \tag{4.8}$$

where as previously one can without loss of generality regard θ_m as positive. In view of the properties of the solution considered, integration of equation (4.5) leads to

$$F(\theta)\theta'^2 = \chi_a H^2(\sin^2\theta_m - \sin^2\theta), \tag{4.9}$$

which gives in turn using the boundary conditions (4.6)

$$\chi_a^{1/2}hH = \int_0^{\theta_m}[F(\theta)/(\sin^2\theta_m - \sin^2\theta)]^{1/2}d\theta, \tag{4.10}$$

this providing a relationship between the parameters θ_m, h and H. Alternatively employing the substitution

$$\sin\theta = \sin\theta_m \sin\lambda, \tag{4.11}$$

the above takes the more convenient form

$$\chi_a^{1/2}hH = \int_0^{\pi/2}(F(\theta)/\cos^2\theta)^{1/2}d\lambda, \tag{4.12}$$

from which one readily deduces that the expression on the right is a monotonic increasing function of θ_m in the range $(0,\pi/2)$. Consequently, this solution first becomes available at a critical field strength H_c, where

$$hH_c = (K_1/\chi_\alpha)^{1/2}\pi/2, \qquad (4.13)$$

and, moreover, H becomes infinitely large as θ_m approaches the value $\pi/2$.

To compare the overall energies of the two solutions, let ΔE denote the difference between the energies per unit area of plate of the distorted and unperturbed configurations, and from equations (4.3), (4.4), (4.8), (4.9) and (4.11) one finds that

$$\Delta E = 1/2 \int_{-h}^{h} (F(\theta)\theta'^2 - \chi_a H^2 \sin^2\theta)dz$$

$$= \chi_a^{1/2} H \int_0^{\theta_m} (\sin^2\theta_m - 2\sin^2\theta)\left[\frac{F(\theta)}{\sin^2\theta_m - \sin^2\theta}\right]^{1/2} d\theta$$

$$= \chi_a^{1/2} H \sin^2\theta_m \int_0^{\pi/2} \cos 2\lambda [F(\theta)/\cos^2\theta]^{1/2} d\lambda$$

$$< 0, \quad \theta_m \neq 0, \qquad (4.14)$$

given the behaviour of $\cos 2\lambda$ and the monotonicity of the function $F(\theta)/\cos^2\theta$ in the relevant interval. Thus the distorted solution when available represents a state of lower energy than the initial uniform alignment. Other more complex solutions of equation (4.5) are possible, but they all require a larger field strength and have higher energies than the simple symmetric distortion. From these results one concludes that the initial parallel alignment is unaffected by sufficiently weak magnetic fields, but as the field strength increases beyond the critical value given by equation (4.13) the alignment begins to tilt towards the field direction.

The above conclusions are in agreement with observations first made by Freedericksz and Zolina [26]. They also conducted two similar experiments with the direction of the magnetic field parallel to the plates but again perpendicular to the initial alignment, in one case this being parallel and in the other normal to the plates. In each experiment their observations were similar, namely that no distortion of the initial uniform alignment occurs until a critical field strength that varies with the particular experiment, but in each case is inversely proportional to the thickness of the layer. Analyses of these other two arrangements are equally feasible, following almost identical lines to that just given with similar conclusions. The critical field strengths for the initial parallel and perpendicular alignments are given by the expression (4.13) but with K_2 and K_3 replacing K_1, respectively. Zocher [7] was the first to obtain such results, and subsequently Saupe [27] and Dafermos [28] have given more complete analyses. Such changes in orientation induced by an external field normal to the initial alignment are commonly called Freedericksz transitions after their discoverer.

Such transitions are also possible when the initial alignment is not uniform. As in the preceding section, consider a nematic with uniform parallel alignment confined between two parallel plates, and suppose that one of them is rotated about a normal to produce a twisted configuration in the sample. As Leslie [29] shows, a magnetic field applied perpendicular to the plates can lead to a Freedericksz transition. To see this consider solutions in which the Cartesian components of the director take the forms

$$n_x = \cos\theta(z)\cos\phi(z), \quad n_y = \cos\theta(z)\sin\phi(z), \quad n_z = \sin\theta(z), \tag{4.15}$$

the z-axis again normal to the plates, and

$$H_x = H_y = 0, \; H_z = H, \tag{4.16}$$

H as before a constant. The related energy density follows from the expression (2.6) and is

$$\begin{gathered} 2W = F(\theta)\theta'^2 + G(\theta)\phi'^2, \\ F(\theta) = K_1\cos^2\theta + K_3\sin^2\theta, \quad G(\theta) = (K_2\cos^2\theta + K_3\sin^2\theta)\cos^2\theta, \end{gathered} \tag{4.17}$$

the prime again denoting a derivative, and as above

$$2\psi_{m} = \chi(\theta)H^2, \quad \chi(\theta) = \chi_\perp + \chi_a\sin^2\theta. \tag{4.18}$$

Here the equations (2.35) yield

$$\begin{gathered} 2[F(\theta)\theta']' - \frac{dF(\theta)}{d\theta}\theta'^2 - \frac{dG(\theta)}{d\theta}\phi'^2 + \frac{d\chi(\theta)}{d\theta}H^2 = 0, \\ [G(\theta)\phi']' = 0, \end{gathered} \tag{4.19}$$

which integrate to give

$$G(\theta)\phi' = a, \quad F(\theta)\theta'^2 + G(\theta)\phi'^2 + \chi(\theta)H^2 = b, \tag{4.20}$$

a and b constants of integration. Once again with the strong anchoring assumption, a suitable choice of axes leads to boundary conditions

$$\theta(\pm h) = 0, \quad \phi(h) = -\phi(-h) = \phi_0, \tag{4.21}$$

with ϕ_0 a given positive constant.

One solution of the equations (4.19) meeting the required boundary conditions is simply

$$\theta = 0, \quad \phi = \phi_0 z/h, \tag{4.22}$$

the unperturbed twisted alignment. However, another possibility is a symmetric distortion with

$$\begin{aligned} \theta(z) &= \theta(-z), \quad \phi(z) = -\phi(-z), \\ \theta'(0) &= 0, \quad \theta(0) = \theta_m, \quad \phi(0) = 0\ , \end{aligned} \tag{4.23}$$

θ_m again considered positive without loss of generality. In this case, the equations (4.20) soon yield

$$F(\theta)\theta'^2 = \chi_a H^2(\sin^2\theta_m - \sin^2\theta) + a^2(1/G(\theta_m) - 1/G(\theta)), \tag{4.24}$$

and with the substitution (4.11) and the conditions (4.21) and (4.23) one finally obtains the relations

$$\begin{aligned} h &= \int_0^{\pi/2} \left[\frac{F(\theta)}{\chi_a H^2 - a^2 P(\theta,\theta_m)} \right]^{1/2} / \cos\theta \, d\lambda, \\ \phi_0 &= \int_0^{\pi/2} \left[\frac{aF(\theta)}{\chi_a H^2 - a^2 P(\theta,\theta_m)} \right]^{1/2} / \cos\theta \, G(\theta) d\lambda, \end{aligned} \tag{4.25}$$

where

$$P(\theta,\theta_m) = (1/G(\theta) - 1/G(\theta_m))/(\sin^2\theta_m - \sin^2\theta). \tag{4.26}$$

The equations (4.25) determine the tilt parameter θ_m and the constant a as functions of the field strength H and the gap parameter h. From these one can show that this second solution becomes a possibility under certain conditions once the field strength exceeds a critical value given by

$$\chi_a h^2 H_c^2 = K_1(\pi/2)^2 + (K_3 - 2K_2)\phi_0^2, \tag{4.27}$$

and, moreover, that the distorted solution has a lower overall energy than the unperturbed configuration. Here again the critical field strength is inversely proportional to the sample thickness, a prediction confirmed experimentally by Gerritsma, de Jeu and Van Zanten [30]. Confirmation of this scaling in these transitions lends strong support to the choice of a quadratic dependence in the energy function (2.6).

In the absence of external fields, cholesteric liquid crystals between parallel plates commonly adopt twisted configurations of the type described by equations (4.15) when the angles θ and ϕ take the forms (4.22), particularly if the gap between the plates is not too large. Consequently, for this type of liquid crystal one examines solutions referred to Cartesian axes of the form

$$n_x = \cos\phi(z), \quad n_y = \sin\phi(z), \quad n_z = 0, \tag{4.28}$$

the z-axis again normal to the plates, and straightforwardly it follows from the expression (2.5) that

$$2W = K_2(\phi' - \tau)^2, \tag{4.29}$$

the prime again denoting differentiation with respect to z. Hence here equation (2.35) reduces to

$$\phi'' = 0. \tag{4.30}$$

For cholesterics it is perhaps better to write the strong anchoring boundary condition in the form

$$\phi(h) = \phi_0 + n\pi, \quad \phi(-h) = -\phi_0, \tag{4.31}$$

where n is an integer and ϕ_0 a given acute angle, and solutions of equation (4.30) subject to this condition are

$$\phi = (2\phi_0 + n\pi)z/2h + n\pi/2. \tag{4.32}$$

The member of this family of solutions most likely to be relevant in practice is that which minimizes the energy (4.31), this having its gradient most nearly equal to the parameter τ. Barratt and Fraser [31] investigate Freedericksz transitions connected with such a twisted configuration, the magnetic field being normal to the plates, and their analysis is rather similar to that for the twisted nematic described above, but includes a discussion of both strong and weak anchoring boundary conditions. Earlier Leslie [29] also examined the same topic, but he employed a mixed boundary condition that assumes parallel alignment with the normal component of couple stress zero. On the other hand, de Gennes [32] and Dreher [33] consider the case when the magnetic field is parallel to the plates, and seek to determine how the twisted structure untwists in such a field. While de Gennes ignores the influence of the boundaries, Dreher employs the strong anchoring condition. Both conclude that a field of strength

$$H = (K_2/\chi_a)^{1/2}\pi\tau/2 \tag{4.33}$$

completely untwists the sample, and de Gennes [1] cites experimental results in good agreement with this prediction.

We conclude this section by indicating the modifications that arise when an electric field replaces the magnetic field. As before consider a uniformly aligned layer of nematic liquid crystal confined between parallel plates, but which are now subject to a voltage difference, and examine solutions of the form (4.1), the Frank energy again given by equation (4.3). Here the electric field is of the form

$$E_x = E_y = 0, \quad E_z = E, \tag{4.34}$$

with the associated energy given by

$$2\psi_e = \underline{D} \cdot \underline{E} = \varepsilon(\theta)E^2, \quad \varepsilon(\theta) = \varepsilon_\perp + \varepsilon_a \sin^2\theta. \tag{4.35}$$

In this case the equation (2.35) yields

$$2[F(\theta)\theta']' - \frac{dF(\theta)}{d\theta}\theta'^2 + \frac{d\varepsilon(\theta)}{d\theta}E^2 = 0, \tag{4.36}$$

exactly as before. However, as Deuling [34] notes, the dielectric anisotropy ε_a is not in general small for liquid crystals, and one must therefore allow for the presence of the material interacting with the applied field. To do this, Deuling employs the reduced Maxwell equations

$$\text{curl } \underline{E} = 0, \quad \text{div } \underline{D} = 0, \tag{4.37}$$

the latter assuming that no free charge is present. The former justifies the introduction of an electrostatic potential V, which here can depend solely upon z, and the relation (2.26) therefore yields

$$D_z = \varepsilon(\theta)E = \varepsilon(\theta)V', \tag{4.38}$$

the prime denoting differentiation, and D_z is of course constant in view of the second of equations (4.37). Thus one has

$$E = D_z/\varepsilon(\theta) \tag{4.39}$$

with the value of D_z ultimately obtained from

$$\Delta V = D_z \int_{-h}^{h} \frac{dz}{\varepsilon(\theta)}, \tag{4.40}$$

where ΔV denotes the voltage difference between the plates, again a distance $2h$ apart. Consequently, for an electric field the integral of equation (4.36) is

$$F(\theta)\theta'^2 - D_z^2/\varepsilon(\theta) = a, \tag{4.41}$$

where a is a constant of integration. Deuling [34] provides an analysis of this particular case, and gives results for the classical Freedericksz transition discussed at the beginning of this section. Also, Barratt and Fraser [31] give examples of such transitions in cholesterics in the presence of electric fields.

One topic of current interest in this area of research, which employs the various ideas developed above, is the possibility of switching between configurations of almost equal energy by using electric fields, this being of particular relevance with regard to applications in display devices. Thurston [35] discusses one aspect of this question for nematics, while Berreman and Heffner [36] do likewise for cholesterics, each citing references to earlier work on their respective topics.

References

1. P.G. de Gennes, "The Physics of Liquid Crystals". Oxford: Clarendon Press (1974).
2. S. Chandrasekhar, "Liquid Crystals". Cambridge University Press (1977).
3. M.J. Stephen and J.P. Straley, Rev. Mod. Phys. 46, 617-704 (1974).
4. J.L. Ericksen, Adv. Liq. Cryst. 2, 233-298 (1976).
5. H.J. Deuling, Solid State Phys. Suppl. 14, 77-107 (1978).
6. C.W. Oseen, Ark. Mat. Astr. Fys. A 19, 1-19 (1925).
7. H. Zocher, Trans. Faraday Soc. 29, 945-957 (1933).
8. F.C. Frank, Discuss. Faraday Soc. 25, 19-28 (1958).
9. J.L. Ericksen, Arch. Rat. Mech. Anal. 9, 371-378 (1962).
10. J.L. Ericksen, Phys. Fluids 9, 1205-1207 (1966).
11. J.T. Jenkins, J. Fluid Mech. 45, 465-475 (1971).
12. J.L. Ericksen, Trans. Soc. Rheol. 5, 23-34 (1961).
13. J.T. Jenkins and P.J. Barratt, Quart. J. Mech. Appl. Math 27, 111-127 (1974).
14. J.L. Ericksen, Quart. J. Mech. Appl. Math. 29, 203-208 (1976).
15. J.L. Ericksen, J. Fluid Mech. 27, 59-64 (1967).
16. F.M. Leslie, Pramana Suppl. 1, 41-55 (1975).
17. G. Porte and J.P. Jadot, J. Phys. (Paris) 39, 213-223 (1978).
18. R. Thurston, J. Phys. (Paris) 42, 419-425 (1981).
19. F. Fischer, Z. Naturforsch. A 31, 302-305 (1976).
20. J.L. Ericksen, Trans. Soc. Rheol. 11, 5-14 (1967).
21. J.L. Ericksen, "Liquid Crystals and Ordered Fluids." Eds. J.F. Johnson and R.S. Porter, Plenum: New York, 181-193 (1970).

22. P.E. Cladis and M. Kleman, J. Phys. (Paris) 33, 591-598 (1972).

23. R.B. Meyer, Phil. Mag. 27, 405-424 (1973).

24. P.J. Barratt, Mol. Cryst. Liq. Cryst. 24, 223-237 (1973).

25. P.J. Barratt, Quart. J. Mech. Appl. Math. 27, 505-522 (1974).

26. V. Freedericksz and V. Zolina, Trans. Faraday Soc. 29, 919-930 (1933).

27. A. Saupe, Z. Naturforsch. A 15, 810-814 (1960).

28. C.M. Dafermos, SIAM J. Appl. Math. 16, 1305-1318 (1968).

29. F.M. Leslie, Mol. Cryst. Liq. Cryst. 12, 57-72 (1970).

30. C.J. Gerritsma, W.H. de Jeu and P. Van Zanten, Phys. Lett. A 36, 389-390 (1971).

31. P.J. Barratt and C. Fraser, Proc. Edin. Math. Soc. 26, 319-332 (1983).

32. P.G. de Gennes, Solid State Commun. 6, 163-165 (1968).

33. R. Dreher, Solid State Commun. 13, 1571-1574 (1973).

34. H.J. Deuling, Mol. Cryst. Liq. Cryst. 19, 123-131 (1972).

35. R. Thurston, J. Phys. (Paris) 43, 117-128 (1982).

36. D.W. Berreman and W.R. Heffner, J. Appl. Phys. 52, 3032-3039 (1981).

THEORY OF FLOW PHENOMENA IN NEMATIC LIQUID CRYSTALS

Frank M. Leslie

Department of Mathematics
University of Strathclyde
Livingstone Tower
26 Richmond Street
Glasgow GI 1XH, Scotland

1. Introduction

There has been a remarkable increase in liquid crystal research during the last twenty years, much of it concerned with the material response to flow, electric and magnetic fields, surface alignments and thermal gradients, particularly with regard to liquid crystals of the nematic type. This has led to a marked improvement in our understanding of various known macroscopic effects, and also of striking new ones discovered within this period. A major factor contributing to this expansion of research activity has undoubtedly been the realisation of possible applications in displays and other devices, and this has certainly stimulated interest in macroscopic properties of these strange liquids. A further factor, however, has been the emergence of a continuum theory capable of describing many flow phenomena in nematics satisfactorily, and also the fact that its development coincided with the growth in interest in applications. This theory, due to Ericksen [1] and Leslie [2], successfully predicted several effects and analysed many others rather well, with the result that theoretical and experimental activity both supported and stimulated each other.

My aim here is twofold, firstly to present briefly the continuum equations proposed by Ericksen and Leslie, and secondly to give some impression of the types of problems that arise. The former does merit some attention given the nature of liquid crystals, particularly the fact that these liquids are anisotropic, but equally the occurrence of body couples and couple stresses in the equations does make the theory somewhat unusual in continuum mechanics. Given the constraints, the latter aim is perhaps the more difficult, the number of topics discussed being of necessity few and their description brief, but our selection represents an attempt to give some indication of the novelty and variety of effects that can arise. Greater detail is of course available in the original papers, and also in two relatively recent reviews, one by Jenkins [3] devoted to viscometry of nematics, and the second by Leslie [4] attempting to give a more comprehensive account of flow phenomena in both nematic and cholesteric liquid crystals.

It appears reasonable in such an account to assume some familiarity with physical properties of liquid crystals, particularly of the nematic type, and also with continuum concepts. For the former the books by de Gennes [5] and Chandrasekhar [6] give excellent introductions, including some account of continuum theory. Reviews dealing exclusively with continuum theory are one on static aspects by Ericksen [7] and the two cited above by Jenkins [3] and Leslie [4], but a useful shorter account of both static and dynamic theory is given by Ericksen [8].

2. Continuum Equations

Our aim here is to present a brief account of the continuum theory proposed by Ericksen [1] and Leslie [2] to describe flow effects in nematic liquid crystals. For this purpose it is useful to employ Cartesian tensor notation, and therefore the summation convention applies to repeated indices and a comma preceding a suffix implies partial differentiation with respect to the corresponding spatial variable. Also, a superposed dot denotes a material time derivative, and δ_{ij} and e_{ijk} the familiar Kronecker delta and the alternating tensor, respectively.

Continuum theory for liquid crystals requires two independent vector fields to describe flow satisfactorily, the familiar velocity vector $\underline{v}$ and an additional kinematic variable $\underline{n}$, a unit vector field describing the alignment of the anisotropic axis in these transversely isotropic liquids, which is commonly referred to as the director. With the customary assumption of incompressibility, both are subject to constraints, namely

$$v_{i,i} = 0, \qquad n_i n_i = 1. \tag{2.1}$$

In contrast with continuum theories for isotropic liquids, the balance laws for both linear and angular momentum play a prominent role in liquid crystal theory. Here, these take the respective forms

$$\rho\dot{v}_i = F_i + t_{ij,j}\ , \qquad 0 = K_i + e_{ijk}t_{kj} + \ell_{ij,j}\ , \tag{2.2}$$

when ρ is density assumed constant, $\underline{F}$ the external body force, $\underline{t}$ the stress tensor, $\underline{K}$ the external body couple, and $\underline{\ell}$ the couple stress tensor. No inertia term is included in angular momentum, since it is generally considered negligible in flow problems.

For nematic liquid crystals, the stress and couple stress tensors are given by

$$\begin{aligned} t_{ij} &= -p\,\delta_{ij} - \frac{\partial W}{\partial n_{k,j}}\, n_{k,i} + \tilde{t}_{ij}\ , \\ \ell_{ij} &= e_{ipq}n_p \frac{\partial W}{\partial n_{q,j}}\ , \end{aligned} \tag{2.3}$$

p being an arbitrary pressure arising from the assumption of incompressibility, W the stored energy function of static theory, and $\underline{\tilde{t}}$ the dissipative part of the stress tensor; the energy function is that due to Oseen [9] and Frank [10]

$$\begin{aligned} 2W = {} & k_1(n_{i,i})^2 + k_2(n_i e_{ijk} n_{k,j})^2 + k_3 n_i n_j n_{k,i} n_{k,j} \\ & + (k_2 + k_4)(n_{i,j}n_{j,i} - (n_{i,i})^2), \end{aligned} \tag{2.4}$$

and $\underline{\tilde{t}}$ takes the form first discussed by Ericksen [11],

$$\begin{aligned} \tilde{t}_{ij} = {} & \alpha_1 A_{kp}n_k n_p n_i n_j + \alpha_2 N_i n_j + \alpha_3 N_j n_i + \alpha_4 A_{ij} \\ & + \alpha_5 A_{ik}n_k n_j + \alpha_6 A_{jk} n_k n_i, \end{aligned} \tag{2.5}$$

with

$$2A_{ij} = v_{i,j} + v_{j,i}\ , \qquad 2N_i = 2\dot{n}_i - (v_{i,k} - v_{k,i})n_k. \tag{2.6}$$

If one ignores thermal effects, the coefficients in (2.4) and (2.5) are simply constants. These equations are insensitive to the replacement of $\underline{n}$ by $-\underline{n}$, this reflecting the absence of polarity in nematic liquid crystals.

In cases of interest the external body force and couple arise from the presence of a magnetic or electric field, and in this event they can be expressed as

$$F_i = \frac{\partial\psi}{\partial x_i} \qquad K_i = e_{ipq} n_p \frac{\partial\psi}{\partial n_q}, \tag{2.7}$$

where for a magnetic field

$$2\psi = \chi_\perp H_i H_i + \chi_a (H_i n_i)^2, \quad \chi_a = \chi_\parallel - \chi_\perp, \tag{2.8}$$

the $\chi_\parallel$ and $\chi_\perp$ denoting the magnetic susceptibilities for alignment parallel or perpendicular to the magnetic field, respectively. Clearly analogous forms are available for the case of an electric field.

Given the above constitutive assumptions, it is possible to express the balance laws (2.2) in forms more convenient for specific problems. However, this requires the following identity

$$e_{ipq}\left(n_p \frac{\partial W}{\partial n_q} + n_{p,k} \frac{\partial W}{\partial n_{q,k}} + n_{k,p} \frac{\partial W}{\partial n_{k,q}}\right) = 0, \tag{2.9}$$

which one can deduce either by the argument given by Ericksen [1], or directly from the expression (2.4). Using this result, conservation of angular momentum (2.2b) can be expressed in the form

$$\left(\frac{\partial W}{\partial n_{i,j}}\right)_{,j} - \frac{\partial W}{\partial n_i} + \tilde{g}_i + \frac{\partial\psi}{\partial n_i} + \gamma n_i = 0, \tag{2.10}$$

where γ is an arbitrary scalar and the vector $\tilde{\underline{g}}$ is related to $\tilde{\underline{t}}$ by

$$e_{ijk} n_j \tilde{g}_k = e_{ijk} \tilde{t}_{kj}, \tag{2.11}$$

so that

$$\tilde{g}_i = -\gamma_1 N_i - \gamma_2 A_{ik} n_k, \quad \gamma_1 = \alpha_3 - \alpha_2, \ \gamma_2 = \alpha_6 - \alpha_5. \tag{2.12}$$

Moreover, using equations (2.3a), (2.7a) and (2.10), the balance law (2.2a) may be recast as

$$\tilde{t}_{ij,j} + \tilde{g}_k n_{k,i} - \tilde{p}_{,i} = \rho\dot{v}_i, \quad \tilde{p} = p + W - \psi. \tag{2.13}$$

The equations (2.10) and (2.13) are somewhat more convenient for practical purposes than the versions that follow directly by substitution of the constitutive equations into (2.2).

It is possible to place some restrictions on the possible values of the coefficients in the above constitutive equations by different arguments. Ericksen [12] employs a stability argument to limit those in the energy (2.4) by

$$k_1 \geq 0, \quad k_2 \geq 0, \quad k_3 \geq 0, \quad |k_4| \leq k_2, \; k_2 + k_4 \leq 2k_1, \tag{2.14}$$

and Leslie [13] restricts the viscosity coefficients through the thermodynamic inequality

$$\tilde{t}_{ij} A_{ij} - \tilde{g}_i N_i \geq 0. \tag{2.15}$$

Appealing to another thermodynamic argument, Parodi [14] proposes the relationship

$$\gamma_2 = \alpha_3 + \alpha_2, \tag{2.16}$$

a result also obtained by Currie [15] by an independent stability argument.

At a solid boundary it is customary to assume that the liquid crystal adheres to the solid surface, and therefore that the velocity vector is subject to the familiar no-slip boundary condition. The corresponding boundary condition for the director is the so-called strong anchoring condition discussed by de Gennes [5], which assumes that the prior treatment of the solid surface determines the particular alignment of the director at the interface, and that the alignment so produced is insensitive to the flow. While occasionally questioned, virtually all analyses of flow problems employ this assumption, and no obvious conflict arises between theory and experiment.

3. Scaling Properties

Ericksen [16] points out a rather surprising property of the above equations that leads to relatively straightforward tests of the relevance of the theory to experimental observations. Essentially he notes that a transformation of the Cartesian coordinates x_i and time t by

$$\underline{x} = k\,\underline{x}^*, \quad t = k^2 t^*, \tag{3.1}$$

where k is some constant, leaves the equations completely unchanged provided that one sets

$$\tilde{p} = k^{-2}p^*, \quad \gamma = k^{-2}\gamma^*, \quad \psi = k^{-2}\psi^*, \tag{3.2}$$

this requiring suitable scaling of any magnetic or electric fields when included. In this transformation the velocity and stress transform according to

$$\underline{v}^* = k\,\underline{v}, \quad \underline{t}^* = k^2\underline{t}, \tag{3.3}$$

but the director being a unit vector is unaltered.

To investigate the implications of the above for simple shear flow, one naturally chooses the constant k equal to the gapwidth h, and the above transformation reduces the original problem with relative shear velocity V and gap h to one with unit gap and relative velocity V^*, where

$$V^* = Vh, \tag{3.4}$$

but the strong anchoring condition on the director is unchanged. Given a unique solution to the problem, this presumably involving stability and other arguments, it therefore follows that

$$\underline{n} = \underline{n}^* = \underline{n}^*(\underline{x}^*, V^*) = \underline{n}^*(h^{-1}\,\underline{x}, Vh), \tag{3.5}$$

when electric and magnetic fields are absent. In this way one concludes that in simple shear flow an optical property varies with shear through the product Vh, whatever the wall alignment. Also, defining an apparent viscosity η in the usual way by

$$\eta = \sigma h/V\ , \tag{3.6}$$

where σ is the shear stress, it quickly follows that

$$\eta = \eta^* = F(V^*) = F(Vh), \tag{3.7}$$

the function of course unknown. Hence one predicts the same simple scaling in viscosity measurements, again irrespective of the particular wall alignment. Wahl and Fischer [17] confirm such scaling in optical measurements on a flow rather similar to simple shear, but no direct experimental evidence is available with which to compare the viscosity prediction.

For flow down a circular capillary a natural choice of the scaling parameter k is the radius R of the capillary cross-section. In this event, the transformation (3.1) reduces the problem of flow down a capillary of radius R under a pressure gradient a to that of flow down a capillary of unit radius under a pressure gradient a^*, where

$$a^* = aR^3. \tag{3.8}$$

In this transformation, the flux per unit time Q from the capillary transforms according to

$$Q = RQ^*. \tag{3.9}$$

Granted that a unique solution exists, one finds when electric and magnetic fields are absent that

$$Q^* = H(a^*) \quad \text{or} \quad Q = R\,H(aR^3), \tag{3.10}$$

the function unknown. For this flow, one usually defines an apparent viscosity by

$$\eta = \pi a R^4/8Q, \tag{3.11}$$

this motivated by the Hagen-Poiseuille law for Newtonian liquids, and in this way Atkin [18] predicts for Poiseuille flow of a nematic that

$$\eta = G(Q/R), \tag{3.12}$$

the function unknown. This differs from the corresponding result for an isotropic liquid where the dependence is upon the quotient of the flux and the radius cubed. However, this novel prediction has been confirmed experimentally by Fisher and Frederickson [19].

As Leslie [4] discusses, confirmation of such scaling largely vindicates the various assumptions behind the theory described in the preceding section. For example, any generalisation of the constitutive assumptions must upset this scaling property and therefore lead to predictions in disagreement with experimental observations. Similar remarks apply equally to other proposals for the boundary conditions for the director. However, it would be of interest to determine how widely such scaling occurs in other nematic systems.

4. Flow Alignment

Many novel flow effects in nematic liquid crystals stem from competition between flow and other influences such as boundaries or external fields to dictate particular alignments to the anisotropic axis. Consequently an understanding of the likely response of alignment to flow is a useful starting point in any discussion of such phenomena, and therefore initially we ignore all other influences and consider solely that of shear flow upon alignment. To this end one examines solutions of the foregoing equations in which the Cartesian components of velocity and the director take the forms

$$v_x = \kappa z, \qquad v_y = v_z = 0,$$
$$n_x = \cos\theta\cos\phi, \qquad n_y = \cos\theta\sin\phi, \qquad n_z = \sin\theta \tag{4.1}$$

with κ a positive constant, and θ and ϕ depending only upon time t. This choice automatically satisfies equations (2.1) and (2.13), but conservation of angular momentum (2.10) ultimately yields the pair of equations

$$\gamma_1 \frac{d\theta}{dt} + \kappa m(\theta)\cos\phi = 0, \qquad \gamma_1 \cos\theta \frac{d\phi}{dt} - \kappa \ell(\theta)\sin\phi = 0, \tag{4.2}$$

where

$$m(\theta) = \alpha_3 \cos^2\theta - \alpha_2 \sin^2\theta, \qquad \ell(\theta) = \alpha_2 \sin\theta, \tag{4.3}$$

the latter employing the Parodi relation (2.16). These equations describe the response of the director in simple shear flow, when surface effects may be ignored and external fields are absent, and, not surprisingly, the conclusions that follow are similar to those deduced by Ericksen [20] in the context of a simpler theory of anisotropic liquids.

Naturally it is of interest to determine first of all any steady solutions that correspond to uniform alignments possible in shear flow. In doing so, one must bear in mind that the theory does not distinguish between orientations $\underline{n}$ and $-\underline{n}$, and therefore disregard certain solutions physically equivalent to others. One steady solution is clearly

$$\theta = 0, \qquad \phi = \pi/2, \tag{4.4}$$

this representing alignment everywhere normal to the plane of shear. In addition, if the viscosity coefficients α_2 and α_3 have the same sign, there are a further two steady solutions corresponding to uniform alignment in the plane of shear, namely

$$\theta = \pm\theta_0 , \qquad \phi = 0, \tag{4.5}$$

the acute angle θ_0 being defined by

$$\tan^2\theta_0 = \alpha_3/\alpha_2, \qquad \alpha_2\alpha_3 > 0. \tag{4.6}$$

Since the thermodynamic inequality (2.15) requires that the coefficient γ_1 be positive, one must have either

$$\alpha_2 < \alpha_3 < 0 \Rightarrow 0 < \theta_0 < \pi/4, \tag{4.7}$$

or

$$\alpha_3 > \alpha_2 > 0 \Rightarrow \pi/4 < \theta_0 < \pi/2, \tag{4.8}$$

or lastly α_3 positive and α_2 negative, in which case the solution (4.4) is the only steady solution.

To examine the stability of the above solutions one first notes an integral of equations (4.2) relating the angles θ and ϕ. By elimination of time between these equations, it follows that

$$\frac{d\theta}{d\phi} = (\alpha_2\sin^2\theta - \alpha_3\cos^2\theta)/\alpha_2\tan\theta\tan\phi, \tag{4.9}$$

which promptly yields the integral

$$\sin^2\phi = c(\tan^2\theta - \alpha_3/\alpha_2), \tag{4.10}$$

c a constant. When solutions of the form (4.5) exist, it is possible to show using the above that all time dependent solutions tend to one of this type, the other steady solutions being unstable. When α_2 and α_3 are both negative, the stable solution is that with the positive sign, but, when these viscosity coefficients are positive, the other solution in the plane of shear is the stable one. However, even when the shear plane solutions do not exist, the out-of-plane solution remains unstable.

As Leslie [21] discusses, a similar analysis is also possible under certain assumptions when the Cartesian components for the velocity and the director take the forms

$$v_x = \kappa z, \quad v_y = \tau z, \quad v_z = 0, \qquad n_x = \cos\theta\cos\phi, \quad n_y = \cos\theta\sin\phi, \quad n_z = \sin\theta, \tag{4.11}$$

with the coefficients κ and τ, and the angles θ and ϕ, all functions of time t. In this case, the counterparts of equations (4.2) are

$$\gamma_1\frac{d\theta}{dt} + m(\theta)\xi = 0, \qquad \gamma_1\cos\theta\frac{d\phi}{dt} - \ell(\theta)\zeta = 0 \tag{4.12}$$

where

$$\xi = \kappa\cos\phi + \tau\sin\phi, \qquad \zeta = \kappa\sin\phi - \tau\cos\phi. \tag{4.13}$$

For flow in relatively small gapwidths, Pieranski, Brochard and Guyon [22] argue that one may also neglect the inertia term in the equation for conservation of linear momentum (2.13). In addition, there are situations in which the transverse component of shear stress is zero, or at least appears to be negligible, as for example in the experiments by Pieranski and Guyon [23]. With these assumptions, the equation (2.13) essentially reduces to

$$\tilde{t}_{xz} = \sigma, \quad \tilde{t}_{yz} = 0, \tag{4.14}$$

where σ is some function of time assumed positive, and employing the relationship (2.5) one quickly obtains the equations

$$g(\theta)\xi + m(\theta)\frac{d\theta}{dt} = \sigma\cos\phi, \quad h(\theta)\zeta - \ell(\theta)\cos\theta\frac{d\phi}{dt} = \sigma\sin\phi, \tag{4.15}$$

where

$$2g(\theta) = \alpha_4 + (\alpha_5 - \alpha_2)\sin^2\theta + (\alpha_6 + \alpha_3)\cos^2\theta + 2\alpha_1\sin^2\theta\,\cos^2\theta,$$
$$2h(\theta) = \alpha_4 + (\alpha_5 - \alpha_2)\sin^2\theta, \qquad (4.16)$$

both positive functions on account of the inequality (2.15). Elimination of the velocity gradients between equations (4.12) and (4.15) leads to

$$\gamma_1 G(\theta)\,\frac{d\theta}{dt} + \sigma m(\theta)\cos\,\phi = 0, \quad \gamma_1 H(\theta)\cos\,\theta\,\frac{d\phi}{dt} - \sigma \ell(\theta)\sin\,\phi = 0, \qquad (4.17)$$

with

$$G(\theta) = g(\theta) - m^2(\theta)/\gamma_1, \quad H(\theta) = h(\theta) - \ell^2(\theta)/\gamma_1. \qquad (4.18)$$

Again using the inequality (2.15), Leslie [4] shows that the functions $G(\theta)$ and $H(\theta)$ are both positive for all values of θ, and, given the close similarities between the equations (4.2) and (4.17), one can clearly repeat the earlier discussion regarding steady solutions and their stability.

Equally, however, one may obtain an integral of equations (4.17) analogous to the result (4.9) which proves useful in the analysis of a problem discussed later. As above, it follows directly from equations (4.17) that

$$\frac{d\theta}{d\phi} = -\,\frac{m(\theta)\;H(\theta)}{\alpha_2\tan\,\theta\,\tan\,\phi\;G(\theta)}, \qquad (4.19)$$

or equivalently

$$2\cot\,\phi\,\frac{d\phi}{du} = \frac{[(u+1)^2 + 2\beta u]}{(u+\varepsilon)(u+\lambda)(u+1)}, \qquad u = \tan^2\theta, \qquad (4.20)$$

where the parameters occurring are given by

$$\varepsilon = -\alpha_3/\alpha_2, \quad \lambda = \alpha_4/2G(0), \quad 2\gamma_1 G(0)\beta = \gamma_2^2 + \alpha_1\gamma_1. \qquad (4.21)$$

When λ and ε are unequal, and neither is equal to unity, one finds

$$\sin^2\phi = c(1 + \tan^2\theta)^a(\varepsilon + \tan^2\theta)^b(\lambda + \tan^2\theta)^{1-a-b},$$
$$a = 2\beta/(1-\varepsilon)(\lambda-1), \quad b = [(1-\varepsilon)^2 - 2\beta\varepsilon]/(1-\varepsilon)(\lambda-\varepsilon) \qquad (4.22)$$

and c a constant of integration. Results for special cases are readily obtainable, and are given by Leslie [21].

It is also possible to obtain solutions of the equations of section two for shear flow demonstrating flow alignment and taking account of boundary conditions for both the velocity and the director. Leslie [21] discusses such solutions in some detail summarising progress. These solutions of course exhibit explicitly the scaling noted by Ericksen [16]. In addition there are solutions that describe competition between flow, surface alignment and external fields, but in such cases the analysis becomes rather complex; Jenkins [3] and Leslie [4] give references to such studies.

5. Induced Flow

A feature of continuum theory for liquid crystals is the intimate coupling between flow and alignment of the director. From the previous section, it is clear that flow of a nematic influences the orientation of the director, either tending to align it uniformly, or simply producing a torque that contributes to its ultimate configuration. However, rather less obviously, changes in alignment of the anisotropic axis can induce flow, which in turn can influence the transient behaviour of the director. This induced flow is frequently referred to as backflow, and it can lead to some unexpected behaviour.

An interesting illustration of this phenomenon occurs in a problem first noted by Gerritsma et al. [24] in connection with display devices. In its simplest form this involves a thin layer of nematic confined between two stationary, parallel plates with its initial alignment uniform and parallel to the plates, due to appropriate pretreatment of the bounding surfaces. The application of the sufficiently strong magnetic or electric field perpendicular to the plates rotates the anisotropic axis almost normal to the plates in the greater part of the sample, and its subsequent removal allows the alignment to return to its initial configuration. However, this relaxation turns out to be less straightforward than one might initially imagine. To examine this problem, it proves necessary to consider solutions of the continuum equations in which the Cartesian components of the velocity and the director take the forms

$$v_x = v(z,t)\ ,\ v_y = v_z = 0,$$
$$n_x = \cos\ \theta(z,t),\quad n_y = 0,\quad n_z = \sin\ \theta(z,t), \tag{5.1}$$

where z denotes distance normal to the plates and t time. These forms immediately satisfy the constraints (2.1), and the equations (2.10) and (2.13) reduce to

$$2f(\theta)\frac{\partial^2\theta}{\partial z^2} + \frac{d}{d\theta}f(\theta)\left(\frac{\partial\theta}{\partial z}\right)^2 - 2\gamma_1\frac{\partial\theta}{\partial t} - 2m(\theta)\frac{\partial v}{\partial z} = 0,$$
$$\rho\frac{\partial v}{\partial t} = \frac{\partial}{\partial z}\left[g(\theta)\frac{\partial v}{\partial z} + m(\theta)\frac{\partial\theta}{\partial t}\right], \tag{5.2}$$

with the functions $m(\theta)$ and $g(\theta)$ as defined in equations (4.3a) and (4.16a), and

$$f(\theta) = k_1\cos^2\theta + k_3\sin^2\theta. \tag{5.3}$$

The need to include a flow component is now clear from the above, since otherwise the problem is over-determined. If the layer thickness is $2h$, an obvious choice of origin leads to the boundary conditions

$$\theta(\pm h,t) = 0,\quad v(\pm h,t) = 0,\quad t > 0, \tag{5.4}$$

and the initial conditions are

$$\theta(z,0) = \Theta(z),\quad v(z,0) = 0,\quad |z| < h, \tag{5.5}$$

the function $\Theta(z)$ known from the static version of the theory.

Van Doorn [25] integrates equations (5.2) numerically neglecting the inertial term in the latter, but here we briefly describe an approximate analysis of this problem by Clark and Leslie [26] that also neglects this term. Since the distorted alignment is approximately normal to the plates in the greater part of the sample, it does not appear unreasonable in order to describe the initial stages of the relaxation to replace the equations (5.2) by

$$k_3\frac{\partial^2\theta}{\partial z^2} - \gamma_1\frac{\partial\theta}{\partial t} + \alpha_2\frac{\partial v}{\partial z} = 0,\quad \eta\frac{\partial^2 v}{\partial z^2} - \alpha_2\frac{\partial^2\theta}{\partial z\,\partial t} = 0, \tag{5.6}$$

where

$$2\eta = \alpha_4 + \alpha_5 - \alpha_2 = 2g(\pi/2), \tag{5.7}$$

these being the outcome of replacing the coefficient functions by their values when θ is equal to $\pi/2$, and setting ρ equal to zero. With these approximations, a solution of the resulting equations satisfying the boundary conditions (5.4) is

$$\begin{aligned} \theta &= \alpha A(\cos(qz/h) - \cos q)e^{-q^2\mu t/\alpha} \\ v &= -(\alpha_2\mu/\eta)\, qA(h\sin(qz/h) - z\sin q)e^{-q^2\mu t/\alpha}, \end{aligned} \tag{5.8}$$

with A a constant, and α and μ defined by

$$\alpha = 1 - \alpha_2^2/\gamma_1\eta\,, \qquad \mu = k_3/\gamma_1 h^2, \tag{5.9}$$

provided that the parameter q satisfies the equation

$$(1 - \alpha)\tan q = q. \tag{5.10}$$

Rather clearly, the parameter α is less than unity, since γ_1 and η are both positive, but it is also positive, being in the notation of the previous section $G(\pi/2)$ divided by η. In addition, given the inequalities (2.14), μ is positive. By superposition, therefore, one obtains solutions that decay in time of the form

$$\theta = \alpha \sum_{n=1}^{\infty} A_n(\cos(q_n z/h) - \cos q_n)e^{-q_n^2\mu t/\alpha}, \tag{5.11}$$

$$v = -(\alpha_2\mu/\eta) \sum_{n=1}^{\infty} q_n A_n(h\sin(q_n z/h) - z\sin q_n)e^{-q_n^2\mu t/\alpha}$$

where q_n are roots of equation (5.10), and the coefficients A_n are chosen to meet the initial condition (5.5a), so that

$$\alpha \sum_{n=1}^{\infty} A_n(\cos(q_n z/h) - \cos q_n) = \theta(z). \tag{5.12}$$

However, since

$$\int_{-h}^{h} (\cos(q_n z/h) - \cos q_n)\cos(q_m z/h)dz = 0, \quad n \neq m,$$
$$\int_{-h}^{h} (\cos(q_n z/h) - \cos q_n)\cos(q_n z/h)dz = (q_n - \sin q_n \cos q_n)h/q_n, \tag{5.13}$$

it quickly follows that

$$\alpha A_n = [2q_n/h(q_n - \sin q_n \cos q_n)] \int_0^h \theta(z)\cos(q_n z/h)dz, \tag{5.14}$$

$\theta(z)$ being an even function. Having neglected the fluid inertia, it is impossible to satisfy the second initial condition.

For very strong fields a reasonable approximation to the function $\theta(z)$ is simply

$$\theta(z) = \pi/2, \tag{5.15}$$

and in this event it follows straightforwardly from the expression (5.14) that

$$\alpha A_n = \pi \cos q_n/(\sin^2 q_n - \alpha). \tag{5.16}$$

However, since the roots of equation (5.10) are

$$q_n = (n - \tfrac{1}{2})\pi - \delta_n, \tag{5.17}$$

where δ_n tend to zero as n becomes large, one finds that

$$\alpha A_n \sim (-1)^{n-1}/(n - \tfrac{1}{2}) \tag{5.18}$$

for large n. Thus it is evident that the convergence of the series is rather slow, and therefore that one must take a high number of terms to compute accurate values. However, from these results, Clark and Leslie find that the value of θ increases initially, this increase being a maximum in the centre of the gap. With their choice of material parameters, this maximum increase reaches approximately 25° before the angle ultimately begins to decrease. This "kickback", as it is called, can have serious implications in twisted nematic devices as Gerritsma et al [24] found.

6. Instabilities

Given the variety of configurations possible with liquid crystals, one anticipates a correspondingly wider range of phenomena in these anisotropic liquids than in isotropic liquids. This proves to be the case with regard to the different instabilities that can occur in these materials. For example, consider simple shear of a nematic between parallel plates with the initial alignment uniform and parallel to the plates. Clearly, there is a non-trivial choice of direction in which to shear the layer, and one may select this so that the anisotropic axis is everywhere perpendicular to the plane of shear. In this case, no distortion of the alignment occurs at sufficiently low shear rates, this being in agreement with theory, since there is a solution of the continuum equations with the director normal to the plane of shear. However, at a critical shear rate distortion commences, and for nematics that align in shear this perturbation is homogeneous, being uniform in the plane of the plates. Also, the critical shear rate is inversely proportional to the square of the layer thickness in agreement with the scaling discussed in section three. Thus an appropriate choice of the shear direction produces a flow instability, termed the Pieranski-Guyon instability. A further variant is possible if one employs a magnetic field parallel to the initial alignment to inhibit the onset of the instability. As anticipated, the critical shear rate increases with field strength, the distortion remaining homogeneous, but at a critical field strength the nature of the instability changes to one with a roll structure, the roll axes coincident with the direction of shearing, and remains of this type with increasing field strength. This instability also occurs in plane Poiseuille flow, but in the experiments it is always homogeneous irrespective of the field strength. Essentially all analyses of this instability employ the appropriate linearised version of the above equations, and describe these different aspects rather well, as Dubois-Violette et al [27] and Dubois-Violette et al [28] discuss citing relevant references. Subsequently, however, Manneville [29] makes some progress with a non-linear analysis of the homogeneous instability.

A somewhat similar situation prevails with regard to thermally induced or Benard convection in the sample of nematic between two horizontal plates. If the alignment is uniform and parallel to the plates, Benard convection occurs at a much lower threshold than for a similar isotropic liquid, a result predicted initially by Dubois-Violette [30] using continuum theory. For example, such convection can occur in a layer of nematic so aligned with thickness 1 mm. However, if the anisotropic axis is everywhere perpendicular to the plates, it is impossible to induce convection in such a thin layer of nematic by heating from below, but, if one heats from above, convection occurs, as predicted by Currie [31] from an analysis based on the continuum equations. Such novel experimental results naturally stimulated further studies, with virtually all analyses employing linearised theory, but again there is acceptable agreement between theory and experiment, as Dubois-Violette et al [28] describe. References to more recent studies of this topic are given by Hodson et al [32] in a recent non-linear analysis of aspects of this problem.

Dubois-Violette et al. [28] and Goossens [33] review similar instabilities induced by application of an electric field to a layer of nematic. However, rather than discuss these reasonably well documented instabilities further we turn instead to some more recent work that illustrates other aspects, and also demonstrates the relevance of flow phenomena to certain device applications.

Once again consider a layer of nematic confined between parallel plates with its initial alignment uniform and parallel to the bounding surfaces, but here subject to an oscillatory shear produced by the upper plate executing sinusoidal oscillations of amplitude a and frequency ω parallel to the initial alignment, the lower plate remaining at rest. In a preliminary investigation, it appears reasonable to concentrate upon the influence of the oscillatory flow, assuming that surface and inertial effects are of secondary importance. Consequently, with an appropriate choice of Cartesian axes, we discuss solutions of the continuum equations in which

$$\begin{aligned} v_x &= \kappa(t)z, \quad v_y = v_z = 0, \quad \kappa(t) = (a\omega/h)\cos \omega t \\ n_x &= \cos \theta(t), \quad n_y = 0, \quad n_z = \sin \theta(t), \end{aligned} \tag{6.1}$$

h denoting the gapwidth. As in section four, the equations reduce to

$$\gamma_1 \frac{d\theta}{dt} + \kappa(t)\, m(\theta) = 0, \tag{6.2}$$

with $m(\theta)$ again defined by equation (4.3a), or equivalently one has

$$\frac{d\theta}{dt} = -(\sin^2\theta + \varepsilon\cos^2\theta)A\omega \cos \omega t, \tag{6.3}$$

where

$$\varepsilon = -\alpha_3/\alpha_2, \quad A = |\alpha_2|a/\gamma_1 h, \tag{6.4}$$

this assuming that α_2 is negative. Two possibilities occur depending upon the sign of ε. If the nematic aligns in shear so that ε is negative, one obtains

$$\tan\theta = \delta \tanh(\delta A \sin\omega t), \quad \delta^2 = |\varepsilon|, \tag{6.5}$$

but otherwise

$$\tan\theta = -\delta \tan(\delta A \sin\omega t), \quad \delta^2 = \varepsilon, \tag{6.6}$$

in each choosing the origin of time to coincide with θ being zero, consistent with the initial parallel alignment. In the former, the oscillations of the angle θ remain bounded, but in the latter they increase without bound. However, if δ is small compared with unity, even in this latter case the amplitude remains small until that of the plate approaches a critical value given by

$$a_c = \pi\gamma_1 h/2\delta|\alpha_2|, \tag{6.7}$$

when it increases sharply.

To examine the stability of the above, Clark et al. [34] essentially consider solutions of the form (4.11), but regard the angle ϕ and the transverse flow gradient τ as infinitesimal, in which event equation (4.12a) reduces to (6.2), or equivalently (6.3). Also, they assume that the transverse component of stress is zero, and therefore obtain the integral (4.22). Consequently, having found the angle θ as above, they deduce the angle ϕ from this integral, and thus determine the behaviour of small deviations of the director from the plane of shear during a cycle. If the nematic aligns in shear, such small perturbations remain small, but, for nematics that do not so align, they find that such small departures can deviate significantly from the shear plane as the amplitude of the oscillations approaches the critical value given by the expression (6.7). In this way, they conclude that oscillatory shear flow of certain nematics becomes unstable to small perturbations of the alignment out of the plane of shear, and they describe experimental results in agreement with this conclusion. As Clark et al. discuss, the problem does have relevance to transient behaviour in certain display devices.

References

1. J.L. Ericksen, Trans. Soc. Rheol., 5, 22-34, 1961.
2. F.M. Leslie, Arch. ration. Mech. Analysis, 28, 265-283, 1968.
3. J.T. Jenkins, Ann. Rev. Fluid Mech., 10, 197-219, 1978.
4. F.M. Leslie, Adv. Liq. Cryst., 4, 1-81, 1979.
5. P.G. de Gennes, 'The Physics of Liquid Crystals', Oxford, Clarendon Press, 1974.
6. S. Chandrasekhar, 'Liquid Crystals', Cambridge University Press, 1977.
7. J.L. Ericksen, Adv. Liq. Cryst., 2, 233-298, 1976.
8. J.L. Ericksen, 'The Mechanics of Viscoelastic Fluids', ed. R.S. Rivlin, Amer. Soc. Mech. Eng., New York, 1977, pp. 47-58.
9. C.W. Oseen, Ark. Mat. Astr. Fys., A19, 1-19, 1925.
10. F.C. Frank, Discuss. Faraday Soc., 25, 19-28, 1958.
11. J.L. Ericksen, Arch. ration. Mech. Analysis, 4, 231-237, 1960.
12. J.L. Ericksen, Phys. of Fluids, 9, 1205-1207, 1966.
13. F.M. Leslie, Quart. J. Mech. Appl. Math., 19, 357-370, 1966.
14. O. Parodi, J. Phys. (Paris), 31, 581-584, 1970.
15. P.K. Currie, Mol. Cryst. Liq. Cryst., 28, 335-338, 1974.
16. J.L. Ericksen, Trans. Soc. Rheol., 13, 9-15, 1969.
17. J. Wahl and F. Fischer, Mol. Cryst. Liq. Cryst., 22, 359-373, 1973.
18. R.J. Atkin, Arch. ration. Mech. Analysis, 38, 224-240, 1970.

19. J. Fisher and A.G. Fredrickson, Mol. Cryst. Liq. Cryst., 8, 267-284, 1969.

20. J.L. Ericksen, Kolloidzeitschrift, 173, 117-122, 1960.

21. F.M. Leslie, Mol. Cryst. Liq. Cryst., 63, 111-128, 1981.

22. P. Pieranski, F. Brochard and E. Guyon, J. Phys. (Paris), 34, 35-48, 1973.

23. P. Pieranski and E. Guyon, Phys. Lett. A, 49, 237-238, 1974.

24. C.J. Gerritsma, C.Z. Van Doorn, and P. Van Zanten, Phys. Lett. A, 48, 263-264, 1974.

25. C.Z. Van Doorn, J. Phys. (Paris), 36, C1, 261-263, 1975.

26. M.G. Clark and F.M. Leslie, Proc. R. Soc. Lond. A, 361, 463-485, 1978.

27. E. Dubois-Violette, E. Guyon, I. Janossy, P. Pieranski and P. Manneville, J. Mecanique, 16, 733-767, 1977.

28. E. Dubois-Violette, G. Durand, E. Guyon, P. Manneville and P. Pieranski, Solid State Physics, Supplement 14, 147-208, 1978.

29. P. Manneville, J. Phys. (Paris), 39, 911-925, 1978.

30. E. Dubois-Violette, C.R. Hebd. Seances Acad. Sci. Ser. B 273, 923-926, 1971.

31. P.K. Currie, Rheol. Acta, 12, 165-169, 1973.

32. D.A. Hodson, P.J. Barratt and D.M. Sloan, J. Non-equilib. Thermodyn., 9, 107-130, 1984.

33. W.J.A. Goossens, Adv. Liq. Cryst., 3, 1-39, 1978.

34. M.G. Clark, F.C. Saunders, I.A. Shanks and F.M. Leslie, Mol. Cryst. Liq. Cryst., 70, 195-222, 1981.

A Model for Disclinations in Nematic Liquid Crystal

J.H.Maddocks *
Department of Mathematics
University of Maryland
College Park, Md. 20742

§1 Introduction

Nematic liquid crystals typically have many threadlike regions, called disclinations, along which the optic axis displays singular behaviour. The standard Frank-Oseen model for nematic liquid crystals uses a field of unit-vectors to describe the equilibrium configurations, and line singularities in the vector field are interpreted as disclinations. The simplest possible line singularity has cylindrical symmetry, with a planar vector field that is everywhere radial and orthogonal to the cylinder axis. Such a disclination is a (formal) solution to the Frank-Oseen equilibrium equations, but the theory associates an infinite potential energy with this solution.

Physical arguments indicate that the difficulty of an infinite energy can be circumvented if the Frank-Oseen theory is modified by permitting variation of the degree of orientation in the order parameter. The purpose of this note is to discuss some simple solutions of an extended model. In particular, it is shown that there is a solution described by a planar vector field with cylindrical symmetry that approximates a disclination, and which has finite energy.

It should be stated at the outset that the Frank-Oseen theory does admit a nonplanar singular solution with cylindrical symmetry and finite energy. The vector field corresponding to this configuration is parallel to the cylinder axis in the centre

---

* This work was supported in part by the Air Force Office of Scientific Research under Contract Number 86-0097.

and then splays out to become a radial planar field. Moreover, there is considerable experimental evidence to suggest that amongst cylindrically symmetric singularities, the nonplanar configurations are the physically relevant ones (Williams et al, 1973). Nevertheless, there are more complicated singularities observed in experiment that cannot be fully described within the framework of the Frank-Oseen theory. Accordingly, an extended model is required, and this analysis can be regarded as a first simple investigation of one physically plausible modified theory.

In this study we assume that any configuration of a nematic liquid crystal is described by a pair of functions (n(x), S(x)) defined over a region $\Omega \subset R^3$ containing the fluid, and that the equilibrium configurations are the extremals of a variational problem:

$$\min \int_{\Omega} W(n,S)\, dx\,, \qquad n\big|_{\partial\Omega} = n_0. \tag{2.1}$$

Here n(x) is a unit vector called the director, and the scalar S(x) is the degree of orientation. The boundary condition that prescribes n(x) on the surface $\partial\Omega$ is known as strong anchoring; other conditions are sometimes relevant. Physically, it is unclear what boundary conditions should be imposed on S. Implicit in (2.1) is one choice which seems reasonable.

Variational principle (2.1) reduces to the standard Frank-Oseen model if the variable S is assumed to be constant, that is $W(n,S) = W_1(n)$, and $W_1(n)$ is defined to be

$$\begin{aligned} 2W_1(n) \equiv\ & \kappa_1(\nabla\cdot n)^2 + \kappa_2(n\cdot \mathrm{curl}\, n)^2 \\ & + \kappa_3\|n\times \mathrm{curl}\, n\|^2 + (\kappa_2+\kappa_4)[\mathrm{tr}(\nabla n)^2 - (\nabla\cdot n)^2] \end{aligned} \tag{2.2}$$

Here κ_1, κ_2, κ_3, and κ_4 are material constants.

The Frank-Oseen model gives excellent agreement with experiment, but it gives rise to the paradox concerning singularities that was described in §1. Several workers, for example Fan & Stephen (1970) and Fan (1971), argue that close to such singularities the scalar S cannot be accurately modelled as being constant. Ericksen (1985) has proposed that W should take the form:

$$2W(n,S) = 2W_0(S) + S^2W_1(n) + \kappa_5\|\nabla S\|^2 + \kappa_6\|n\cdot\nabla S\|^2 . \qquad (2.3)$$

Here $W_0(S)\in C^2(-1/2,1)$ has the qualitative features of the function sketched in Figure1, $W_1(n)$ is the Frank-Oseen energy defined in (2.2), and κ_5 and κ_6 are additional material constants. In the sequel the material constants κ_1, κ_5, and κ_6 will be assumed to satisfy the constitutive hypothesis

$$\kappa_5 + \kappa_6 > \kappa_1 . \qquad (2.4)$$

The role of this hypothesis is discussed in §8.

An <u>extremal</u> of (2.1) is a pair $(n(x), S(x))$ that satisfy the pertinent Euler-Lagrange equations and boundary conditions. With the further assumption that the "potential barrier" or "spinoidal" region of the function W_0 is sufficiently small (in a sense to be made precise), I shall show that when Ω is a circular cylinder, there is an extremal of (2.1) with W of the form (2.3), that has a line singularity in the vector field n(x), but has finite energy. Moreover, the singular vector field is everywhere radial and orthogonal to the cylinder axis.

§3 Cylindrical symmetry

The domain Ω is now taken to be a circular cylinder of radius R. The unit-vector n can be written in terms of the two scalar variables θ, and φ:

$$n(x) = \sin\theta\cos\varphi\, e_1 + \cos\theta\, e_2 + \sin\theta\sin\varphi\, e_3. \qquad (3.1)$$

Here $\{e_1,e_2,e_3\}$ is the orthonormal basis for cylindrical coordinates (r,ψ,z) that is defined in terms of the usual Cartesian orthonormal basis $\{i,j,k\}$ by $e_1 = \cos\psi\, i + \sin\psi\, j$, $e_2 = -\sin\psi\, i + \cos\psi\, j$, and $e_3 = k$.

In principle, problem (2.1) can be stated in terms of the three scalar variables θ, φ, and S, which are each functions of the variables (r,ψ,z). We shall simplify our

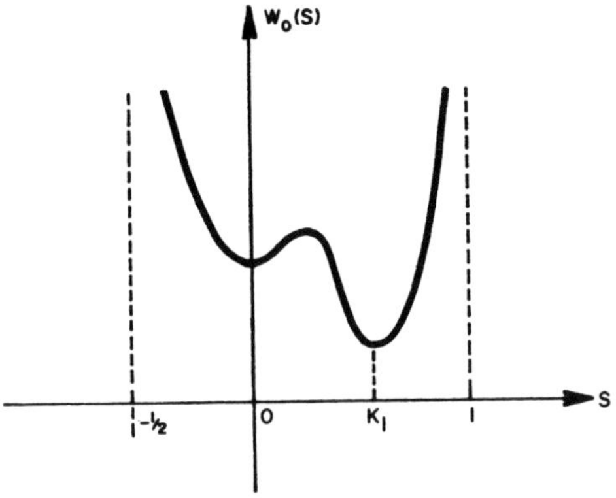

Figure 1 The qualitative form of potential $W_0(S)$. There is a global minimum at $S=K_1$, $0 < K_1 < 1$, and either a local minimum or inflexion point at $S = 0$.

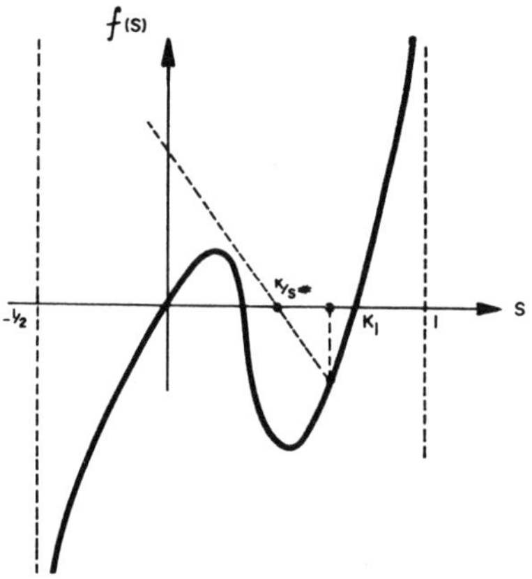

Figure 2 The qualitative form of the function $f(S)$. It is assumed that for a given $S^{\#} > 1$, there exists a constant K such that the line through the points $(K/S^{\#}, 0)$ and $(K, f(K))$ lies above the graph of f for $S < K$.

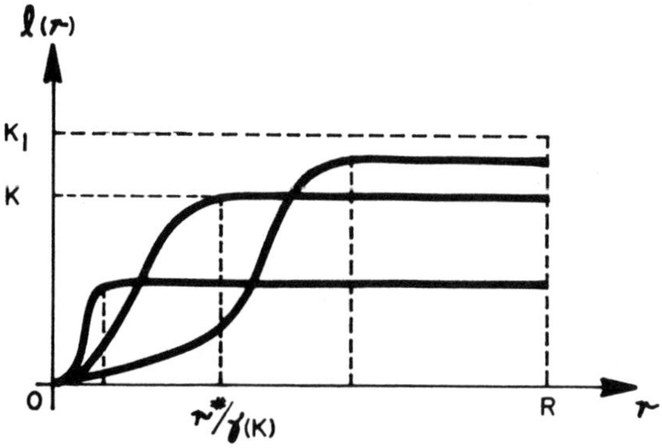

Figure 3 The qualitative form of the lower solution $\ell(n)$ defined in (6.8). K may assume a range of values, and three lower solutions are illustrated. The upper envelope of the lower solutions is a lower bound for a solution of integral equation (5.7).

task by considering solutions that are both symmetric under rotations about the z-axis, and symmetric under translations along the z-axis; that is θ, φ, and S are assumed to be functions of r only. The following expressions can then be calculated:

$$\nabla\cdot n = r^{-1}\{r\sin\theta\cos\varphi\}' , \tag{3.2}$$

$$n\cdot \mathrm{curl}\, n = -r\cos^2\theta\ \{r^{-1}\tan\theta\sin\varphi\}' , \tag{3.3}$$

$$\mathrm{tr}(\nabla n)^2 = \{\sin\theta\cos\varphi\}'^2 - r^{-1}\{\cos^2\theta\}' + r^{-2}\sin^2\theta\cos^2\varphi , \tag{3.4}$$

$$\begin{aligned}\|n\times \mathrm{curl}\, n\|^2 = {} & \sin^2\theta\cos^2\varphi\,\theta'^2 + \sin^4\theta\cos^2\varphi\,\varphi'^2 \\ & -2\sin^3\theta\cos\theta\sin\varphi\cos\varphi\,\theta'\varphi' \\ & +\cos^2\theta\ \{r^{-1}\sin^2\theta\sin^2\varphi\}',\end{aligned} \tag{3.5}$$

$$\|\nabla S\|^2 = S'^2, \qquad \text{and} \qquad \|n\cdot\nabla S\|^2 = \sin^2\theta\cos^2\varphi\, S'^2. \tag{3.6}$$

With the above expressions the energy density (2.3) can be expressed as a function $W(\theta(r),\varphi(r),S(r))$. Of course, this assumption of cylindrical symmetry, is only realistic if the imposed strong anchoring boundary conditions have the same symmetry.

The equilibrium equations are now a set of three coupled nonlinear ordinary differential equations that are the Euler-Lagrange equations of

$$\int_0^R r\ W(\theta(r),\varphi(r),S(r))\ dr. \tag{3.7}$$

However the calculation of these equations is still a considerable task. Instead, symmetry is used to further simplify the problem.

§4 Solutions with No Twist

I define an equilibrium solution with no twist to be one for which $\theta(r)\equiv\pi/2$. On such solutions n(r) has no azimuthal component. Consideration of (3.2) - (3.6) demonstrates that:

$$W(\pi/2-\theta,\varphi,S) = W(\pi/2+\theta,\varphi,S).$$

As a consequence, the Euler-Lagrange equation corresponding to θ must be satisfied for $\theta \equiv \pi/2$, and the remaining two Euler-Lagrange equations can be determined after θ is set equal to $\pi/2$. Of course, this procedure can only be adopted if such solutions are compatible with the boundary condition.

With $\theta \equiv \pi/2$, the following expressions are obtained:

$$2W(\varphi,S) = 2W_0(S) + S^2 W_1(\varphi) + \{\kappa_5 + \kappa_6 \cos^2\varphi\} S'^2, \tag{4.1}$$

where

$$W_1(\varphi) = \{\kappa_1 \sin^2\varphi + \kappa_3 \cos^2\varphi\}\varphi'^2 + r^{-2}\kappa_1 \cos^2\varphi. \tag{4.2}$$

Accordingly the Euler-Lagrange equation with respect to φ is

$$\begin{aligned} -\{(\kappa_1 \sin^2\varphi + \kappa_3 \cos^2\varphi) S^2 r \varphi'\}' + \\ + \{(\kappa_1 - \kappa_3)\sin\varphi\cos\varphi\,\varphi'^2 r - \kappa_1 \cos\varphi\sin\varphi\, r^{-1}\} S^2 \\ - \kappa_6 \cos\varphi\sin\varphi\, S'^2 r = 0, \end{aligned} \tag{4.3}$$

and with respect to S is

$$-\{r(\kappa_5 + \kappa_6 \cos^2\varphi) S'\}' + rSW_1(\varphi) + rW_0'(S) = 0. \tag{4.4}$$

§5 A Planar Singular Solution of Finite Energy

When $\varphi(r) \equiv 0$, equation (4.3) is satisfied identically. The associated planar vector field has n everywhere purely radial, with no z-component, and the solution has a disclination on the z-axis. Equation (4.4) then becomes

$$-\{rS'\}' + \mu^2 r^{-1} S + r f(S) = 0, \tag{5.1}$$

where

$$\mu = \{\kappa_1/(\kappa_5 + \kappa_6)\}^{1/2}, \tag{5.2}$$

and

$$f(S) \equiv (\kappa_5 + \kappa_6)^{-1} W_0'(S). \tag{5.3}$$

The issue to be addressed is whether there is a nontrivial solution of (5.1) that satisfies the natural boundary condition

$$S'(R) = 0, \tag{5.4}$$

and which has finite energy.

Inspection of integral (3.7) and energy density (4.2) demonstrates that S must vanish at r=0 in order that the solution have finite energy. Moreover a series expansion about r=0 shows that the boundary condition

$$S(0) = 0 \tag{5.5}$$

is also sufficient to guarantee that the extremal in question has finite energy.

Existence of a nontrivial solution to boundary-value problem (5.1), (5.4) and (5.5) can be demonstrated by application of the theory of upper and lower solutions associated with monotone operators. One elementary, concise account of this technique is given by Hutson & Pym (1980, p211).

Equation (5.1) can be rewritten as

$$-\{rS'\}' + \{\mu^2/r + r\omega^2\}\, S = r\{\omega^2 S - f(S)\} \tag{5.6}$$

where ω^2 is a (large) constant chosen such that the right hand side is a monotone increasing function of S for $-1/2 < l \leq S \leq u < 1$, for given constants l and u. The assumed smoothness of f(S) guarantees that such an ω^2 exists.

The left hand side of (5.6) is a modified Bessel operator which can be inverted with a positive Green's function $k(r,\rho)$, to obtain the integral equation

$$S = A[S], \tag{5.7}$$

where the nonlinear operator A[.] is defined by

$$A[S](r) = \int_0^R \rho\, k(r,\rho)\ \{\omega^2 S(\rho) - f(S(\rho))\}\ d\rho. \tag{5.8}$$

If $\mu^2 < 1$ (cf. constitutive hypothesis (2.4) and definition (5.2)), the operator A is continuous from $L^2_r \rightarrow L^2_r$ (see, for example, Stakgold 1979 p.439) where the space L^2_r is the l^2-space on the interval [0,R] with weight function r. Moreover, by the choice of ω^2 and because the Green's function is positive, the operator A is <u>monotone</u> in the following sense:

if $v,w \in L^2_r$ satisfy $v(r) \leq w(r)$, a.e. , then $Av(r) \leq Aw(r)$, a.e. .

Upper and lower solutions of equation (5.7), say $l(r)$, $u(r) \in L^2_r$, are defined by the property $l(r) \leq Al(r)$, or $Au(r) \leq u(r)$, respectively. The abstract theory mentioned above exploits the monotonicity and continuity of the operator A to prove that if an upper and a lower solution satisfying $l(r) \leq u(r)$ can be found, then the sequences $A^n l$ and $A^n u$ both converge (in L^2_r) to solutions $\underline{S}(r)$, and $\overline{S}(r)$ of integral equation (5.7). Furthermore, the following bounds hold

$$l(r) \leq \underline{S}(r) \leq \overline{S}(r) \leq u(r). \tag{5.9}$$

A standard argument can be used to demonstrate that the solutions of the integral equation are actually classical solutions of the pertinent boundary value problem.

In §6, I shall exhibit upper and lower solutions in C[0,R] that satisfy $-1/2 < l(r) \leq u(r) < 1$, and $l(r)$ is not less than the zero function. By the above development this is sufficient to guarantee the existence of a nontrivial solution to boundary value problem (5.1), (5.4) and (5.5).

Two remarks should be made. First, the above arguments provide a constructive iteration procedure by which solutions could be found. Second, there is often a uniqueness result associated with monotone iteration schemes (e.g. Stakgold, op.cit. p.612), but for the particular form of nonlinearity arising in this example the standard argument breaks down. Accordingly, it appears that the two solutions $\underline{S}(r)$ and $\overline{S}(r)$ could be distinct.

§6 Upper and lower solutions

Upper solutions of (5.7) are easily obtained. The one that will be used here is:

$$u(r) \equiv K_1 , \tag{6.1}$$

where the constant K_1 is the largest root of $f(S) = 0$ (cf. Figure 2).

Construction of lower solutions is more intricate, and several preliminary steps are necessary. Consider the nonhomogeneous Bessel equation:

$$-\{rv'\}' + \{\mu^2/r - r\}\, v = -r, \qquad v(0) = v'(0) = 0, \tag{6.2}$$

and define S^* and r^* to be the ordinate and abscissa of the first local maximum of the function $v(r)$ defined by (6.2). In the case $\mu = 1$, the function $v(r)$ is a Struve function, which functions are tabulated. In the more general case, simple contradiction arguments demonstrate that S^* and r^* satisfy the estimates

$$S^* > 1 + \{\mu/r^*\}^2, \tag{6.3}$$

and

$$\rho_1 < r^* \le \rho_2\,, \tag{6.4}$$

where ρ_1 is the abscissa of the first local maximum of $J_\mu(r)$ (i.e. the Bessel function of order $\mu>0$), and ρ_2 is the first nontrivial zero of $J_\mu(r)$.

The smallness condition on the nonlinear term can now be introduced. The assumption that will be made is that there exists a constant K with the property

$$f(x) \le \gamma^2(K/S^* - x), \qquad -1/2 < x \le K\,. \tag{6.5}$$

Here γ^2 is defined to be

$$\gamma^2 = \{-f(K)/K\}\,\{S^*/(S^* - 1)\}. \tag{6.6}$$

Thus, condition (6.5) is the requirement that the straight line through the points $(0, K/S^*)$ and $(K, f(K))$ lies above the graph of f for $x \le K$ (cf. Figure 2).

If smallness requirement (6.5), and the further condition

$$\gamma R \ge r^*, \tag{6.7}$$

are satisfied, then the following function is a lower solution of integral equation (5.7) (cf. Figure 3):

$$l(r) = \begin{cases} (K/S^*)\, v(\gamma r), & 0 \le r \le r^*/\gamma, \\ K\,, & r^*/\gamma \le r \le R. \end{cases} \tag{6.8}$$

Recall that the function $v(r)$ was defined by (6.2). It is straightforward to verify that (6.8) represents a lower solution once it has been observed that, by (6.5) and (6.8), the function $l(r)$ satisfies (in a piecewise sense) the differential inequality:

$$-(r\,l')' + (\mu^2/r - \gamma^2 r)\,l + \gamma^2 K/S^*\, r \le -r\{\, f(l) + \gamma^2(l - K/S^*)\}. \tag{6.9}$$

Inequality (6.9) can be rewritten in the form

$$-(r l')' + (\mu^2/r + \omega^2 r) l \leq r(\omega^2 l - f(l)), \tag{6.10}$$

and provided that (6.7) is valid, multiplication of (6.10) by the Green's function followed by integrations by parts over the intervals $[0, r^*/\gamma]$ and $[r^*/\gamma, R]$ yield the desired result, namely

$$l(r) \leq Al(r). \tag{6.11}$$

It should be remarked that if condition (6.7) does not hold then boundary terms of the wrong sign appear in the integrations by parts, and (6.11) no longer follows from (6.10).

The lower solution $l(r)$ is sketched in Figure 3. In particular the following estimates hold

$$0 \leq l(r) \leq K \leq K_1 = u(r). \tag{6.12}$$

Consequently, all the requirements of the previous section are satisfied, and the existence of a solution is guaranteed. Moreover $l(r)$ vanishes only at $r = 0$, which, with the left hand inequality of (6.12), implies that the solution is nontrivial.

§7 Estimates on the solution

A major benefit of the existence result described in §§5 and 6 is that the solutions $\overline{S}$ and $\underline{S}$ satisfy bounds (5.9). Of course, when different upper and lower solutions are used in the existence proof, different estimates are obtained. In this section the parameters arising during the construction of the lower solution are varied, and improved estimates are obtained. Throughout, the upper solution is taken to be that given by (6.1). Consequently, because the sequence $A^n u(r)$ is independent of the choice of lower bound, the solution $\overline{S}$ is actually bounded below by the upper envelope of all lower solutions that are found. It should be reiterated that there is no guarantee that the solutions $\overline{S}$ and $\underline{S}$ coincide, and, as $\underline{S}$ depends upon the choice of lower bound, there is no estimate on $\underline{S}$ in terms of the envelope of lower solutions. Nevertheless, estimates can be obtained on one solution, namely $\overline{S}$.

A function of the general form (6.8) must satisfy inequalities (6.5) and (6.7) in order that it be a lower solution, and the existence result was predicated on there being one value of K that satisfies the inequalities. But typically one expects there to be either no such values of K, or a range of admissible values. In the latter case K can be regarded as a parameter in the problem.

If K is chosen to be as large as is permissible, the lower solution obtained eventually reaches a relatively high value, namely K, but has slow growth near $r=0$. This last conclusion arises because the assumed form of f (cf. Figure 2) associates a large value of K with a small value of γ. Contrariwise, when K is chosen to maximize $-f(K)/K$ amongst admissible K, the lower solution grows relatively quickly close to the origin, but is eventually smaller than lower solutions corresponding to larger values of K. The best of both worlds is obtained when the envelope of lower solutions is constructed.

A few remarks concerning the range of values of K are in order. First, it is pointless to consider values of K below that which maximizes $-f(K)/K$; the corresponding lower solutions lie entirely below some other lower solution. Thus maximization of $-f(K)/K$ provides an effective lower bound. Inequality (6.5) implies another lower-bound, and one of these two bounds determines the best growth estimate for small r.

Upper bounds on the range of admissible K are determined by (6.5) and (6.7). The domain size R appears only in (6.7), and this condition can be expected to be the active one in small domains. Notice that (6.5) is in some sense a measure of the "hump" region of f, and if this region is high (6.5) will determine the optimal bound. However there are cases of interest in which the energy density $W_0(S)$ has only an inflexion point at $S=0$, with the associated function $f(S)$ being non-positive for $x \leq K_1$. In this circumstance (6.5) is irrelevant, and the upper bound is provided by (6.7). Notice that inequalities (6.3) and (6.7) imply that admissible K satisfy

$$-f(K)/K > \mu^2/R^2. \tag{7.1}$$

Consequently, the lower solution on any finite domain is strictly less than K_1.

More refined estimates can be obtained if S^* is also treated as a parameter. This strategy can be adopted if the inner-part of lower solution (6.8) is redefined in terms of the function $w(r) = \alpha J_\mu(r) + v(r)$, which is a solution of

$$-\{rw'\}' + \{\mu^2/r - r\} w = -r, \qquad w(0) = 0. \tag{7.2}$$

Here the constant α can be chosen such that the first local maximum of w is achieved at any r^* satisfying (6.4). Moreover, the corresponding value of S^* is given by the formula

$$S^*(r^*) = -\{r^* J_\mu'(r^*)\}^{-1} \int_0^{r^*} r J_\mu(r)\, dr. \tag{7.3}$$

The function $S^*(r^*)$ so defined is monotone decreasing in the interval $(\rho_1, \rho_2]$ with a local minimum at ρ_2. More importantly, inequality (6.3) remains valid when the left hand side is regarded as a function of r^*. All of the above arguments and estimates hold when S^* and r^* are interpreted in the sense of (7.3). The one added complication is that account must be taken of the requirement

$$l(r) > -{}^1/_2\,, \tag{7.4}$$

that was used in §5 to guarantee the monotonicity of the operator A. Inequality (7.4) did not arise explicitly in §6 because it was satisfied trivially, the lower solution under consideration being positive. It appears that the values of α that arise are either positive, or negative and of small magnitude, so that (7.4) is not an active constraint, but I have not been able to prove this.

The more general interpretation of S^* strengthens the existence result described previously; the hypothesis that there exist a pair S^* and K satisfying the pertinent inequalities is a less stringent requirement than the hypothesis that there exist some K for a fixed value of S^*. However, when μ is close to one little benefit is gained. This is apparent from the tabulated values of the Struve function, which show that the simple definition of S^* given in §6 happens to be close to optimal. A numerical study for the case of small μ would be of some interest.

The main benefit of allowing S^* (or, equivalently, α) to vary, lies in the estimates that can be obtained. When α is positive the associated lower solutions have a positive derivative at $r=0$. Consequently, although these lower solutions provide no better estimate at large values of r, the envelope of lower solutions obtained by varying both S^* and K is a significantly improved bound for small values of r.

The following less general version of the existence and estimate results is perhaps the most practical result described here. When μ is set equal to 1 in either the definition (6.2) of v(r), or (7.3) of w(r), it can be verified that the function l(r) that is constructed as before, is still a lower solution. Consequently, if inequalities (6.3), (6.5), and (7.4) hold for some S^* (defined with $\mu=1$) and K, then there is a nontrivial extremal of finite energy for all values of $\mu<1$. This result has associated estimates that are not as sharp as those described previously, but its main significance is that it encompasses the physically realistic case in which there is uncertainty in the value of μ.

§8 Discussion

The objective of this work was to model disclinations in nematic liquid crystals by a solution of Ericksen's model that has a planar vector field and finite energy. That goal has been attained under various simplifying assumptions and restrictions. The most restrictive condition is that of cylindrical symmetry, which reduces the governing equations to ordinary differential equations. This simplification has two justifications, it is required to make the problem tractable, and there is a reasonable expectation that singular solutions for more general domains and boundary conditions are a smooth perturbation of a radially symmetric singular solution.

The mathematical development is based on three hypotheses that are encapsulated in inequalities (2.4), (6.5) and (6.7). If any of these conditions fail, the method employed to prove existence breaks down. It has not been proven that the problem has no solution when the inequalities are invalid. However, each of the conditions has a plausible physical interpretation, so there is some possibility that the inequalities are not entirely technical artifices.

Inequality (2.4) is a constitutive hypothesis on various material constants. The condition is used to guarantee that the nonlinear operator A is continuous. To my knowledge there is no direct experimental evidence either to support or to refute the assumption. Indirect support for the validity of the condition is lent by the following argument. The Frank-Oseen model assumes the variable S to be constant, and the theory provides good agreement with experiment away from singularities. Accordingly, any modification of the Frank-Oseen theory should heavily penalize the appearance of gradients in S. Inequality (2.4) is a requirement of this type.

Inequalities (6.5) and (6.7) are coupled conditions of a somewhat complicated geometrical type, that are used in the construction of the lower solutions. Inequality (6.5) is primarily a restriction on the shape of the potential function $W_0(S)$, and (6.7) is primarily a restriction on the domain size (cf. Figures 1 & 2). Inequality (6.5) will be satisfied unless the barrier region of W_0 is too large in comparison to the depth of the wells, and this seems to be a physically reasonable requirement. Inequality (6.7) will be satisfied unless the domain size is too small. This is also a reasonable condition because the imposed boundary conditions are radial strong anchoring, and as the domain size approaches zero it is to be expected that the only solution is $S \equiv 0$, in which the molecules of the liquid crystal have no coherence.

Two issues have so far been ignored, namely: is the solution of the Euler-Lagrange equations that has been found either a global or local minimum of the potential? Neither of these questions has an obvious answer. As was mentioned in the Introduction, the Frank-Oseen model has a nonplanar singular extremal with finite energy that satisfies the radial boundary conditions described in §3. This solution has no twist ($\theta \equiv \pi/2$), and $\varphi(r)$ is defined implicitly by

$$r^2 \varphi'^2 = \cos^2\varphi / \{\kappa_1 \sin^2\varphi + \kappa_3 \cos^2\varphi\}.$$

The function $\varphi(r)$ corresponds to a segment of the separatrix in the phase-plane diagram that arises after the change of variable $r = Re^t$. Presumably Ericksen's model has an analogous nonplanar singular solution, and, a priori, either of the two singular solutions could have the lesser energy. Verification of the conditions for a local minimum are also nontrivial. In particular, it appears that the second variation of (4.1) with respect to perturbations in S need not be positive definite. An analysis based on bifurcation theory might be profitably pursued here.

Acknowledgements This research was initiated whilst I was a Senior Fellow at the Institute for Mathematics and Its Applications, University of Minnesota, during the program year *ContinuumPhysicsand PartialDifferentiaEquations*. I am happy to acknowledge the support and hospitality of that institution. I also wish to thank Professor Ericksen for suggesting this problem to me.

References

J.L.Ericksen, 1985, Private communication.

C.Fan & M.J.Stephen, 1970, Isotropic nematic phase transitions in liquid crystals, Phys. Review Letters V.25 pp.500-503.

C.Fan, 1971 Disclination lines in liquid crystals, Phys. Letters V.34A pp.335-336.

V.Hutson & J.S.Pym, 1980, Applications of Functional Analysis and Operator Theory, Academic Press, New York.

I.Stakgold, 1979, Green's Functions and Boundary Value Problems, Wiley-Interscience, New York.

C.E.Williams, P.E.Cladis & M.Kleman, (1973), Screw disclinations in nematic samples with cylindrical symmetry, Mol. Cryst. & Liquid Cryst. V.21 pp.355-373.

SOME REMARKS ABOUT A FREE BOUNDARY TYPE PROBLEM

Mario Miranda

Dipartimento di Matematica
Università di Trento
38050 Povo
Trento, ITALY

Let me call a "free boundary type problem" any problem consisting of trying to get information about the boundary of an open set of R^n, through calculations carried on functions defined in the open set itself.

Measuring the boundary of an open set, can be seen as a free boundary type problem, whose solution is given by the formula

$$\text{(1)} \qquad \text{meas}_{n-1}\partial\Omega = \left\{\sup\{\textstyle\int_\Omega \text{div}\phi(x)dx \mid \phi \in [C_0^1(R^n)]^n,\ |\phi(x)| \leq 1\ \forall\ x\}\right\} .$$

Actually (1) should be seen as a general definition of $\text{meas}_{n-1}\partial\Omega$, that does not require regularity assumptions about $\partial\Omega$. For (1) to make sense, Ω only needs to be Lebesgue measurable. Regularity assumptions about $\partial\Omega$ make meas_{n-1} coincide with other classical definitions of $(n - 1)$-dimensional measures.

A local version of (1) is given by

$$\text{(2)} \qquad \text{meas}_{n-1}(\partial\Omega \cap K) = \left\{\sup\{\textstyle\int_\Omega \text{div}\phi(x)dx \mid \phi \in [C_0^1(K)]^n,\ |\phi(x)| \leq 1\ \forall\ x\}\right\}$$

where K is any given open set of R^n.

Definition (1) was introduced by E. De Giorgi in 1954 for treating the isoperimetric problem. De Giorgi himself studied the Plateau Problem in R^n by means of (2) in 1960.

Formula (2) shows how the decision about the minimal character of $\partial\Omega$ can be seen as a free boundary type problem. Let divide this problem in two parts: minimal character of $\partial\Omega$ with respect to compact modifications of $\partial\Omega$ interior to Ω and exterior to it.

In [8] a criterium for deciding whether $\partial\Omega$ is minimal or not, with respect to compact modifications of Ω was given in the general case and then in the special case of cones. In order to make the statements simpler, the following definition of M-concave sets was given.

DEFINITION: A measurable set Ω is said to be M-concave if: $\forall$ K open bounded, $\forall$ measurable $E \subset\subset K$ ($E \subset\subset K$ means that $\bar{E}$ is a compact subset of K)

$$\text{meas}_{n-1}(\partial\Omega \cap K) \leq \text{meas}(\partial[\Omega - E] \cap K).$$

The general criterion of M-concavity proven in [8] is the following

THEOREM 1. If Ω is an open set of R^n and f_j is a sequence of subsolutions to the minimal surface equation in Ω, i.e.

$$\text{div}\left(\frac{\text{grad } f_j(x)}{\sqrt{1 + |\text{grad } f_j(x)|^2}}\right) \geq 0, \quad \forall x \in \Omega,$$

$$\lim_j f_j(x) = +\infty, \quad \forall x \in \Omega,$$

$$\limsup_{x \to y} f_j(x) \leq 0, \quad \forall y \in \partial\Omega, \forall j,$$

then Ω is M-concave.

When Ω is an open cone, that is Ω satisfies

(3) $$x \in \Omega, \lambda > 0 \Rightarrow \lambda x \in \Omega,$$

a very simplified version of theorem 1 is given by the following:

THEOREM, 2. If Ω is an open cone of R^n and f is a positive subsolution to the m.s.e. in Ω, vanishing at all points of $\partial\Omega$, if f is homogeneous of degree $\neq 1$, then Ω is M-concave.

We shall consider now a special class of cones with zero mean curvature at all regular points. The class, firstly studied by H.B. Lawson, is the following

$$\Omega = \{(x,y) | x \in R^{k+1}, y \in R^{h+1}, hx^2 > ky^2\},$$

where h and k are integers greater or equal to one.

In order to make the calculations simpler we introduce the variables

$$s = hx^2 \quad , t = ky^2$$

through which the inequality defining Ω becomes

$$s > t. \tag{4}$$

We shall try to decide whether the set Ω is M-concave or not, through the use of theorem 2 and subsolutions f depending on (s,t).

Consider the function $F(s,t)$ of the two variables s and t and the function

$$f(x,y) = F(hx^2, ky^2) \tag{5}$$

of the $h + k + 2$ variables $x_1, x_2, \dots, x_{k+1}, y_1, \dots, y_{h+1}$.

Consider the $(h + k + 2)$-dimensional minimal surface operator in the form

$$Mf = \sqrt{1 + |\text{grad } f|^2}^{\,3} \ \text{div}\Big(\frac{\text{grad } f}{\sqrt{1 + |\text{grad } f|^2}} \Big) = \tag{6}$$

$$= (1 + |\text{grad } f|^2)\, \Delta f - \sum_{i,j=1}^{h+k+2} D_i f D_j f D_i D_j f.$$

In the case f is depending on (s,t), i.e. f is given by (5), also Mf depends on (s,t) only and the following identity holds

$$\begin{aligned} \frac{1}{2} Mf = {} & h(2sF_{ss} + F_s) + k(2tF_{tt} + F_t) + hk(F_s + F_t) + \\ & + 4hk\{tF_t^2(2sF_{ss} + F_s) + sF_s^2(2tF_{tt} + F_t) - 4stF_sF_tF_{st} + \\ & + (F_s + F_t)(hsF_s^2 + ktF_t^2)\}. \end{aligned} \tag{7}$$

In the symmetric case, $h = k$, let us consider the function

$$F(s,t) = \frac{1}{2}(s^2 - t^2). \tag{8}$$

We have

$$F_s = s,\ F_t = -t,\ F_{ss} = 1,\ F_{st} = 0,\ F_{tt} = -1, \tag{9}$$

therefore, through (9) and (7) we obtain

$$\frac{1}{2}Mf = (3 + k)k(s - t) + 4k^2\{3st^3 - 3s^3t + (s - t)k(s^3 + t^3)\} =$$

$$= (k^2 + 3k)(s - t) + 4k^2(s - t)\{k(s^3 + t^3) - 3st(s + t)\} =$$

$$= (k^2 + 3k)(s - t) + 4k^2(s - t)\{(k - 3)(s^3 + t^3) +$$
$$+ 3(s + t)(s^2 - st + t^2 - st)\} =$$
$$= (k^2 + 3k)(s - t) + 4k^2(s - t)\{(k - 3)(s^3 + t^3) + 3(s + t)(s - t)^2\}.$$

This formula implies that $fMf > 0$ for $k > 3$, so f is a subsolution where it is positive and a supersolution where it is negative.

Since f is a homogeneous function of degree 4 in (x,y), we obtain, from Theorem 2, that

$$\Omega = \{(x,y) | x \in R^{k+1}, y \in R^{k+1}, x^2 > y^2\}$$

is M-concave for $k > 3$, and the same is true for the set

$$R^{2k+2} - \overline{\Omega} = \{(x,y) | x \in R^{k+1}, y \in R^{k+1}, x^2 < y^2\}.$$

We can conclude that the singular surface

$$\partial\Omega = \{(x,y) | x \in R^{k+1}, y \in R^{k+1}, x^2 = y^2\}$$

is a minimal surface if $k > 3$. This result was known since 1969. A non elementary proof of it was firstly presented in [6]. The simple proof we have given here was already presented in [4].

We should remark that for $h + k < 6$ the surfaces $\partial\Omega$ cannot be minimal, see [5].

For the general case $h \neq k$, following an idea of Sassudelli and Tamanini, consider the function

(10) $$F(s,t) = s(s - t) = s^2 - st$$

for which

(11) $$F_s = 2s - t, \; F_t = -s, \; F_{ss} = 2, \; F_{st} = -1, \; F_{tt} = 0.$$

Therefore, through (11) and (7) we obtain

$$\frac{1}{2}\,Mf = h(k + 1)(s - t) + (5h - k)s +$$

$$+ 4hk(s - t)s\{h(2s - t)^2 + (k - 2)st - 4s^2\}.$$

It is obvious that we can assume

(12) $$1 \leqslant k \leqslant h.$$

So, in order to have

(13) $$f\,Mf \geqslant 0 \quad \text{for} \quad s > t,$$

as required by Theorem 2, we only need to have

(14) $$h(2s - t)^2 + (k - 2)st - 4s^2 \geqslant 0, \quad \text{for} \quad s > t.$$

Putting $u = s - t$ in (14) we get

(15) $$4(h - 1)u^2 + (3h + h + k - 10)ut + (h + k - 6)t^2 \geqslant 0$$

which is obviously satisfied for all $u \geqslant 0$, $t \geqslant 0$ as soon as (12) and

(16) $$h + k \geqslant 6$$

are assumed to be valid. We can then conclude that the open set

$$\Omega = \{(x,y)\,|\,x \in R^{k+1},\ y \in R^{h+1}, hx^2 > ky^2\}$$

is M-concave if h and k satisfy (12) and (16).

In order to study the set

(17) $$\Omega^* = \{(x,y)\,|\,x \in R^{k+1}, y \in R^{h+1}, hx^2 < ky^2\}$$

consider the function

(18) $$F(s,t) = t(s - t) = st - t^2$$

for which

(19) $$F_s = t,\ F_t = s - 2t,\ F_{ss} = 0,\ F_{st} = 1,\ F_{tt} = -2.$$

Through (19) and (17) we obtain

$$\frac{1}{2} Mf = (hk + k)(s - t) + (h - 5k)t + 4hk\{t^2(s - 2t)^2 +$$

$$+ st^2(s - 6t) - 4st^2(s - 2t) + (s - t)[hst^2 + kt(s - 2t)^2]\}.$$

The linear term is negative when $s - t < 0$ if

(20) $$5k > h.$$

So, assuming (20), we are reduced to the study of the fourth order term. This can be written as

$$4hkt\{(t(s - t)^2 - 2t^2(s - t) + t^3 + st(s - t) - 5st^2 - 4st(s - t) +$$

(21) $$+ 4st^2 + (s - t)[hst + k(s - 2t)^2]\} =$$

$$= 4hkt(s - t)\{t(s - t) - 3t^2 - 3st + hst + k(s - 2t)^2\}.$$

This term can also be written as

$$4hk(s - t)t\{t(s - t) - 3t(t + s) + hst + k(s - t)^2 - 2kt(s - t) + kt^2\}.$$

We are then reduced to see whether the following inequality holds

(22) $$t(s - t) - 3t(s + t) + hst + k(s - t)^2 + kt^2 - 2kt(s - t) > 0,$$
for $s < t$.

Substituting $u = t - s$, we get

$$-u(s + u) - 3(s + u)(2s + u) + hs(s + u) + ku^2 + k(s + u)^2 +$$

$$+ 2ku(s + u) > 0,$$

or

(23) $$(4k - 4)u^2 + (4k + h - 10)su + (h + k - 6)s^2 > 0.$$

This inequality is sastisfied as soon as $k > 1$, in addition to the assumption

(16). In the case $k = 1$ the (23) is not valid for all positive values of u and s. But when $k = 1$, since we are assuming (16) and (21), the only possible choice for h is the value 5.

We have so proven that the set Ω^* is M-concave for all values of (h,k) satisfying

(24) $$h + k \geq 6,\ 1 < k \leq h \leq 5k.$$

For the same values of (h,k) the surface

$$\partial\Omega = \{(x,y) \mid x \in R^{k+1},\ y \in R^{h+1},\ hx^2 = ky^2\}$$

is minimal.

We remain with the cases

(25) $$h + k \geq 6,\quad h > 5k,\ k \geq 1.$$

To prove the M-concavity of Ω^* in these cases let us go back to the function

(26) $$F(s,t) = (s - t)(as + bt)$$

considered by Concus-Miranda in [8].

Assuming (26) we have

(27) $$\begin{aligned} &F_s = as + bt + a(s - t),\ F_t = b(s - t) - (as + bt),\ F_{ss} = 2a, \\ &F_{st} = b - a,\ F_{tt} = -2b. \end{aligned}$$

Through (27) and (7) we obtain

$$\begin{aligned} \frac{1}{2} Mf = {} & h[5as + bt + a(s - t)] + k[b(s - t) - 5bt - as] + hk(a + b)(s - t) + \\ & + 4hk\{t[b(s - t) - (as + bt)]^2 \cdot [5as + bt + a(s - t)] + \\ & + s[a(s - t) + as + bt]^2 \cdot [b(s - t) - as - bt] - \\ & - 4st[a(s - t) + as + bt] \cdot [b(s - t) - as - bt](b - a) + \\ & + (a + b)(s - t)[hs(2as + bt - at)^2 + kt(bs - as - 2bt)^2]\}. \end{aligned}$$

Let us consider separately the linear term and the term of degree 4. For the former we have

$$(28) \qquad (5h - k)as - (5k - h)bt + (s - t)[ah + bk + hk(a + b)].$$

By making the choice

$$(29) \qquad a = 5k - h, \; b = 5h - k,$$

we obtain, for (28),

$$(30) \qquad (s - t)[(4k - 6)h^2 + (4h - 6)k^2 + 36hk].$$

Observe that with the choice (29) the sign of the function f in Ω^* is the same as the sign of $(s - t)$. Therefore, for the linear term of Mf, what interests us is the inequality

$$(31) \qquad (4k - 6)h^2 + (4h - 6)k^2 + 36hk > 0.$$

The only bad cases correspond to the choice $k = 1$, when the inequality becomes

$$(32) \qquad -2(h^2 + 3 - 20h) > 0,$$

which is false as soon as $h > 20$.

So the linear term of Mf has the right sign except for the choices $k = 1$, $h > 20$, which are all permitted by (25).

The sign of the fourth order term of Mf, when the function f is defined through (26) and (29), is the right one in Ω^*, for the values of (h,k) satisfying (25), according to the calculations carried on by Concus-Miranda.

In order to decide about the set Ω^* in the cases

$$(33) \qquad k = 1, \; h > 20,$$

consider the function

$$(34) \qquad F(s,t) = (s - t)t^{\alpha},$$

where α is a positive real number to be determined. From (34) we get

$$F_s = t^\alpha,\ F_t = -t^\alpha + \alpha t^{\alpha-1}(s-t),\ F_{ss} = 0,\ F_{st} = \alpha t^{\alpha-1},$$

(35)

$$F_{tt} = -2\alpha t^{\alpha-1} + \alpha(\alpha-1)t^{\alpha-2}(s-t).$$

Through the usual computations, for the lower order term of $\frac{1}{2}Mf$ we obtain

(36) $$(s-t)t^{\alpha-1}\alpha[h+2\alpha-1] + t^\alpha[h-4\alpha-1],$$

which is surely non positive for $s < t$ as soon as

(37) $$4\alpha + 1 \geqslant h.$$

For the higher order term we obtain a very simplified expression, if we require α to satisfy the following

(38) $$h\alpha = 2\alpha^2 + 2\alpha + 1,$$

which is compatible with (37) for all $h \geqslant 20$.

The fourth order term becomes

$$4h(s-t)^2 t^{3\alpha-2}\{\alpha^3(s-t) - (\alpha^2-\alpha-1)t\}$$

which has obviously the sign of $s - t$.

We can conclude that Ω^* is M-concave for all values of $h \geqslant 20$ and for $k = 1$. For these same choices the surface $\partial\Omega$ is minimal.

Our conclusion is then:

$$\partial\Omega = \{(x,y) \mid x \in R^{k+1},\ y \in R^{h+1},\ hx^2 = hy^2\}$$

is minimal if $h + k \geqslant 6$, except for the cases

$$(h,k) = (1,5),\quad (h,k) = (5,1).$$

BIBLIOGRAPHY

[1] E. De Giorgi, Su una teoria generale della misura (r - 1)-dimensionale in uno spazio euclideo ad r dimensioni, Ann. Math. Pura e Appl. 36 (1954), 191-213.

[2] ___________, Frontiere orientate di misura minima, Sem. Mat. Scuola Normale Superiore Pisa, 1960-61.

[3] E. De Giorgi, F. Colombini, L.C. Piccinini, Frontiere orientate di misura minima e questioni collegate, Scuola Normale Superiore, Pisa, 1972.

[4] U. Massari, M. Miranda, A remark on minimal cones, Boll. Un. Mat. Ital. Series VI, 2-A (1983), 123-125.

[5] ____________________, Minimal surfaces of codimension one, Notas de Matematica 91, North-Holland, 1984.

[6] E. Bombieri, E. De Giorgi, E. Giusti, Minimal cones and the Bernstein problem, Inv. Math. 7 (1969), 243-268.

[7] H.B. Lawson, The equivariant Plateau problem and interior regularity, Trans. Amer. Math. Soc. 173 (1972), 231-249.

[8] P. Concus, M. Miranda, Macsyma and minimal surfaces, (to appear in Proceedings of A.M.S., Symposia in Pure Mathematics).

[9] G. Sassudelli, I. Tamanini, A remark on minimal cones, (to appear in Boll. U.M.I.)

[10] P.A. Simoes, A class of minimal cones in R^n, $n \geq 8$, that minimize area, Ph.D. thesis, University of California, Berkeley, Calif., 1973.

COMPUTER SIMULATION OF FLOW OF LIQUID CRYSTAL POLYMERS

Gregory Ryskin

Department of Chemical Engineering
Northwestern University
Evanston, Illinois 60201

1. Introduction

In the lecture delivered at the IMA Workshop on Orienting Polymers two years ago, Wissbrun (1984) presented an overview of the rheology of liquid crystal polymers and also emphasized the need for numerical simulations of the flow and orientation development. The great practical importance of such studies stems from the well-known fact that processing of liquid crystal polymers may be used for production of solid materials which have extremely valuable properties (strength, etc.) due to the high degree of molecular orientation. Molecular orientation is thus a very important function of the processing conditions (equipment geometry, flow regime, etc.), and the ability to predict and control this function can be crucial for the successful production of solid materials with required properties.

The continuum theory of liquid crystals due to Ericksen and Leslie provides a natural starting point for the development of numerical algorithms for simulation of the motion and molecular orientation in liquid crystal polymers. It is, of course, possible that the polymer case may require more complicated constitutive description than the Ericksen-Leslie one, which is linear in velocity gradients, may provide (cf. Wissbrun 1984 and references therein). We think, however, that one should thoroughly explore the possibilities afforded by the Ericksen-Leslie theory before trying to include additional effects. Note also that the molecular theory for rod-like polymers due to Doi (1981) leads to constitutive relations which are quite similar in form to the phenomenological relations of Ericksen and Leslie.

We have recently begun to develop a numerical technique for simulation of the flow of liquid crystal polymers. Here we present the general framework and the first results, obtained in collaboration with W.B. VanderHeyden.

2. The Governing Equations

The mechanics of nematic liquid crystals are governed by the Ericksen-Leslie equations for velocity, pressure, and director. These equations are presented, e.g., in deGennes (1974), Leslie (1979), but will not be reproduced here in their full form since for nematic polymers some very significant simplifications can be introduced from the outset. These stem from the fact that, according to the available experimental evidence (cf. Wissbrun 1984), the elastic constants for nematic polymers are of the same order (10^{-6} dyn) as for low-molecular-weight nematics, while the viscous constants are higher by several orders of magnitude. This suggests that the effects associated with the elasticity of the director, that is, forces arising when the director field is distorted from its equilibrium (say, uniform) configuration, are likely to be negligible in comparison with the viscous forces in most realistic flow fields. More precisely, it is known (see, e.g., deGennes 1974) that at high values of the Ericksen number $E \equiv \frac{\mu VL}{K}$ (where μ and K are representative viscous and elastic coefficients respectively, V is a characteristic velocity, and L a characteristic distance) the effects of the director elasticity are confined to the thin <u>transition layers</u> (of thickness $O(LE^{-1/2})$) where the director orientation changes rapidly from, e.g., the one dictated by the walls to the one dictated by the flow. In liquid crystal polymers viscosities range from 10 to 10^5 poise. Taking $L = 1$ cm and $V = 1$ cm/sec, one obtains $O(10^7) < E < O(10^{11})$ and so the thickness of the transition layers is $O(10^{-3}\text{cm}) < \delta < O(10^{-5}\text{cm})$. Clearly, these layers can be neglected for most practical purposes. Note that the no-slip boundary condition for velocity is completely independent of the director in general and the transition layers in particular.

It appears, therefore, that only minor physical effects (orientation inside the transition layers, contribution of the director elasticity to the forces on the boundaries, etc.) will be lost due to the neglect of the transition layers.

The above considerations lead to the main physical assumption of the present work, namely, the assumption of negligible influence of the director elasticity. In other words, we will be studying the $E \to \infty$ limit of the Ericksen-Leslie theory.

In this limit the Ericksen-Leslie equations simplify very substantially. One can formally obtain the simplified form by setting all elastic constants in these equations equal to zero. The result is the following constitutive equation for the stress tensor (which becomes symmetric in this limit):

$$\underline{\underline{\sigma}} = \alpha_4 \underline{\underline{A}} + \beta_1 (\underline{n} \cdot \underline{\underline{A}} \cdot \underline{n}) \underline{n} \otimes \underline{n} + \beta_2 [(\underline{n} \cdot \underline{\underline{A}}) \otimes \underline{n} + \underline{n} \otimes (\underline{\underline{A}} \cdot \underline{n})] \qquad (1)$$

Here $\underline{\underline{A}}$ is the rate-of-strain tensor, $A_{ij} = 1/2(u_{i,j} + u_{j,i})$ where $\underline{u}$ is the velocity, α_4 is one of the Leslie coefficients, and the unit vector $\underline{n}$ is the director. The viscosity coefficients β_1 and β_2 in (1) are simple combinations of the standard Leslie coefficients, viz.

$$\beta_1 \equiv \alpha_1 - \lambda(\alpha_2 + \alpha_3); \qquad \beta_2 \equiv \alpha_5 + \lambda\alpha_2;$$
$$\lambda \equiv -\gamma_2/\gamma_1.$$

The stress enters the usual equation of motion

$$\rho\left(\frac{\partial u}{\partial t} + \underline{u} \cdot \nabla \underline{u}\right) = -\nabla p + \nabla \cdot \underline{\underline{\sigma}} \qquad (2)$$

where ρ is the density and p the pressure (modified in the usual way to take account of the hydrostatic pressure due to gravity.)

As usual, the liquids will be considered incompressible, i.e.,

$$\text{div } \underline{u} = 0 \qquad (3)$$

The evolution of the director $\underline{n}$ is described by

$$\frac{\partial n}{\partial t} + \underline{u} \cdot \nabla \underline{n} = \underline{\underline{W}} \cdot \underline{n} + \lambda(\underline{\underline{A}} \cdot \underline{n} - (\underline{n} \cdot \underline{\underline{A}} \cdot \underline{n})\underline{n}) \qquad (4)$$

Here $\underline{\underline{W}}$ is the vorticity tensor, $W_{ij} \equiv \frac{1}{2}(u_{i,j} - u_{j,i})$; the last term $-\lambda(\underline{n} \cdot \underline{\underline{A}} \cdot \underline{n})\underline{n}$ is required solely in order to keep the director length always equal to 1.

The boundary conditions for velocity and pressure are of the usual type, including no-slip on solid boundaries. The full Ericksen-Leslie theory would include an elliptic equation for the director and require boundary conditions for it on all boundaries; however, in the simplified form given above only entrance conditions can be specified for the director, that is, the orientation of the director must be given at each streamline as it enters the domain of solution.

Considering for clarity the steady case, one observes that for a given velocity distribution the full director field can be found once entrance conditions for the director have been specified. This does not mean, of course, that the director field decouples from the velocity and pressure fields -- the latter fields depend on the director through the constitutive equation (1).

Equations (1)-(4) are the same as the equations of the theory of anisotropic fluids developed by Ericksen (1960a,b) before the emergence of the complete mechanical theory of liquid crystals. It is possible, therefore, to start the investigation of liquid crystal polymers directly on the basis of this anisotropic fluid theory, without appeal to the Ericksen-Leslie theory of liquid crystals. We feel, however, that it is preferable to follow the above route of simplifying the equations of the latter theory, since in this way the physical nature of our assumptions becomes transparent, and some otherwise surprising results receive an easy explanation. For example, the orientation of the director near the centerline of symmetry in plane Poiseuille flow is discontinuous in the solution obtained on the basis of the anisotropic fluid theory (1)-(4), which could be a cause for some concern. Physically, of course, no discontinuity is present: a solution on the basis of the complete (i.e., including the director elasticity) Ericksen-Leslie theory of nematics would show a very thin transition layer instead of a discontinuity (see Fig. 5.6b in deGennes 1974).

Equations (1)-(4), which include three viscosity coefficients $\alpha_4, \beta_1, \beta_2$ and one dimensionless ratio λ, need to be solved in order to predict the flow field and director orientation. We will be interested mainly in steady solutions, but retain the time derivatives in (2) and (4) since the time-dependent form of the equations provides a natural starting point for developing an iterative numerical technique. Since the liquid crystal polymers are generally quite viscous as mentioned above, we will assume that inertial effects can be neglected and consider the limit of zero Reynolds number, so that the equation of motion takes the form

$$\rho \frac{\partial u}{\partial t} = -\nabla p + \nabla \cdot \underline{\underline{\sigma}}$$

(as just mentioned, the time derivative is retained for essentially numerical reasons). Note that the assumption of zero Reynolds number is less crucial that the assumption of infinite Ericksen number; solutions for Reynolds numbers up to $0(10^2)$ could probably be obtained using the same technique without much additional effort.

Finally, to keep the computer time within reasonable limits, we will deal only with two-dimensional flow fields. Axisymmetric flows can also be considered, but three-dimensional computations will have to await the next reduction in the cost of computations.

3. Numerical Technique

Before discussing the actual numerical scheme for solving the Equations (1)-(4), we have to consider the choice between finite-elements and finite-difference approaches. The main advantage of the finite element method is its ability to fit boundaries of arbitrary shape; finite-differences, on the other hand, are easier to code and usually require less computer time. Here we use a finite-difference technique which is suitable for domains of arbitrary shape since these domains are mapped on a standard unit square domain before actual solution of equations of the physical problem. Such an approach is, of course, well known in mathematical physics, conformal mapping being the foremost example. For numerical solution, however, conformal mapping is not convenient because of the absence of control over the distribution of the grid lines. We use, therefore, the technique of orthogonal mapping (Ryskin & Leal 1983, 1984) which is capable of producing orthogonal curvilinear coordinates but allows for considerable control over the density of the grid lines.

The essence of the orthogonal mapping can be summarized as follows. Given a two-dimensional domain of general shape in x,y coordinates in physical space, we need mapping functions $x(\xi,\eta)$, $y(\xi,\eta)$ which would map this domain on the unit square $0 < \xi,\eta < 1$ in the computational plane. This can be done in a variety of ways, ranging from interpolation to solution of some elliptic equations; however, most of these ways would not guarantee the resulting grid (i.e., curvilinear coordinate system) in the physical domain to be orthogonal. To obtain orthogonality, the equations for $x(\xi,\eta)$ and $y(\xi,\eta)$ are derived via the following very short argument: (a) consider the Cartesian coordinate x(or y) as a scalar function of position in physical space, then its gradient is a constant vector field whose divergence is zero, i.e.,

$$\text{divgrad } x = 0 \quad \text{or} \quad \nabla^2 x = 0$$

where ∇^2 is the covariant Laplace operator; (b) this operator can be written in any coordinate system, including ξ,η, if its metric tensor is known; (c) if the coordinate system is to be orthogonal its metric tensor must be diagonal ($g_{11} = h_\xi^2$; $g_{12} = 0$; $g_{22} = h_\eta^2$ where h_ξ, h_η are the scale factors); (d) only the ratio of the scale factors $f(\xi,\eta) \equiv \frac{h_\xi}{h_\eta}$ is needed to write out the equations $\nabla^2 x = 0$; $\nabla^2 y = 0$, and this ratio, called distortion function, can be specified at will.

The resulting equations take the form

$$\frac{\partial}{\partial\xi}\left(\frac{1}{f}\frac{\partial x}{\partial\xi}\right) + \frac{\partial}{\partial\eta}\left(f\frac{\partial x}{\partial\eta}\right) = 0; \qquad \frac{\partial}{\partial\xi}\left(\frac{1}{f}\frac{\partial y}{\partial\xi}\right) + \frac{\partial}{\partial\eta}\left(f\frac{\partial y}{\partial\eta}\right) = 0$$

The conformal mapping is, clearly, a special case of an orthogonal mapping, obtained by setting the distortion function $f(\xi,\eta)$ equal to 1 everywhere. Generally, the distortion function for given ξ,η specifies the ratio of the sides of a small rectangle in the x,y plane which is an image of a small square in the ξ,η plane.

The main potential of the orthogonal mapping is due to the fact that, unlike in the case of conformal mapping, one can construct an orthogonal mapping with prescribed boundary correspondence.

It is obviously very convenient to be able to prescribe the distribution of the grid points on the boundary, and also in this way to control the distribution of the grid points throughout the domain. Orthogonal mapping has this capability of prescribing the boundary points because of the added degree of freedom associated with the existence of the distortion function, which is free for the user to specify. In the so called "weak constraint" sub-method of orthogonal mapping, this available degree of freedom is utilized in order to specify the boundary correspondence. One divides this available degree of freedom into two "parts", the boundary part and the domain part. The boundary part is used to specify the prescribed boundary correspondence; the domain part is used to specify the distortion function inside the domain, while its values at the boundaries are obtained through its definition via the scale factors as indicated above, the scale factors being computed from the mapping, using the prescribed boundary correspondence and the mapping functions at the current iteration. All the details of the weak constraint method of orthogonal mapping are given in the paper by Ryskin and Leal (1983).

There exist some intriguing mathematical questions associated with existence and uniqueness of orthogonal mappings. In the case of the so called "strong constraint" method of orthogonal mapping, existence and, probably, uniqueness are fairly plausible because conformal mapping is one of the special cases of orthogonal mapping in this sub case, and for conformal mapping the existence and uniqueness are well known. However, for the case of the weak constraint method of orthogonal mapping, which is the one necessary to produce a mapping with prescribed boundary correspondence, no theorems, of course, have been proved, and the question remains open of whether the mapping with prescribed boundary correspondence can be found for any domain. The question is not only interesting mathematically, but also important in practice. So far, numerically we have been able to find orthogonal mappings for the domains which we use, but it should be noted that these domains are fairly smooth, and also the distribution of the grid points on the boundaries was reasonable in some sense. It is not clear if an arbitrary distribution of the boundary points could be used to construct an orthogonal mapping. In fact, it is our suspicion that it could not.

While the questions of the existence and uniquness of the orthogonal mappings await attention of mathematicians, we will use orthogonal mappings in the practical sense to construct numerical grids which are suitable for domains of different shapes and at the same time are sufficiently orthogonal. It should be noted that orthogonality of a numerical grid is important not only because the differential equations take simple form in orthogonal coordinates, but also because it is desirable in a numerical grid to have its lines crossing at an angle which is as close to 90° as possible; if the angle between the coordinate lines is small, large numerical errors will occur.

The above differential equations for the mapping functions are solved by finite differences in an iterative manner, the distortion function being updated at each step. We will not go into details of how the distortion function is being specified at the boundaries and inside the domain because they have been presented in the paper already mentioned (Ryskin and Leal 1983).

The finite difference technique which we use, ADI, or Alternating Directions Implicit, is well known among people who do numerical computations, and I will only very briefly review the main ideas of this technique here. Suppose one needs to solve the equation

$$\frac{\partial^2 w}{\partial \xi^2} + \frac{\partial^2 w}{\partial \eta^2} = 0,$$

then one introduces an artificial time derivative into the equation so it becomes parabolic, then the equation is discretized on a grid and iterated to convergence so the resulting steady state solution is the solution of the original equation without the time derivative. Now, the iteration consists of two steps. In one step, the spatial derivatives with respect to one spatial direction are taken explicitly, which means they can be computed from the values of functions which are already known, while the derivatives in the second spatial direction are taken implicitly, which means they depend on the values of functions which are to be computed at this step. This results in a system of linear equations for each grid line, which are interconnected, but the connection exists only between three neighboring points, so we obtain a system of linear equations with a tri-diagonal matrix which can be very easily solved by the so called Thomas algorithm. In fact, the solution of each of these systems of equations is only by a constant factor more time consuming than simple evaluation of the functions at each point. Such systems are solved for each grid line in turn, and this constitutes the first half of the iteration step. In the second half of the iteration step, the order is reversed in the sense that now the derivatives along the spatial directions which were previously taken explicitly will be taken implicitly, and vice versa. As a result, each of the spatial directions is treated explicitly at one half of the iteration and implicitly at the other, and the resulting numerical scheme is extremely stable. The following equations represent the ADI scheme symbolically:

$$\frac{\tilde{w} - w^n}{(\Delta t/2)} = \frac{\delta^2 w^n}{\delta\xi^2} + \frac{\delta^2\tilde{w}}{\delta\eta^2}$$

$$\frac{w^{n+1} - \tilde{w}}{(\Delta t/2)} = \frac{\delta^2 w^{n+1}}{\delta\xi^2} + \frac{\delta^2\tilde{w}}{\partial\eta^2}$$

Here, $\frac{\delta^2}{\delta\xi^2}$ is the difference operator, and superscripts denote the iteration.

Evidently, in order to be able to use the ADI scheme, one has to write equations of the problem in the form which would explicitly separate first derivatives along each direction, second derivatives, and so on, so that each of the derivatives could be evaluated using a three point finite difference formula.

It is usually a trivial matter to write out the equations in the required form, especially if one is using Cartesian coordinates. However, in the case of the equations given in section 2, and especially when using curvilinear coordinates, the task of writing out the required form of the equations becomes extremeley serious. Let us consider, for example, a couple of components of the rate of strain tensor. For example,

$$A_{\xi\xi} = \frac{1}{h_\xi}\frac{\partial u_\xi}{\partial\xi} + \frac{u_\eta}{h_\xi h_\eta}\frac{\partial h_\xi}{\partial\eta} ;$$

$$A_{\xi\eta} = \frac{1}{2}\left(\frac{h_\eta}{h_\xi}\frac{\partial}{\partial\xi}\left(\frac{u_\eta}{h_\eta}\right) + \frac{h_\xi}{h_\eta}\frac{\partial}{\partial\eta}\left(\frac{u_\xi}{h_\xi}\right)\right).$$

These components already look somewhat messy. However, it is only the beginning. Now one has to combine these components with the components of the director, in different combinations as indicated in equation (1), in order to obtain the stress tensor. This will involve a significant number of multiplications and summations in order to perform all the necessary tensor operations. Then, after the stress tensor has been written out, its divergence has to be taken, as specified by the equation of motion (3).

If one wanted to write out all the terms separately in the resulting expression, one would probably need 20 pages of normal size paper. And this is exactly what is apparently needed to use ADI, since ADI is normally applied to an equation with every derivative separated in the following form:

$$q_1 u_{\xi\xi} + q_2 u_{\eta\eta} + q_3 u_{\xi} + q_4 u_{\eta} + q_5 u + q_6 u_{\xi\eta} + q_7 = 0$$

It is extremely unlikely that a human being would be able to write out many pages of the terms and not make a single mistake in them, and, of course, a single mistake will completely nullify the results. So the question is how one can circumvent this problem. One answer which we are looking into at the moment is to use some kind of a symbolic manipulation program, for example, MACSYMA, or SMP. However, these programs are not easily available. At our installation, at least, they were not available at the time we began this research, and therefore we had to invent some other way of dealing with these huge differential expressions.

It is, of course, quite easy to find the result of all these differential operations numerically if operations are not expanded, but each differential operation is carried out in turn. For example, first one computes the components of the rate-of-strain tensor numerically. Then, from these numbers one computes the components of the stress tensor, again numerically. And then, one takes the divergence of it, again numerically. This is not difficult to do. However, obviously this can be done only in an explicit manner, which means the result can be obtained only for the known values of the velocity field. And therefore the stability enhancement due to the usage of the implicit formulation would be lost. The task is, therefore, to try to combine, on the one hand, the ease of the explicit numerical evaluation of the necessary expressions in turn, and on the other hand, the stability enhancements due to the implicit formulation. These seemingly contradictory requirements, it turns out, can be met if one uses the following trick.

Experience with numerical solutions suggests that it is second derivatives which are mainly responsible for the stability enhancement in the implicit formulation. The first derivatives and non differentiated terms do not necessarily perform better if they are taken implicitly as opposed to explicitly. And, in fact, in some cases explicit formulation is preferable (see, for example, Ryskin and Leal 1984). This suggests the following idea. Try to formulate the algorithm in such a way that only the second derivatives of velocity are expanded, and then evaluated implicitly, while the rest of the differential expression is computed explicitly. It is not very difficult to find the coefficients of the second derivatives of velocity, because, when you do that, you can consider components of the director as constant. One can, therefore, write the following expression for a component of the divergence of the stress tensor

$$(\nabla\cdot\underline{\underline{\sigma}})_\xi = (\alpha_4 + \beta_1 n_\xi^4 + 2\beta_2 n_\eta^2)\ \frac{\partial^2 u_\xi}{\partial \xi^2} + \left(\frac{\alpha_4 + \beta_2}{2} + \beta_1 n_\xi^2 n_\eta^2\right)\ \frac{\partial^2 u_\xi}{\partial \eta^2}$$

$$+ \ \ldots$$

where the dots stand for the remaining part, which has not been expanded, in fact. The idea is now to evaluate, say, the second derivative of velocity with respect to the ξ direction implicitly, and all the rest explicitly. The question arises as to how one can evaluate explicitly the rest of the divergence of the stress tensor if it has not been expanded. Now, instead of doing that literally, one can do the following. Evaluate explicitly the whole divergence of the stress tensor, then evaluate explicitly also the term which will be in implicit form in the equation, and subtract it. On the converged solution, the explicit and implicit evaluation should coincide, and therefore the results in the converged solution should satisfy the original equation. The following equations will clarify the meaning of this trick: (here $u \equiv u_\xi$)

$$\frac{\tilde{u} - u^n}{(\Delta t/2)} = a\ \frac{\delta^2\tilde{u}}{\partial \xi^2} + [(\nabla\cdot\underline{\underline{\sigma}})_\xi^n - a\ \frac{\delta^2 u^n}{\partial \xi^2}] - (\nabla p)_\xi^n\ ;$$

$$\frac{u^{n+1} - \tilde{u}}{(\Delta t\ /2)} = b\ \frac{\partial u^{n+1}}{\partial \eta^2} + [(\nabla\cdot\underline{\underline{\sigma}})_\xi^n - b\ \frac{\partial^2 u^n}{\partial \eta^2}] + a\left(\frac{\partial^2\tilde{u}}{\partial \xi^2} - \frac{\partial^2 u^n}{\partial \xi^2}\right) - (\nabla p)_\xi$$

The above description is, of course, very schematic. The detailed implementation involves some additional questions concerning the fact that we use the ADI scheme by Beam and Warming (1980), which is somewhat more complicated than the basic ADI which I presented above. Also, the fact that the divergence of the velocity is not zero before the converged solution is obtained leads to some changes in the above subtraction procedure. The details of the implementation will be reported elsewhere.

To ensure the satisfaction of the incompressibility condition -- that is, $\mathrm{div}\,\underline{u} = 0$ -- we use the well known Chorin (1967) "artificial compressiblity" method, according to which a new pressure at each iteration is computed on the basis of the old pressure and the value of the divergence of the velocity, according to the following equation:

$$p^{n+1} = p^{n} - c \,\mathrm{div}\, \underline{u}^{n}.$$

It should be noted that we are using compact differencing for computation of the divergence of the velocity (see Aubert and DeVille 1983). This allows us to compute both pressure and velocity on the same grid without the usual difficulties which are commonly remedied by a staggered grid approach.

The director is a new variable in this problem, not encountered in the usual fluid dynamics calculations. A straightforward attempt to solve the director equation (4) would be difficult because of its nonlinearity. Again, a trick is needed, and this time it is based on the fact that the sole purpose of the nonlinear term in (4) is to keep the length of the director equal to 1.

Therefore, the following procedure can be implemented. First, one computes the orientation of the director without paying attention to its length. More precisely, one introduces another vector, $\underline{m}$, such that it is always directed along the director, $\underline{n}$, but its length can be variable. Then, for the vector $\underline{m}$, one can write an equation which is analogous to equation (4), but without the terms responsible for keeping its length equal to one. This equation can now be solved by an explicit iteration as shown below, where the velocity field is taken from the last available iteration:

$$\frac{\underline{m}^{k+1} - \underline{m}^{k}}{\Delta t} = -\underline{u} \cdot \underline{\nabla} \underline{m}^k + (\underline{\underline{W}} + \lambda \underline{\underline{A}}) \cdot \underline{m}^k$$

Several iterations on the director orientation can be made for each iteration on the velocity field. Because the director iteration is explicit, it will require a rather small time step. When the vector $\underline{m}$ has been computed throughout the flow field, one can obtain the vector $\underline{n}$ very simply by normalizing the length of vector $\underline{m}$ to one.

The entire algorithm has been coded in Fortran and run on the VAX 11/780. Convergence required about 200 iterations with time steps of 0.1 for the momentum equation, and of .001 for the director equation. The Chorin factor c in the pressure equation was 0.3. The total time required for the computation of a single case was on the order of 20 min. of CPU time on the VAX 11/780 for a 31×31 grid. We have also made runs on a 61×61 grid in order to check the internal consistency of the numerical scheme. The consistency was quite good; in fact, there was very little difference between the solutions. This gave us some confidence in the overall scheme and also allowed us to use a 31×31 grid for the majority of the computations.

4. Numerical Results

To check our numerical procedure, first of all we decided to compute the plane Poiseuille flow for which the analytical solution has been obtained by Leslie (1964). Poiseuille flow, and the rest of our solutions, were obtained using the dimensionless form of the governing equations so that α_4 was used as a scale for all viscosities. As mentioned above, the creeping flow approximation was used. The boundary conditions were the Poiseuille flow solution for the inlet velocity on the boundary, and symmetry conditions for velocity on centerline. The initial guess was a strongly perturbed Poiseuille flow solution. The computed solution was extremely close to the analytical one for the cases of $\lambda > 1$. This is in agreement with an analytical solution which predicts that stable orientation of the director exists only for the values of $\lambda > 1$. Therefore, we are fairly confident that the director orientation and its stability are computed accurately by the numerical procedure. The velocity field and the pressure field were also in good agreement with analytical solution for the Poiseuille flow for those values of "non-Newtonian" viscosities, β_1 and β_2, for which a solution could be obtained. The highest such values were $\beta_1 = 0.5$, $\beta_2 = -0.5$. Note that the signs for the viscosities, β_1 and β_2, were chosen on the basis of two pieces of evidence. One is experimental values of the viscosities for low molecular weight nematics, as given, for example, in the book by deGennes (1974). Another is the evaluation of the Leslie coefficients on the basis of the Doi (1981) theory as done by Marrucci (1982). In both cases, β_1 turns out to be positive, and β_2 negative. For the absolute values of β_1 and β_2 higher than 0.5 the numerical solution would not converge. Also, for values of $\lambda < 1$ no steady orientation of the director was obtained. The director continued rotating, which is what is predicted by the analytical solution also.

Having checked our numerical procedure on the Poiseuille flow, we began an investigation of flows in more complicated geometries. Particularly, we began studying the flow in a smooth contraction. These types of flows may be relevant to many processing operations. Figure 1 presents the computational grid for the domain of solution constructed by the technique of orthogonal mapping, and Figure 2 presents the velocity field, pressure field, and director field for the same domain. It can be seen that the director orientation is very much influenced by the character of the flow. In the contraction part of the flow field, the director assumes an orientation which is almost parallel to the centerline or to the streamlines, while after the contraction part the director quickly restores its orientation to the one characteristic of the Poiseuille flow.

It should be noted that no solutions could be obtained for values of $\lambda < 2$. This is in contrast to the Poiseuille flow where $\lambda > 1$ are known to produce stable solutions. One can understand the influence of λ on the stability of the orientation of the director in the following way. Looking at the equation for director evolution (4), one can see that λ, in essence, describes the relative influence of pure strain and pure vorticity on the motion of the director. In other words, if λ is very small, say 0, the rate of strain does not influence the motion of the director at all, and the director rotates with the local vorticity. On the other hand, if λ is very large, say infinite, the motion of the director is completely determined by the rate of strain, and vorticity is not important, and in this case the director will try to orient itself along the principal axis of the rate of stain. At intermediate values of λ there is a competition between rate of strain and vorticity, and so for each given λ there is some vorticity which is able to overcome the influence of the rate of strain and make the director rotate. Obviously, in order to keep the director in a given orientation and not rotating, λ should be large enough, and the larger is the ratio of vorticity to the rate of strain, the larger should be λ. Apparently, in the contration flow which we are computing here, the relative magnitude of vorticity in comparison to the rate of strain is larger than in the pure Poiseuille flow, and therefore a larger value of λ is needed in order to keep the director in some stable orientation, as opposed to rotation with vorticity.

The next Figure 3 shows the results for the same viscosities, but for a value of λ 10 times higher. It can be seen that the character of the flow is very similar.

The strong reorientation of the director in the area of contraction, so that the director lies almost along the stream lines, is not surprising since we know from the Leslie (1964) solution for the Hamel flow that this should be expected in the converging flow. The velocity field can be seen to be quite similar to the one which would be expected for a Newtonian fluid; and, in fact, the velocity field for the Newtonian fluid (Fig. 4) is very close to the one for the liquid crystal polymer, while the pressure is substantially different. The similarity of the velocity fields again can be expected since the conservation of mass and no-slip boundary conditions provide two very strong constraints on the velocity field in this very simple flow domain. For this flow, again, the computations for the "non-Newtonian" viscosities with absolute values larger than 0.5 did not converge.

Fig. 5 shows the flow in what we call a "dimpled" channel. It can be seen that there is a very interesting restructuring of the director orientation in the area of the dimple. In the initial converging flow, similarly to the contraction flow, the director takes on an orientation which is essentially parallel to the stream lines near the centerline. However, immediately after the dimple, or, let us say, immediately after the minimum cross section, the director reorients itself very abruptly and becomes essentially normal to the streamlines along the centerline of the channel. This is again in general agreement with analytical solution for the flow in the diverging channel, which can be easily obtained along the lines of the Leslie (1964) investigation. It can be easily shown for the analytical case of the divergent channel that the orientation of the director normal to the streamline near the centerline is a stable one. What is, perhaps, a bit surpising is the abruptness of the change of the director orientation along the centerline from the one along the streamline to the one normal to the streamline. This result suggested to us that it might be interesting to investigate what happens when the dimple is very small, which would correspond in some sense to the possible irregularity of the shape of the processing equipment. Next Fig. 6 shows the flow field and the director field computed for the channel with a very slight dimple. It can be seen that even for this extremely small dimple the director orientation along the streamline changes substantially, though, of course, not so strongly as for the case of the large dimple. In practical terms, this means that one should watch very carefully and avoid any irregularities in the shape of the processing equipment because even small such irregularities can cause the director orientation to change in an unfavorable direction. Note that the orientation of the director normal to the streamlines is unlikely to be useful for the production of highly oriented solid materials. This last result also provides an interesting indication of the rather high accuracy of our numerical procedure since the director orientation turns out to be so sensitive to the small changes in the shape of the channel. At the same time, as mentioned above, for the straight channel (Poiseuille flow) the director orientation was everywhere extremely close to the analytical solution. This accuracy is probably due to the fact that we use boundary-fitted orthogonal coordinates for construction of the computational grid

Acknowledgements

I am grateful to J.L. Ericksen and D. Kinderlehrer for the invitation to participate in this workshop. This research is being sponsored by a grant from the National Science Foundation.

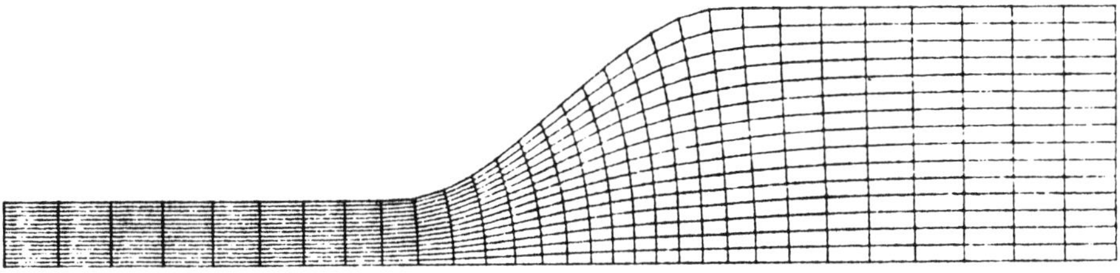

Figure 1 Coordinate system constructed by orthogonal mapping.

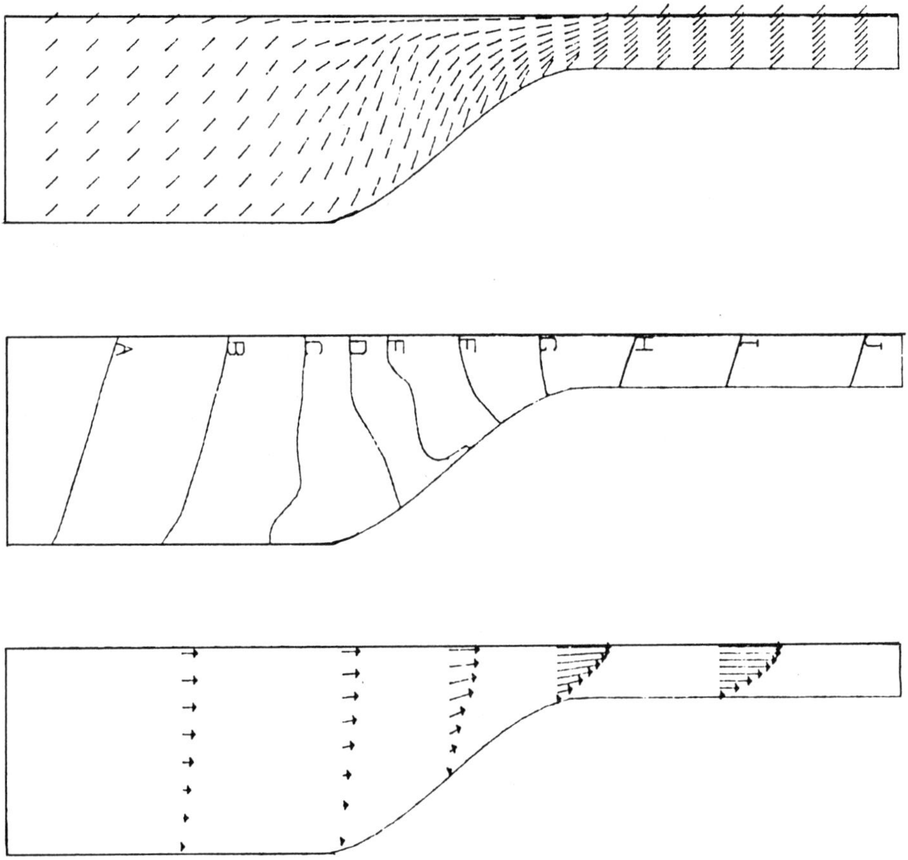

PRESSURE VALUES β_1=0.5, β_2=-0.5, λ= 3.0

A	-1.86
B	-2.75
C	-4.80
D	-5.58
E	-6.58
F	-9.90
G	-24.62
H	-75.62
I	-144.50
J	-223.92

Figure 2 Velocity, pressure, and director fields in a contraction.

$\beta_1 = 0.5,\ \beta_2 = -0.5,\ \lambda = 3.$

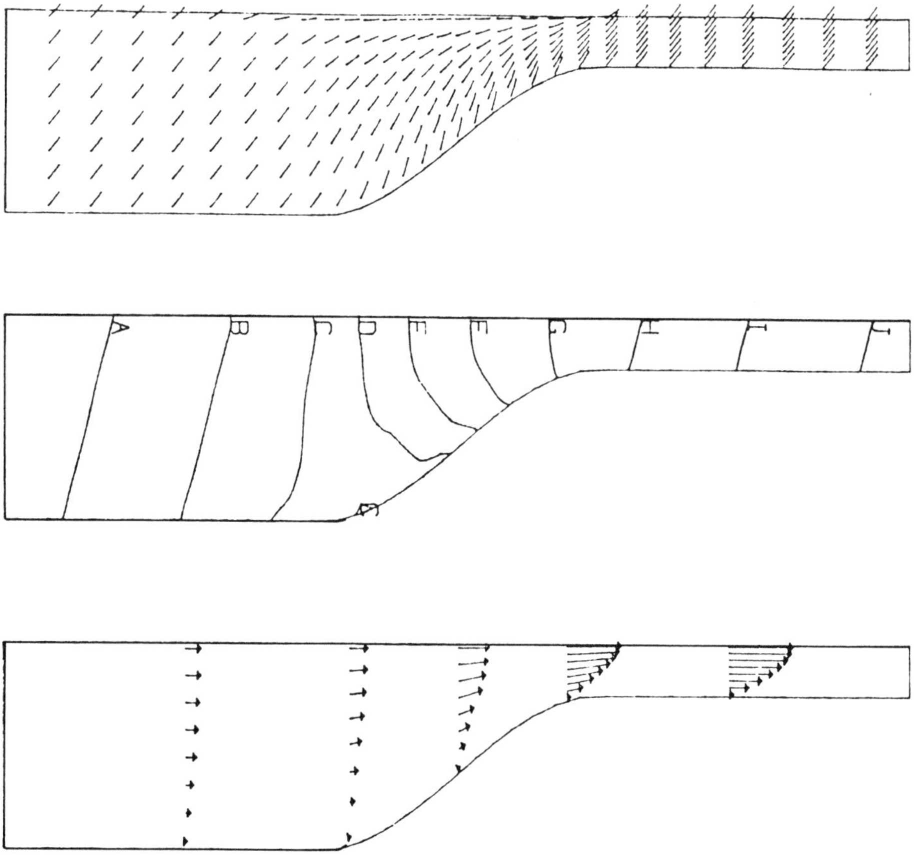

PRESSURE VALUES $\beta_1=0.5$, $\beta_2=-0.5$, $\lambda= 30.0$

A	-1.53
B	-2.41
C	-3.12
D	-3.68
E	-4.57
F	-7.04
G	-18.70
H	-59.20
I	-110.60
J	-169.70

Figure 3 Velocity, pressure, and director fields in a contraction.

$\beta_1 = 0.5$, $\beta_2 = -0.5$, $\lambda = 30$.

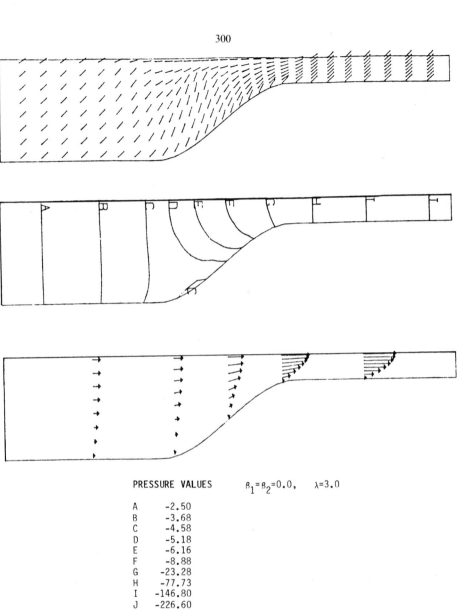

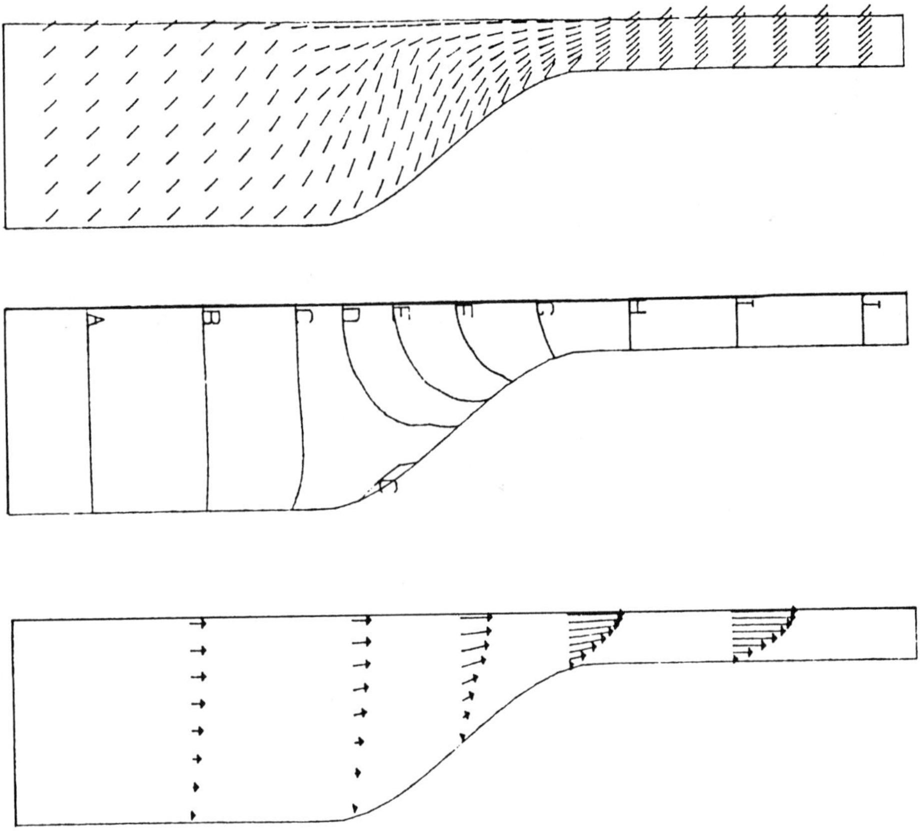

PRESSURE VALUES $\beta_1=\beta_2=0.0$, $\lambda=3.0$

A	-2.50
B	-3.68
C	-4.58
D	-5.18
E	-6.16
F	-8.88
G	-23.28
H	-77.73
I	-146.80
J	-226.60

Figure 4 Velocity, pressure, and director fields in a contraction.

$\beta_1 = \beta_2 = 0.$

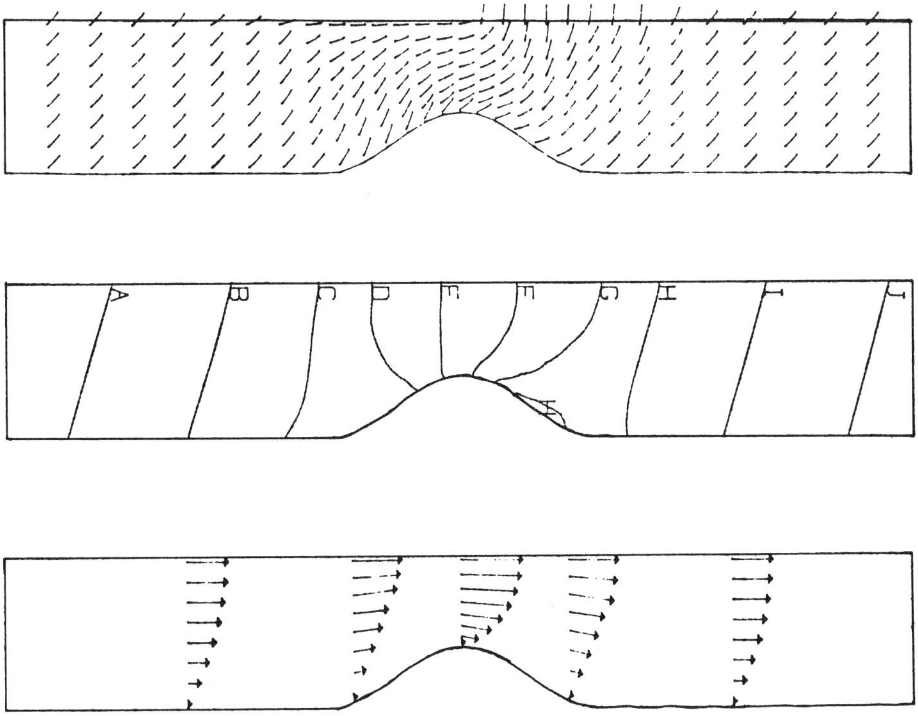

PRESSURE VALUES $\beta_1=0.5$, $\beta_2=-0.5$, $\lambda= 5.0$

A	-4.20
B	-10.46
C	-15.57
D	-20.22
E	-29.70
F	-44.63
G	-52.60
H	-55.54
I	-60.78
J	-67.33

Figure 5 Velocity, pressure, and director fields in a "dimpled" channel

$\beta_1 = 0.5$, $\beta_2 = -0.5$, $\lambda = 5.$

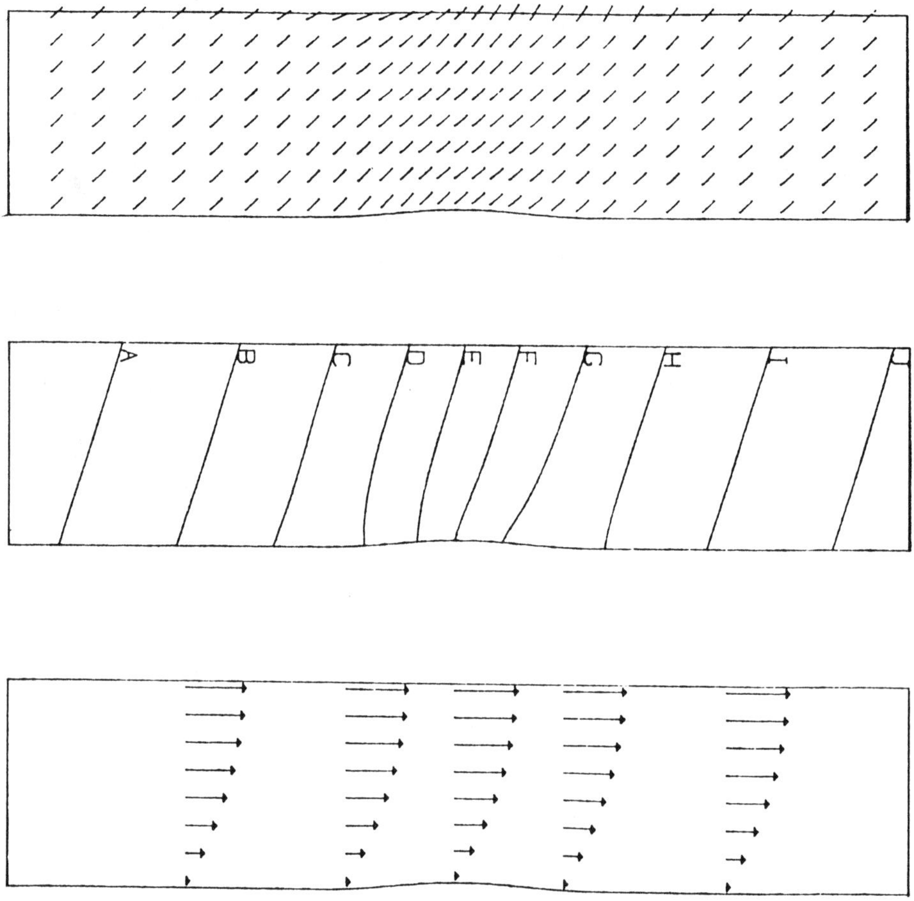

PRESSURE VALUES $\beta_1=0.5,\ \beta_2=-0.5,\ \lambda= 3.0$

A	-2.29
B	-5.71
C	-8.53
D	-10.69
E	-12.46
F	-14.16
G	-16.10
H	-18.45
I	-21.46
J	-25.04

Figure 6 Velocity, pressure, and director fields in a channel with a slight dimple

$\beta_1 = 0.5,\ \beta_2 = -0.5,\ \lambda = 3.$

References

Aubert, X. and M. DeVille (1983). Steady Viscous Flows by Compact Differences in Boundary-Fitted Coordinates. J. Comput. Phys. 49, 490.

Beam, R.M. and R.F. Warming (1980). Alternating Directions Implicit Methods for Parabolic Equations with a Mixed Derivative. SIAM J. Sci. Stat. Compu. 1, 131.

Chorin, A.J. (1967). A Numerical Method For Solving Incompressible Viscous Flow Problems. J. Comput. Phys. 2, 12.

DeGennes, P.G. (1974). The Physics of Liquid Crystals. Oxford Univ. Press.

Doi, M. (1981). Molecular Dynamics and Rheological Properties of Concentrated Solutions of Rodlike Polymers in Isotropic and Liquid Crystalline Phases. J. Polym. Sci.: Polym Phys. 19, 229.

Ericksen, J.L. (1960a). Transversely Isotropic Fluids. Koll. Z. 173, 117.

Ericksen, J.L. (1960b). Anisotropic Fluids. Arch. Rat. Mech. Anal. 4, 231.

Leslie, F.M. (1964). Hammel Flow of Certain Anisotropic Fluids. J. Fluid Mech. 18, 595.

Leslie. F.M. (1979). Theory of Flow Phenomena in Liquid Crystals. Adv. Liq. Cryst. 4, 1.

Marrucci, G. (1982). Prediction of Leslie Coefficients for Rodlike Polymer Nematics. Mol. Cryst. Liq. Cryst. 72, (Lett), 153.

Ryskin, G. and Leal, L.G. (1983). Orthogonal Mapping J. Comput. Phys. 50, 71.

Ryskin, G. and Leal, L.G. (1984). Solution of Free-Boundary Problems in Fluid Mechanics. Part 1. The Finite-Difference Technique. J. Fluid. Mech. 148, 1.

Wissbrun, K.F. 1984. "Orientation Development in Liquid Crystal Polymers," Proc. IMA Workshop "Orienting Polymers", Lect. Notes Math. 1063, Springer-Verlag, p. 1.

THEORY OF THE BLUE PHASES OF CHIRAL NEMATIC LIQUID CRYSTALS

James P. Sethna

Laboratory of Atomic and Solid State Physics
Cornell University
Ithaca, NY 14853

In the condensed matter physics community there has been much recent interest in materials with competing interactions. Because these materials are frustrated (the different terms in the free energy demand incompatible configurations of molecules), these materials form complicated, intricate crystalline phases as compromise structures; also, they often form amorphous, glassy states. Chiral nematic liquid crystals are frustrated, and form up to three blue phases in a small temperature range between the isotropic "melted" phase and the helical phase [1] two exotic crystalline phases and an amorphous phase. The lattice constant in these phases is the wavelength of blue light (hence their brilliant blue colors); because it is large compared to the molecular lengths, the material can be described within a continuum elastic theory. The elastic theory is frustrated, and the ground states contain networks of defect lines to relieve the frustration [2]. The frustration is geometrically described as a curvature in the natural form of parallel transport [3]; indeed, chiral nematics are not frustrated in the unphysical space formed by the surface of a sphere of appropriate radius in four dimensions [4]. (The defect lines are the cuts needed to flatten this sphere.) A total divergence term in the free energy plays a crucial role in keeping the energy of the defect lines finite in the continuum limit, and in stabilizing the blue phases.

1. Frustration

Systems with competing interactions often exhibit rich behaviours, and have become an increasingly important subject in condensed matter physics. They can form glasses, which have an enormous number of metastable configurations that are amorphous (have no long-range order in space), but appear to be frozen (have long-range order in time) [5,6]. They can have quasiperiodic order; highly anisotropic (effectively one dimensional) materials can have incommensurate charge density waves, [7] and metallic alloy systems with three dimensional quasiperiodic icosahedral order have recently been discovered, [8-10]. Finally, they can form complex, exotic crystalline phases. Chiral nematics (also known as cholesteric liquid crystals) form three blue phases. Two of these are crystalline networks of defect lines with a lattice constant (~1000Å) much larger than the molecular size; the last (known as the blue fog) is not understood, but could be glasslike or quasiperiodic.

Crystals have long-range order in space. Given the positions and orientations of a molecule and its neighbors in a cyrstal, the configuration of far distant molecules can be determined within small fluctuations. Liquids lack such order. Although there is some short-range order (e.g. molecules will touch, but won't overlap), no information about distant molecules can be inferred from the local configuration. Liquid crystals are an intermediate form of matter, typically with long-range orientational order, but with incomplete or nonexistent translational order. Nematic liquid crystals, for example, are made of long thin molecules which tend to align with their long axes parallel (but with no preference as to which end is "up"). In the ordinary nematic phase, the molecules in the entire sample choose a common direction in which to point, even though their positions remain disordered and the material remains fluid.

The blue phases† occur in chiral nematics. Chiral molecules lack inversion symmetry; they have a handedness and are distinguishable from their mirror image. The threads on screws are chiral; commercial screws are right-handed. If one

---

† The blue phases have been studied at length. References to the experimental work can be found in the companion article by Cladis [11] and in the forthcoming review article by Wright and Mermin [12]. The director description of the blue phases presented here is based on the traditional description of the cholesteric phases; [2,3,13-15] the Landau quadrapolar description [16] is nicely developed by Wright and Mermin as well [12]. There are closely related theories of metallic alloys [3,15,17-21], which form exotic Frank-Kasper phases [22,23] and (when rapidly cooled from the melt) form metallic glasses. Both the theory of metallic glasses and that of the blue phases are frustrated continuum elastic theories; both have ground states with networks of disclination lines; both are stabilized by the addition of a total divergence term to the free energy which acts to lower the core energy of the defect lines. The blue phase theory presented here is mathematically somewhat simpler, and experimentally is perhaps on a sounder footing.

matches the grooves and threads on neighboring screws, they will align at an angle twice that of the threading angle. The molecules in chiral nematics also preferentially align at a slight angle (figure 1). The alignment angle is small; the pitch (over which the molecules would rotate 360°) is typically hundreds or thousands of Ångstroms.

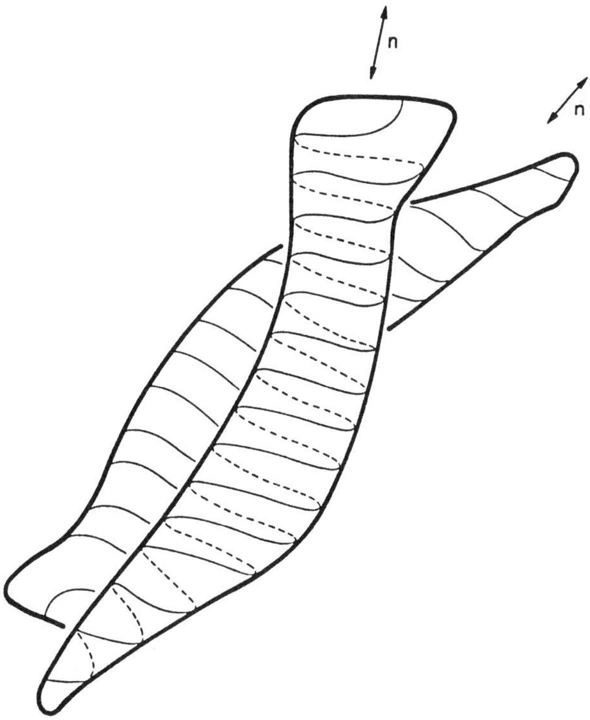

Figure 1

Neighboring chiral nematic molecules sit at a slight angle. Because the molecules are chiral, the pair interaction energy between two molecules is minimized when they are slightly misaligned. Here the forward molecule is twisted slightly counterclockwise (a right-handed twist).

This is not to say that the orientations of the molecules in the nematic or blue phases are rigidly fixed. Typical molecules will fluctuate large distances about their equilibrium orientations; we can define a mean orientation at a point $\overleftrightarrow{n}$ only by averaging over several molecular volumes. (Since there is no long range order in which end of the molecule points along the preferred axis, we can ignore the sign of $\vec{n}$.) We must develop an elastic theory for fields of headless unit vectors (elements of RP^2, also known as "directors").

We can now express the preference for the alignment of chiral nematic molecules to twist in terms of the director field $\overleftrightarrow{n}$. As one moves radially outward from a central molecule, the molecular orientations (in the local low energy configuration) will gradually twist to the right (figure 2).

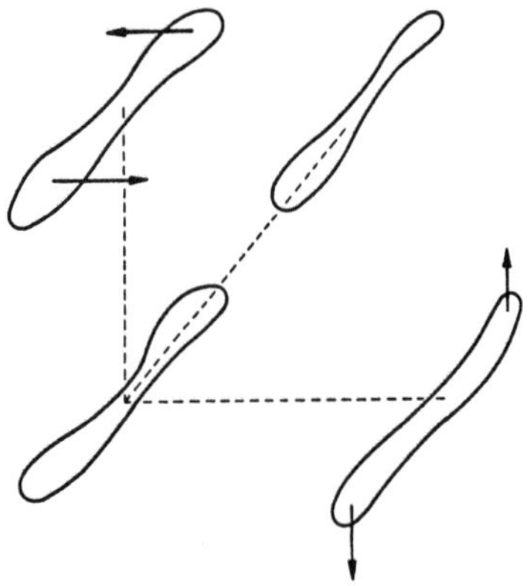

Figure 2

Parallel transport in chiral nematics. As one moves radially outward, the preferred orientation spirals to the right; as one moves along the molecular axis the orientation is unchanged.

As one moves along the long axis of the molecule, the orientations will not change. The optimal change in orientation of the molecule is perpendicular both to the direction of motion and to the molecule:

$$\partial_i n_j = -q(\hat{x}_i \times \mathbf{n})_i = -q\varepsilon_{ijk} n_k. \tag{1}$$

Here $2\pi/q$ is the pitch and ε_{ijk} is the totally antisymmetric three tensor.

At low temperatures, chiral nematics form a helical (cholesteric) phase (figure 3), of the form

$$\mathbf{n}(x) = \hat{x}\cos(qz) + \hat{y}\sin(qz). \tag{2}$$

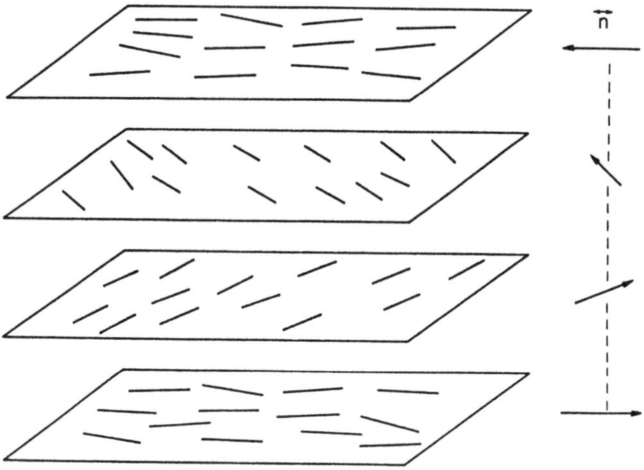

Figure 3

Low temperature cholesteric helical phase twists in one direction. (The planes are to aid visualization; there is no density variation in the vertical direction.)

This phase twists only in the $\hat{z}$ direction; it does not twist in the other diretion perpendicular to $\overleftrightarrow{n}$, and locally is not the lowest energy state. In particular, it does not satisfy (1).

The minimum energy double-twist configuration can be achieved along the center of a tube (figure 4):

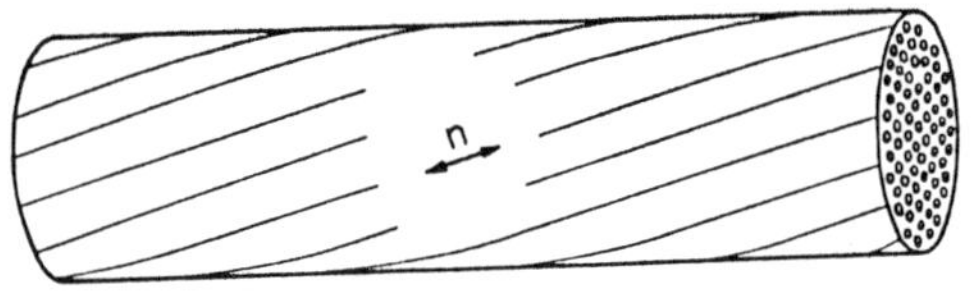

Figure 4

Double twist tube is local low energy structure. Along the center line of the cylinder the chiral nematic free energy is minimized; moving radially outward the energy density increases.

$$\mathbf{n}(\mathbf{x}) = \hat{z}\cos(qr) - \hat{\theta}\sin(qr). \tag{3}$$

The energy density at the center of this tube will be substantially lower than that of the helical phase. The energy density grows with radius; strains build up as the double-twist relation (1) is violated more and more. By the time $\mathbf{n} = -\hat{\theta}$ (90° tubes) the molecules only twist in the $\hat{\mathbf{r}}$ direction, while they bend in the $\hat{\theta}$ diretion: the energy density has risen over that of the helical phase.

Indeed, it is not possible to satisfy the double-twist condition (1) in any finite volume. This can be seen by considering parallel transport around an infinitesimal closed loop. (Figure 5 illustrates transport around a larger loop.) Equation (1) can be used to define a covariant derivative

$$(D_i n)_j = \partial_i n_j + q\varepsilon_{ijk} n_k \tag{4}$$

which is zero for the twisted parallel transport characteristic of the blue phase. If one parallel transports $\vec{n}$ an infinitesimal distance $\vec{a}$, then a distance $\vec{b}$, then back along $-\vec{a}$ and $-\vec{b}$, it will be rotated to a new vector n', given to lower order in $\vec{a}$ and $\vec{b}$ by

$$n_i' = n_i - R_{ijk\ell} a_k b_\ell n_j = n_i - q^2(\delta_{jk}\delta_{i\ell} - \delta_{\ell j}\delta_{ik}) a_k b_\ell n_j. \tag{5}$$

(In particular, an ideal low energy state would have

$$\partial_\ell \partial_k n_i - \partial_k \partial_\ell n_i = q^2(\delta_{i\ell} n_k - \delta_{ik} n_\ell) \neq 0.)$$

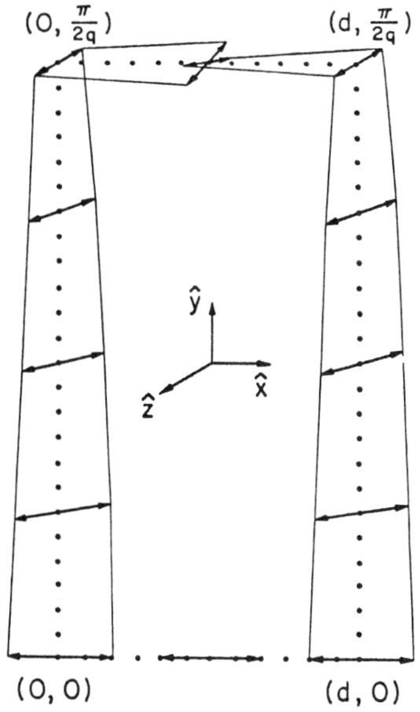

Figure 5

(5) **Frustration in the blue phase.** Parallel transport in chiral nematics is frustrated. The local low energy condition (1) cannot be everywhere satisfied. Consider the closed loop above. Start with a director $\overleftrightarrow{n}$ at the origin pointing in the $\hat{x}$ direction. As we move in the $\hat{y}$ direction, (1) dictates that $\overleftrightarrow{n}$ rotates, until at $(0,\pi/2q)$ it lies in the $\hat{z}$ diretion. Along the $\hat{x}$ axis, the low energy state allows the direction of $\overleftrightarrow{n}$ to remain unchanged; thus if we parallel transport $\overleftrightarrow{n}$ using (1) first to $(d,0)$ and then to $(d,\pi/2q)$, $\overleftrightarrow{n}$ will again point in the $\hat{z}$ direction. Finally, if we transport these two to meet at $(d/2,\pi/2q)$, they will point in different directions, with angular separation $q\,d$.

The tensor $R_{ijk\ell}$ is the analogue of the curvature tensor in differential geometry, and of the field strength tensor $F^{\mu}{}_{\nu}$ in gauge theories. Chiral nematics are frustrated. The local low energy configuration (1) cannot be used to fill space, and the curvature tensor quantifies this frustration. The local low energy state can be realized in the center of the double-twist tubes (3) (figure 4); we shall build the blue phases out of these tubes.

2. Blue Phases

Experimentally, the blue phases occupy a small (~1°K) temperature range between the isotropic (melted) phase and the helical (cholesteric) phase.† The two lower temperature phases are crystalline; blue phase I has a bcc translation group and blue phase II is simple cubic. The lattice constants (roughly the pitch of the helical phase) are wavelengths of blue light, and Bragg reflections cause the brilliant colors. The high temperature phase is called the blue fog. It is largely not understood; it shows no Bragg peaks, and is thought to be amorphous.††

We will build a model for Blue Phase II from double twist tubes (figure 4), whose radius is chosen so that the surface director makes a 45° angle with the axis of the tube. Two 45° tubes will naturally sit at right angles. (If the molecules on the surface of a tube are aligned at an angle θ with respect to the axis, then two tubes will align at an angle of 2θ to keep the molecular orientations continuous at the point of tangency. See figure 1 depicting the similar behavior of two molecules with threads.) Three such tubes can form a corner; this corner can be either right-handed (figure 6A) or left-handed (figure 6B). If the cholesteric pitch is right-handed ($q > 0$, as shown in the figure), then the left-handed corners (B) contain no singularity; nematic fluid can fill the region between the tubes, smoothly pointing out of the corner.

---

† The small temperature range can be explained in terms of the energetics of the defect lines, [2] which we will discuss in section 4. Briefly, at low temperatures, you spend more energy in the "isotropic" core of the defect than you gain from the regions of double-twist.

†† (It is appealing to speculate that the phase might be quasiperiodic; the crystalline blue phase II only has one visible Bragg peak. We are trying to use Landau theory to dignify this speculation. [24])

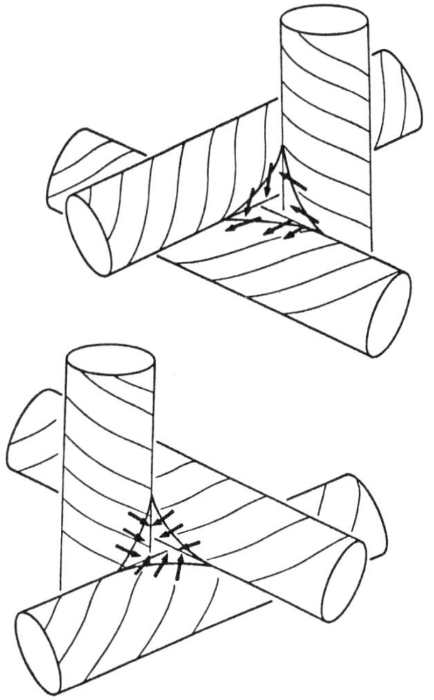

(A)

(B)

Figure 6

(6A) **Right-handed corners contain s = -1/2 defect lines.** Three double-twist tubes forming a right-handed corner. Note that the director rotates 180° as one moves in a closed path around the corner. There is no way to fill the corner smoothly with directors.

(6B) **Left-handed corners have no singularity.** Three double twist tubes in a right-handed chiral nematic forming a left-handed corner. Note that the director rotates 360° as one moves around the corner. The corner can be filled smoothly with arrows pointing into the paper.

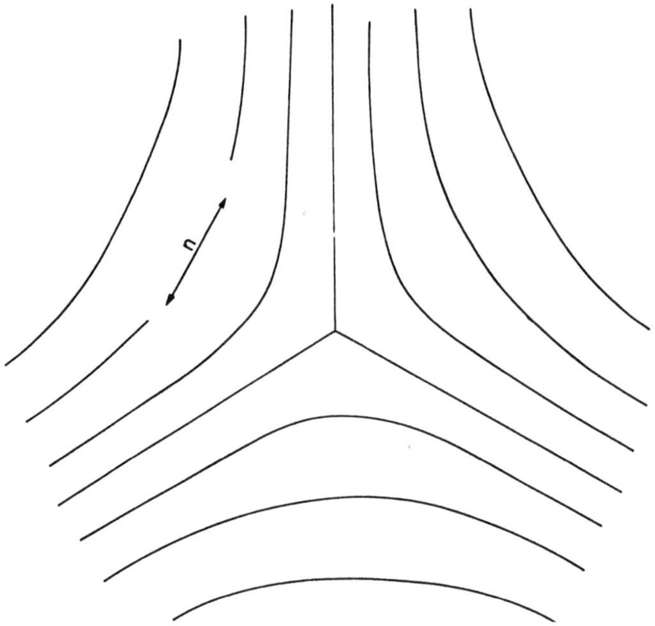

Figure 7

s = -1/2 Disclination line. A cross section of a s = -1/2 disclination line. The local molecular orientation $\overleftrightarrow{n}$ lies tangent to the curves in the figure. As one travels around the defect, the director $\overleftrightarrow{n}$ rotates backward halfway (180°).

Right-handed corners (A) must contain a topological singularity, as the director rotates 180° around a closed path on the surfaces of the cylinders enclosing the corner. The singularity is an $s = -\frac{1}{2}$ disclination line (figure 7). These lines are characteristic of the blue phases. They relax the frustration imposed by the double twist.

We can form a model blue phase[2] based on an array of 45° tubes (figure 8A). This is currently the most successful model of Blue Phase II; the best model for Blue Phase I can be represented as a somewhat more complicated array of tubes. Defect lines ($s = -\frac{1}{2}$ disclination lines) pass through the right-handed corners, and meet in the center of the unit cell (figure 8B). Each corner of the unit cell also starts four defect lines; the defect lines form two interpenetrating diamond lattices.

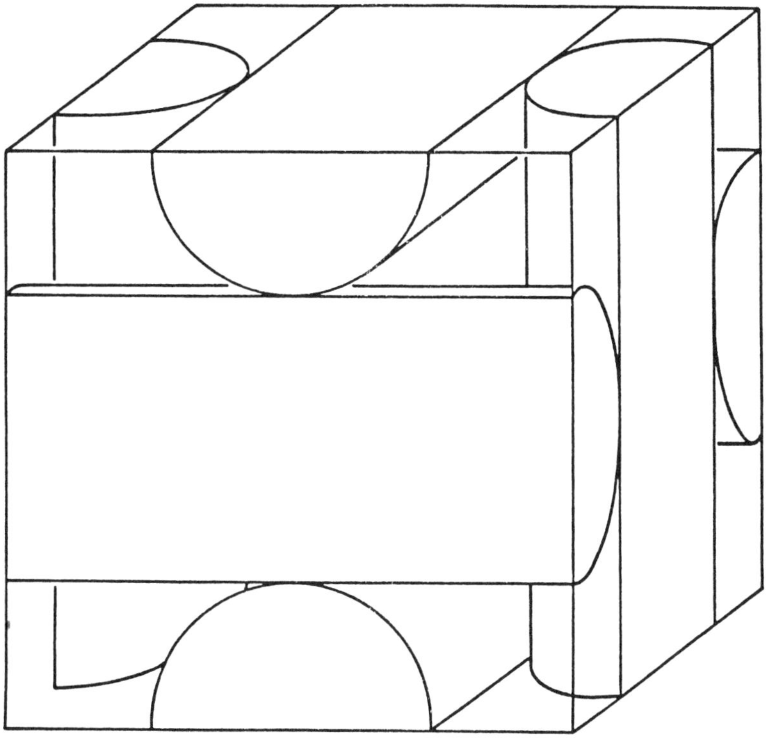

Figure 8A

(8A) **Blue phase ii; tube structure.** Proposed model for Blue Phase II, shown as an array of 45° tubes.

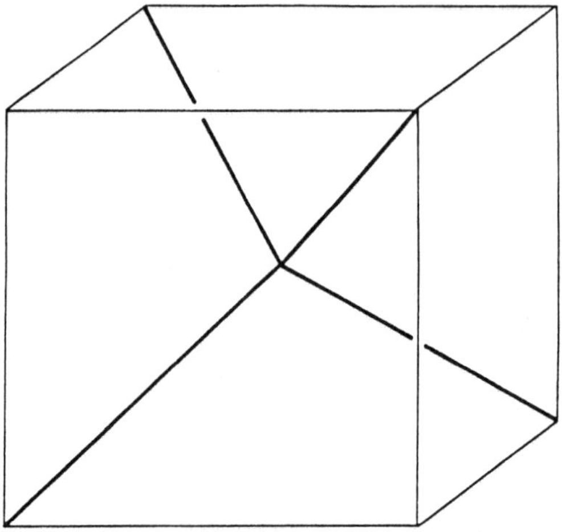

Figure 8B

(8B) **Blue phase II: defect structure.** The same model blue phase, showing the $s = -1/2$ defect lines in the unit cell.

3. Curvatures

In equation (5) we found that chiral nematics are curved in flat space. Indeed, we can contract the curvature tensor (5) to find

$$R_{ijij} = -6q^2 ; \tag{6}$$

that is, the chiral nematics have a negative scalar curvature. This paradox is purely linguistic. The curvature of a manifold is equal to that of the affine connection defining its covariant derivative, so long as the connection is symmetric (has no torsion) [25]. This is normally a sensible assumption, so the word curvature is used for both. In chiral nematics, the connection coefficients

$q\varepsilon_{ijk}$ (equation 4) are not symmetric: the covariant derivative has torsion $T^i_{\ jk} = q\varepsilon_{ijk}$.

Chiral nematics are flat in curved space. They can be made unfrustrated in an unphysical space: the surface S^3 of the sphere in four dimensions of radius q^{-1}. That is, there is an "ideal template" for the blue phase with double-twist everywhere.† Consider the vector field given at a point $\mathbf{x} = (x_0, x_1, x_2, x_3)$ by

$$\mathbf{n}(\mathbf{x}) = q(-x_1, x_0, x_3, -x_2). \tag{7}$$

This is a unit vector field $(n^2 = 1)$ everywhere tangent to the sphere $(n \cdot x = 0)$. Most important, it has double twist everywhere [4]; $n_{i;j} = -q\varepsilon_{ijk}n_k$. Let's check this explicitly at the top of the sphere. At $\mathbf{x}_0 = (q^{-1}, \delta_x, \delta_y, \delta_z)$

$$\mathbf{n}(\mathbf{x}_0) = (-q\delta_x, 1, q\delta_z, -q\delta_y). \tag{8}$$

The first component represents the change in $\vec{n}$ necessary for it to remain tangent to the sphere ($\vec{n}$ points in the $\hat{x}$ direction, and must shift downward if moved forward off the top of the sphere). The last two represent a right-handed double-twist of pitch $2\pi/q$; if I advance in the $\hat{z}$ direction $\vec{n}$ twists in the $+\hat{y}$ direction, if I advance in the $\hat{y}$ direction $\vec{n}$ twists in the $-\hat{z}$ direction.

Notice that the circumference of the sphere equals the pitch of the cholesteric, so for example along the circle $x_1 = x_2 = 0$, $\mathbf{n}(\mathbf{x})$ rotates by 2π. Along the circles $x_2 = x_3 = 0$ and $x_0 = x_1 = 0$, $\mathbf{n}(\mathbf{x})$ always is tangent to the circle; one can think of the template (7) as two 45° double twist tubes centered at these circles and glued together along the surface $x_0^2 + x_1^2 = x_2^2 + x_3^2 = \frac{1}{2}$ †† [26].

---

† This template arises naturally from the identification of the manifold S^3 with the (topologically equivalent) group SU(2). Right-handed blue phase templates are formed by multiplying elements of S^3 on the right by infinitesimal generators in **su**(2).

†† This decomposition is not unique; any left-handed rotation will give another separation of (7) into tubes.

4. Defect Lines

The defect lines now have a geometrical interpretation: they are the disclination lines which form the edges of the cuts needed to flatten the sphere S^3. Consider the problem in one lower dimension - making a carpet out of orange peel. To flatten a piece of peel, one can ease the strains by cutting into the center of the piece. Naturally one must fill the empty wedge with more peel (figure 9).

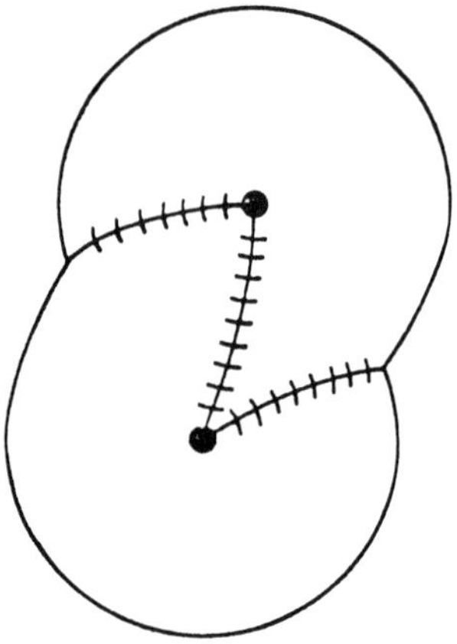

Figure 9

Orange peel carpet

This can be done so as to leave no seam; however, the vertex of the wedge remains a point of high strain energy, and the resulting carpet will be a lattice of interacting point disclinations. In the same way, the line disclinations in the blue phases are the edges of the cuts in the ideal chiral nematic template on S^3. In our problems of course the size of the wedge of material we may add is restricted. For chiral nematics we must add a full half-plane of fluid in order to keep the molecular orientations smooth at the boundaries (the "seams"); the defect formed is the $s = -\frac{1}{2}$ defect line (figure 7).

One problem with discussing defect lines within continuum field theories is the eventual necessity of studying the core of the defect. Since the strains diverge at the defect, linear elasticity breaks down and a more complete theory is necessary. Since the energy diverges only logarithmically as the core radius (cutoff length) goes to zero, this study is not very rewarding. In this section we will develop a method for avoiding the detailed study of the defect cores. In the continuum limit, only the core energy per unit length will matter - all other properties of the core will be irrelevant.†

We can form a sensible energy density for the blue phases (the "one constant approximation") by summing the squares of the components of the covariant derivative (5):††

$$F = \frac{K}{2}(D_i\mathbf{n})_j(D_i\mathbf{n})_j \tag{9}$$

$$= \frac{K}{2}[(\text{div}\,\mathbf{n})^2 + (\mathbf{n}\times\text{curl}\,\mathbf{n})^2 + (\mathbf{n}\cdot\text{curl}\,\mathbf{n} + q)^2] \tag{10}$$

$$+ \frac{K}{2}\,\text{div}\,(\mathbf{n}\cdot\nabla\mathbf{n} - \mathbf{n}\nabla\cdot\mathbf{n}) + \frac{K}{2}q^2.$$

---

† The fact that we can avoid studying the $s = \frac{1}{2}$ defect cores does not mean that they are not a significant feature of the blue phases. First, the ground states have a finite density of defect lines. The magnitude of $\vec{n}$ is fixed in our theory; near the defect line we must either allow the magnitude to vary to zero (as in Landau theory) [12,16] or put the theory on a lattice.[14] Secondly, the core size must be on the order of the pitch for the blue phases to be stable. In avoiding the study of the cores, we ignore the effects of this second length scale; only those properties that are cutoff independent ("universal") will be correctly described with our theory. (Which properties will be universal is not clear yet; nor is it clear what significance the work presented in this section will have.)

†† In the most general form, each of the four gradient terms in (10) can have an independent elastic constant.

Since the covariant derivative D has nonzero curvature, F cannot be made zero in any finite volume. The helical phase just lives with this frustration; the blue phases locally avoid it, but at the cost of the introduction of defect lines.

The total divergence term in (10) cannot be dropped, even though it is a surface energy. In the blue phases there is a finite density of disclination lines. To integrate by parts, one must exclude the core of each defect line. The boundaries of these cores give the blue phases a finite "internal surface area" per unit volume. Indeed, to take the continuum limit (where the core size $\Lambda \to 0$) we shall add a multiple of the total divergence term to the free energy:

$$F_D(\Lambda) = \frac{\overline{K}(\Lambda)}{2}\,\mathrm{div}(n\cdot\nabla n - n\nabla\cdot n). \tag{11}$$

The first term in brackets in equation 10 is a sum of squares, and is zero only for the helical phase (equation 2, figure 3). The last term is an unimportant constant. The second term, together with F_D, can stabilize the blue phase; as a surface term it will add an effective negative energy per unit length to the defect lines. We wish to define $\overline{K}(\Lambda)$ so as to keep the energy of an $s = -1/2$ defect line fixed as $\Lambda \to 0$.

Consider a defect line of length L in the blue phase. Let C be a cylinder of radius Λ about the defect (figure 10).
For simplicity, [15] we consider a straight defect. We assume the free energy outside C is given by the sum $F + F_D$, and we ignore any contributions to the energy from inside C.

Assume the defect lies along the $\hat{z}$ axis. F in equation 10 contributes energy

$$F_\Lambda = \frac{K}{2}\int dz \int_\Lambda dr \int_0^{2\pi} r\,d\theta \left[\left(\frac{\partial n_i}{\partial z}\right)^2 + \left(\frac{\partial n_i}{\partial r}\right)^2 + \frac{1}{r^2}\left(\frac{\partial n_i}{\partial \theta}\right)^2\right] \tag{12}$$

Since we are studying an $s = -1/2$ defect, $n(r,0,z) = -n(r,2\pi,z)$. Clearly

$\frac{\partial n}{\partial z} = \frac{\partial n}{\partial r} = 0$ for minimum energy; in the θ direction $\int_0^{2\pi} (\frac{\partial n_i}{\partial \theta})^2 \, d\theta$ is stationary for $\frac{\partial^2 n_i}{\partial \theta^2} = \frac{1}{4} n_i$. Since at some point $n_z(\theta) = 0$, we may pick coordinates to have $\vec{n}(0) = (1,0,0)$. Then the stationary solutions for F_Λ are

$$\vec{n}(\theta) = (\cos \theta/2,\ A \sin \theta/2,\ \sqrt{1 - A^2} \sin \theta/2), \tag{13}$$

so n moves along great circles on the sphere $n^2 = 1$. The energy from F_Λ is

$$F_\Lambda = \frac{\pi KL}{4} \ln(\Lambda^{-1}). \tag{14}$$

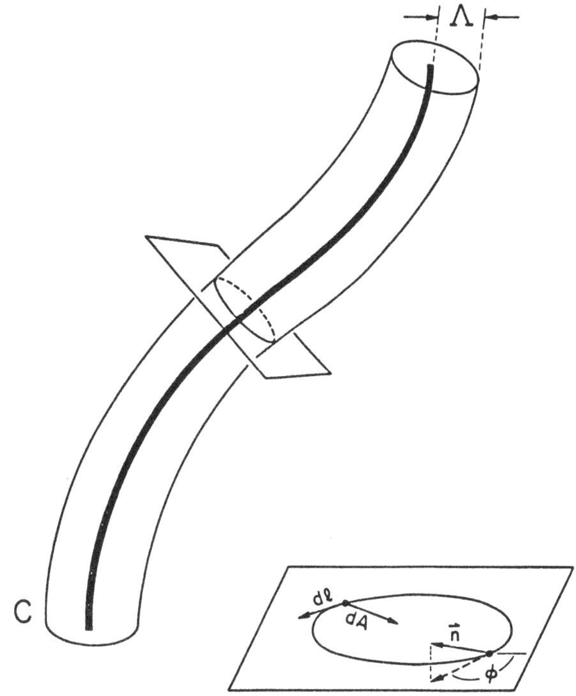

Figure 10

Core of defect line. The cylinder C is of radius Λ about the defect line. The differential surface area $d\vec{A}$ points perpendicular to the cylinder, into it. Consider a plane perpendicular to the defect. The differential arc length $d\vec{\ell}$ measures length along the circular intersection of C with the plane; $d\vec{\ell} \cdot d\vec{A} = 0$. The blue phase director $\vec{n}$, when projected onto the plane, makes an angle ϕ with respect to some fixed axis.

The rest of the free energy (F_D in equation 11) is a surface term

$$S_\Lambda = \frac{\overline{K}(\Lambda)}{2} \int_C (n\cdot\nabla n - n\nabla\cdot n)\cdot dA. \tag{15}$$

Project $\vec{n}$ onto the plane perpendicular to the defect, and let ϕ be the angle between the projection and some fixed axis in the plane (figure 10): $\phi = \arctan(n_2/n_1)$. S_Λ can be reexpressed in terms of $\vec{\nabla}\phi$ and the arc length $d\ell$ along the cylinder C:

$$\vec{\nabla}\phi = \frac{1}{n_1^2 + n_2^2} \begin{bmatrix} -n_2\partial_1 n_1 + n_1\partial_1 n_2 \\ -n_2\partial_2 n_1 + n_1\partial_2 n_2 \end{bmatrix} \tag{16}$$

and

$$(d\ell_x, d\ell_y) = (dA_y, -dA_x), \tag{17}$$

so

$$(n_1^2 + n_2^2)\vec{\nabla}\phi \cdot d\ell = (n \cdot \nabla n - n\nabla\cdot n)\cdot dA \tag{18}$$

and

$$S_\Lambda = \frac{\overline{K}(\Lambda)}{2} \int dz \int (1 - n_3^2)\ \vec{\nabla}\phi\cdot d\ell. \tag{19}$$

Notice two important features of S_Λ. First, if $n_3 = 0$, it measures the winding number of the defect:

$$s = \frac{1}{2\pi} \int \vec{\nabla}\phi \cdot d\ell. \tag{20}$$

Secondly, if ϕ is monotone decreasing, it is minimized by $n_3 \equiv 0$, so minimizing S_Λ gives

$$S_\Lambda = \pi\overline{K}(\Lambda)Ls. \tag{21}$$

In the continuum limit, $\overline{K}(\Lambda) \to \infty$, so the boundary condition becomes rigidly enforced. Thus the total divergence term S_Λ in the continuum limit forces $\vec{n}$ at the core to lie perpendicular to the defect line. With these boundary conditions, it measures the product of the length and topological strength of the defect lines.

F_Λ and S_Λ can be simultaneously minimized by $\mathbf{n} = (\cos\theta/2, -\sin\theta/2, 0)$ with energy per unit length

$$F_\Lambda + S_\Lambda = \pi L[K\,\ln(\Lambda^{-1})/4 + \overline{K}(\Lambda)/2]. \tag{22}$$

To keep the energy per unit length of the defect lines fixed, we set

$$\overline{K}(\Lambda) = K_0 + \frac{K\,\ln(\Lambda^{-1})}{2} \tag{23}$$

The energy per unit length $-\pi/2\,K_0$ of the $s = -1/2$ defect will depend on the shape of the defect line, the core and surface energies, and the position of neighboring defects. Energies of defects of all other winding numbers will diverge as $\Lambda \to 0$; in the continuum limit only $s = -1/2$ defects are allowed.

In summary, the strain energy near defect lines diverges logarithmically as the cutoff length (core size) goes to zero. To compensate, we can add a total divergence to the free energy, which cannot change the bulk energies of defect-free phases, since it can be written as a surface integral. However, in the phases of interest there is a lattice of defect lines, and therefore a finite density of internal surface area. In the continuum limit, the divergence term enforces boundary conditions at the singularity; with these boundary conditions it is a topological measure of the winding number of the defects.

Acknowledgements

This work was supported by NSF Grant #PHY77-27084 and by the Cornell Materials Science Center, NSF Grant #DMR-8217227A-01.

References

1. R. Barbet-Massin, P.E. Cladis, and P. Pieranski, Phys. Rev. A. **30**, 1161 (1984).

2. S. Meiboom, J.P. Sethna, W.F. Brinkman, and P.W. Anderson, Phys. Rev. Lett. **46**, 1216 (1981).

3. J.P. Sethna, Phys. Rev. Lett. **51**, 2198 (1983).

4. J.P. Sethna, D.C. Wright, and N.D. Mermin, Phys. Rev. Lett. **51**, 467 (1983).

5. G. Toulouse, Comm. Phys. **2**, 115 (1977).

6. P.W. Anderson, in Ill Condensed Matter, Proceedings of the 1978 Les Houches Summer School, Session 31, edited by R. Balian, R. Maynard and G. Toulouse (N. Holland, Amsterdam, 1979).

7. P.A. Lee and T.M. Rice, Phys. Rev. B **19**, 3970 (1979).

8. D. Shechtman, I. Blech, D. Gratias, and J.W. Cahn, Phys. Rev. Lett. **53**, 1951 (1984).

9. D. Levine and P.J. Steinhardt, Phys. Rev. Lett. **53**, 2477 (1984).

10. N.D. Mermin and S.M. Troian. Phys. Rev. Lett. 54, 1524 (1985).

11. P Cladis. (This volume.)

12. D.C. Wright and N.D. Mermin. (Review article to be published.)

13. P.G. deGennes, The Physics of Liquid Crystals (Clarendon Press, Oxford, 1975).

14. S. Meiboom, M. Sammon, and W.F. Brinkman, Phys. Rev. A **27**, 438 (1983).

15. J.P. Sethna (To be published in Phys. Rev. B.)

16. H. Grebel, R.M. Hornreich, and S. Shtrikman, Phys. Rev. A **28**, 1114 (1983). (See also references therein.)

17. J.F. Sadoc and R. Mosseri, Phil. Mag. B **45**, 467 (1982).

18. M. Kleman and J.F. Sadoc, J. Phys. (Paris) Lett. **40**, L569 (1979).

19. D.R. Nelson, Phys. Rev. B **28**, 5515 (1983).

20. D.R. Nelson and M. Widom, Nuclear Physics B **240 [FS 12]**, 113 (1984).

21. S. Sachdev and D.R. Nelson, Phys. Rev. Lett. **53**, 1947 (1984).

22. F.C. Frank and J.S. Kasper, Acta Cryst. **11**, 184 (1958).

23. F.C. Frank and J.S. Kasper, Acta Cryst. **12**, 483 (1959).

24. D. Rokhsar and J.P. Sethna. (To be published.)

25. J. Milnor, Morse Theory (Princeton University Press, Princeton, New Jersey, 1969).

26. D. DiVincenzo. Private Communication.

ON THE GLOBAL STRUCTURE OF SOLUTIONS TO SOME SEMILINEAR ELLIPTIC PROBLEMS

Joel Spruck *

Department of Mathematics and Statistics
University of Massachusetts at Amherst
Lederle Graduate Research Center Towers
Amherst, MA 01003

In this paper we discuss two problems involving positive solutions of semilinear elliptic equations which are global in the sense that the problem requires the knowledge of all possible solutions. In the first problem, we will study the semilinear elliptic equation

$$\Delta u + \lambda \sinh u = 0$$

in bounded domains $D \subset R^2$. This equation has sometimes been called the elliptic Sinh-Gordon equation. Of particular interest is the study of the following boundary value problem of "nonlinear eigenvalue" type:

$$(1) \qquad \begin{aligned} \Delta u + \lambda \sinh u &= 0 \quad \text{in } R \\ u &= 0 \quad \text{on } \partial R \\ u &> 0 \quad \text{in } R \end{aligned}$$

where R is a rectangle in R^2. This problem arises in plasma physics and also statistical mechanics as a way of modeling point vortices. However, it arises in a surprising and central way in the construction of compact surfaces of constant man curvature. This will be explained in the following section. The basic question that we discuss is the behavior of solutions as λ tends to zero.

---

* Research supported in part by NSF grant DMS-8501952

Theorem 1 Let (λ_k, u_k) be a sequence of non trivial solutions to (1) with $\lambda_k \downarrow 0$. Then the u_k tend to the Green's function

$$- 2 \log |g(z)|^2$$

where $g(z)$ is the symmetric conformal map of R onto the unit disk. The convergence is uniform on compact subsets of $R \setminus \{0,0\}$ and in $W^{1,p}(R)$, $p < 2$.

This theorem easily extends to nonlinearities $f(u)$ with f asymptotic to ce^u as $u \to \infty$. It also holds for a class of domains D which have reflective x,y symmetry about the origin. It is a remarkable fact that all solutions of (1) can be represented in terms of elliptic functions and that there is uniqueness; however, we shall not pursue these matters here.

The second problem that we discuss concerns solutions of

$$(1') \qquad \Delta u + u^{\frac{n+2}{n-2}} = 0 \quad \text{in} \quad R^n \qquad n > 2$$
$$u > 0$$

Here we impose no conditions on u at infinity so that infinity may be considered as an isolated singularity. This second problem has been much studied [5] when $u \to 0$ at ∞ sufficienty fast because of its interesting geometrical and physical interpretations.

Theorem 2 [2] Let u be a solution of (1'). Then u is regular at infinity (decays like $\frac{1}{r^{n-2}}$) and radially symmetric about some origin.

Theorem 2 is a special case of the work of [2] which shows how to improve the famous method of moving planes [5] in certain situations so as to include the study of solutions (both local and global) with an isolated singularity at infinity.

The proof of Theorem 1 takes up the majority of the paper and serves to illustrate the interplay of a priori bounds and symmetry arguments in proving global results (in this regard see also [6]). Section 1 develops the connection between problem (1) and toroidal soap bubbles. In Section 2 we prove an a priori bound for all solutions of (1) that is central to the proof of Theorem 1. In section 3 we use this bound and symmetry arguments to complete the proof of Theorem 1.

The proof of Theorem 2 is sketched in Section 4 and serves as an introduction to [2]. The techniques of [2] are especially important in problems involving critical Sobolev growth because of the failure of most analytic techniques.

1. The Construction of Toroidal Soap Bubbles

A classical problem in differential geometry, made famous by H. Hopf [8], is the following: Does there exists a compact surface in R^3 of constant mean curvature (without boundary) other than the standard sphere? Hopf himself showed that such a surface could not be of genus zero. Later, A.D. Alexandrov [1] showed that such a surface could not be embedded, that is, it must have self-intersections if it exists. In fact, Alexandrov invented the method of moving planes to prove that an embedded compact surface of constant mean curvature is a sphere. Finally, in 1984 Wente [11] prove the existence of infinitely many non-congruent compact surfaces of constant mean curvature of genus one. A key point in his construction is the analysis of problem (1) for λ near zero.

To see how problem (1) arises, let $F:D \subset R^2 \to R^3$ be a conformal representation of a surface, that is

$$F_x^2 = F_y^2 = E > 0$$

$$F_x \cdot F_y = 0 .$$

Here, $z = x + iy$ are coordinates in R^2. The metric on the surface is given by $ds^2 = E|dz|^2$ with normal vector field $\xi = \frac{F_x \wedge F_y}{|F_x \wedge F_y|}$. The second fundamental form of the surface is given by $\ell dx^2 + 2mdxy + ndy^2$ where

$$\ell = F_{xx} \cdot \xi \qquad m = F_{xy} \cdot \xi \qquad n = F_{yy} \cdot \xi$$

The mean curvature H and Gauss curvature K are given by

$$H = \frac{\ell + n}{2E}, \qquad K = \frac{\ell n - m^2}{E^2}.$$

The first fundamental equation of surface theory is the Gauss equation

$$(2) \qquad -K = \frac{\Delta \log E}{2E}$$

The other fundamental equations are the Codazzi-Mainardi equations. These can be expressed as follows (see [4].)

Let $\phi = \ell - n - 2im$. Then

$$(3) \qquad |\phi| = 2E\sqrt{h^2 - K} = E(K_2 - K_1)$$

(where $k_2 > k_1$ are the principal curvatures).

The Codazzi-Mainardi equations are given (in complex notation) by

$$(4) \qquad \phi_{\bar{z}} = EH_z .$$

Now suppose F is a doubly periodic immersion of constant mean curvature (genus 1). Then from (4), ϕ is an analytic function which is doubly periodic. Hence

(5) $$|\phi| = \lambda \text{ constant} .$$

By making a rotation of the x,y coordinates we can arrange that ϕ is real, that is, $m = 0$. Geometrically, this means that the coordinate lines are lines of curvature.

From (3) and (5) we find

$$H^2 - K = \frac{\lambda^2}{4E^2} .$$

Substitution into equation (2) gives

$$\frac{\lambda^2}{4E^2} = \frac{\Delta \log E}{2E} + H^2$$

or

$$\Delta \log E + 2EH^2 - \frac{\lambda^2}{2E} = 0 .$$

Finally, we set $E = \frac{\lambda}{2H} e^u$. Then

(6) $$\Delta u + 2\lambda H \sinh u = 0 .$$

For convenience we may assume $H = 1/2$. Then (6) is just the elliptic Sinh-gordon equation. We can also easily compute

$$k_2 = \frac{1 + e^{-u}}{2} \qquad \ell = k_1 E \qquad E = \lambda e^u$$

$$m = 0$$

$$k_1 = \frac{1 - e^{-u}}{2} \qquad n = k_2 E$$

Now we want to reverse this procedure and start with a doubly periodic solution of

$$\Delta u + \lambda \sinh u = 0 \quad \text{in} \quad R^2$$

and construct the map F with first and second fundamental forms given by (7). The simplest way to do this is to solve

$$\text{(8)} \qquad \begin{aligned} \Delta u + \sinh u &= 0 \quad \text{on} \quad R \\ u &= 0 \quad \text{on} \quad \partial R \end{aligned}$$

where $R = [0,a] \times [0,b]$. Then we can extend u by odd reflection to a doubly periodic solution on R^2. The parameter λ is a scale parameter which can be removed by the transformation

$$\omega(x,y) = \frac{1}{2} u \left(\frac{x}{\sqrt{\lambda}} , \frac{y}{\sqrt{\lambda}} \right)$$

Then,

$$\Delta\omega + \sinh \omega \cosh \omega = 0 \quad \text{in} \quad R^\lambda$$

$$\omega = 0 \quad \text{on} \quad \partial R^\lambda$$

where $R^\lambda = [0,\sqrt{\lambda}\, a] \times [0,\sqrt{\lambda}\, b]$.

For later reference, we have $E = e^{2\omega}$ and

$$\begin{aligned} k_2 &= e^{-\omega}\cosh \omega \qquad & \ell &= e^{\omega} \sinh \omega \\ k_1 &= e^{-\omega}\sinh \omega \qquad & m &= 0 \\ & & n &= e^{\omega} \cosh \omega . \end{aligned}$$

The fundamental equations of surface theory for the triple F_x, F_y, ξ are

$$
(10)\qquad
\begin{aligned}
F_{xx} &= \omega_x F_x - \omega_y F_y + e^{\omega}\sin\omega\ \xi\\
F_{xy} &= \omega_y F_x + \omega_x F_y\\
F_{yy} &= -\omega_x F_x + w_y F_y + e^{\omega}\cosh\omega\ \xi\\
\xi_x &= -e^{-\omega}\sinh\omega\ F_x\\
\xi_y &= -e^{-\omega}\cosh\omega\ F_y\ .
\end{aligned}
$$

This is an overdetermined system which can be integrated if u satisfies (8) or equivalently ω satisfies (9). The surface $F(x,y)$ is unique up to a Euclidean motion (see [4].

The difficulty is that F need not be doubly periodic; there is a problem of periods. In order to overcome this difficulty, Wente required u to be non-negative, that is, u solves (1). Then by Theorem (3.2) of [5], u inherits the symmetry of the rectangle about its center. By analyzing the system (1) it is then easy to analyze the inherited symmetries of the surface $F(x,y)$. Wente then shows that it is possible to choose the parameters of the problem $(\lambda, b/a)$ so that "F closes up" (see [11]).

2, An a priori Bound for Positive Solutions

In this section we will prove an a priori sup norm estimate for solutions of (1) that will be central to our analysis.

Let $u \geq 0$ satisfy

$$\Delta u + \lambda f(u) = 0 \qquad \text{in } D \subset R^2$$

and let $w = e^{-\alpha u}$ where $1/2 < \alpha < 1$. Then

$$(11)\qquad \Delta w = \alpha\lambda w f(u) + \frac{|\nabla w|^2}{w}\ .$$

Define a vector field $V(x)$ by

$$V^j = \frac{1}{w}(w_i w_{ij} - 1/2 w_j \Delta w) .$$

We note that V can be expressed as a gradient:

$$(12) \qquad V^j = \left(\frac{|\nabla w|^2}{2w} - \lambda\, G(w)\right)_j \qquad \text{where} \quad G'(t) = tf(\ell n\, \frac{1}{t}) .$$

Using (11), a short computation gives

$$(13) \qquad \operatorname{div} V = J + (2\alpha f(u) - f'(u)) \cdot \frac{\lambda |\nabla w|^2}{2w}$$

where $\qquad J = \frac{1}{w} \sum_{i,j} w_{ij}^2 - \frac{1}{2}(\Delta w)^2 \geq 0 .$

If $f(u) = \sinh u$ then

$$(14) \qquad \operatorname{div} V \geq \frac{\lambda\alpha^2}{2} ((2\alpha - 1)e^{(1-\alpha)u} - (2\alpha + 1)e^{-(1+\alpha)u})|\Delta u|^2 .$$

Let η be a non-negative $C_0^\infty(D)$ function. Then

$$0 = \int_D \operatorname{div} \eta V\, dxdy = \int_D (\eta \operatorname{div} V + \nabla\eta \cdot V) dxdy.$$

Using (12) this gives

$$(15) \qquad \int_D \eta \operatorname{div} V\, dx\, dy = \int_D \Delta\eta \left(\frac{|\nabla w|^2}{2w} - \lambda\, G(w)\right) dxdy .$$

Using (14) for $f(u) = \sinh u$ (15) leads to the estimate

$$(16) \qquad \lambda \int_D \eta\, e^{(1-\alpha)u} |\nabla u|^2 dxdy \leq c_1 + c_2 \int_{\text{supp}\ \eta} e^{\alpha u} |\nabla u|^2 dxdy .$$

Now let D satisfy the conditions of [5]. Then the maximum of u is achieved in a compact subdomain $\overline{D}_{4\varepsilon} = \{(x,y) \in D : \text{dist}((x,y),\partial D) > 4\varepsilon\}$, where ε is independent of u. Choose η such that $\eta \equiv 1$ on $D_{2\varepsilon}$, $\eta \in C_0^\infty(D_\varepsilon)$. Then we can estimate

$$\int_{D_\varepsilon} e^{-\alpha u} |\nabla u|^2 dxdy < C(\varepsilon)$$

$$\lambda \int_{D_\varepsilon} e^{(1-\alpha)u} dxdy < C(\varepsilon) . \tag{17}$$

From (16) and (17) it follows that

$$\int_D (\nabla \eta e^{\frac{1-\alpha}{2} u})^2 \, dxdy < C/\lambda . \tag{18}$$

Using the Sobolev inequality, estimate (18) gives the following L^p estimate for e^u:

$$\| e^u \|_{L^p(D_{2\varepsilon})} < c_1 + \frac{c_2}{\lambda^{\frac{1+\alpha}{1-\alpha}}} . \tag{19}$$

We can now easily prove

Theorem 2.1. Let $u > 0$ satisfy

$$\Delta u + \lambda \sinh u = 0 \quad \text{in} \quad D \subset R^2$$
$$u = 0 \quad \text{on} \quad \partial D .$$

Then $u < \log(c_1 + \frac{c_2}{\lambda})$ where c_1, c_2 depend only on D.

Proof: We compute

$$\Delta\zeta^2 e^u = \zeta^2 \Delta e^u + e^u \Delta\zeta^2 + 4\zeta e^u \nabla\zeta\nabla u$$

$$= \zeta^2 e^u\,(\Delta u + |\nabla u|^2) + e^u(\Delta\zeta^2 + 4\zeta\nabla u\cdot\nabla\zeta)$$

$$\geq -c(\lambda\zeta^2 e^{2u} + (|\nabla\zeta|^2 + \zeta|\Delta\zeta|)e^u)\ .$$

Choosing $\zeta \in C_0^\infty(D_{2\varepsilon})$, $\zeta \equiv 1$ on $D_{4\varepsilon}$, we have from elliptic regularity theory

$$\sup_{D_{4\varepsilon}} e^u \leq \sup_{D} \zeta^2 e^u \leq C(\lambda \| e^{2u} \|_{L^P(D_{2\varepsilon})} + \| e^u \|_{L^P(D_{2\varepsilon})}) \qquad P > 1$$

$$\leq c_1 + \frac{c_2}{\lambda^{\frac{1+\alpha}{1-\alpha}}}$$

by (19). This proves the theorem.

3. The Singular Limit as $\lambda \to 0$

In this section we will discuss the asymptotic behavior of solutions as $\lambda \to 0$. In order to understand what happens consider a very classical example.

Example Let $u > 0$ satisfy $\Delta u + \lambda e^u = 0$ in $B_1 \subset R^2$. By Theorem (1) of [5] u is radial and it is not difficult to show that all solutions are given by

$$u = -2\log\frac{r^2 + \beta}{1 + \beta}\ .$$

Then,

$$\Delta u = \frac{-8\beta}{(r + \beta)^2} = \frac{-8\beta}{(1 + \beta)^2}\, e^u\ .$$

Thus β must satisfy $\dfrac{8\beta}{(1 + \beta)^2} = \lambda$, i.e.

$$\beta^{\pm} = \frac{4}{\lambda} - 1 \pm \sqrt{(\frac{4}{\lambda} - 1)^2 - 1}$$

and the picture is

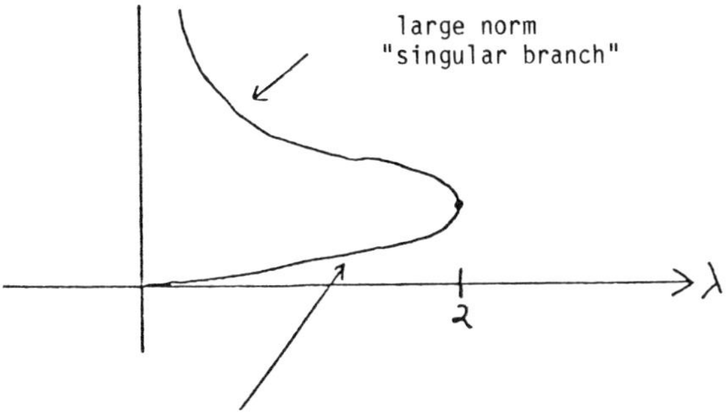

the choice $\beta_- \approx \frac{\lambda}{8}$ for small λ gives the singular branch of solutions which converge to the Green's function $-4 \log r$.

In our situation, the minimal solution is replaced by the trivial solution $u = 0$ and there is a bifurcation from the trivial solution at the first Dirichlet eigenvalue $\lambda_1 = \lambda_1(D)$.

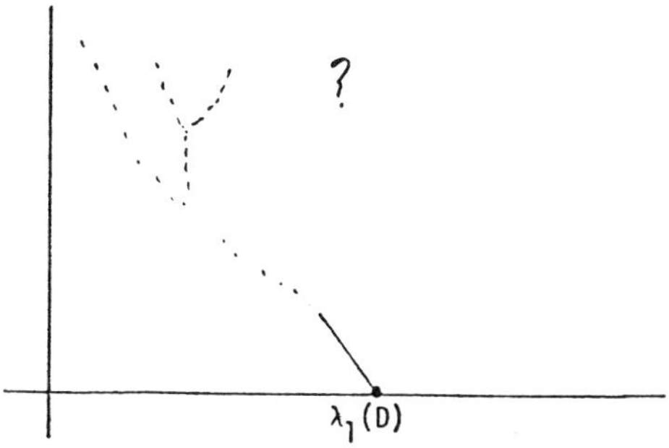

Before we proceed to the study of the singular limit of solutions as $\lambda \to 0$ we summarize the existence and uniqueness situation for a general domain $D \quad R^2$.

Theorem 3.2 For $\lambda > \lambda_1(D)$ there are no nontrivial positive solutions of (1) in D. There exists a maximally connected closed branch of solutions C bifurcating from the trivial solution at $\lambda = \lambda_1$. The projection of C on λ space covers the interval $(0,\lambda_1)$. For λ near λ_1 there is uniqueness.

Proof Since $\frac{\sinh u}{u} > 1$ the first Dirichlet eiigenvalue μ_1 of

$\Delta\phi + \mu \cdot \frac{\sin u}{u} \phi = 0$ satisfies $\mu_1 < \lambda_1(D)$ (with equality if and only if $u \equiv 0$)

by the variational characterization of μ_1 and λ_1. Hence since $\phi = u$, $\mu = \lambda$ there is no solution of (1) if $\lambda > \lambda_1$.

The existence of C near $\lambda = \lambda_1$ $(\lambda < \lambda_1)$ is well known (see [3]). The global existence of C follows from our a priori bound Theorem 3.1 and a theorem of Rabinowitz [9].

Finally, the asserted uniqueness of solutions follows from known results once we show that for λ near λ_1 all solutions of (1) are close to the trivial solution. This again follows from Theorem 3.1. For if (λ_k, u_k) are a sequence of solutions of (1) then by Theorem 3.1 the u_k are uniformly bounded so by elliptic regularity theory we can choose a subsequence which we still call u_k which converge uniformly to a solution of $\Delta u + \lambda_1 \sinh u = 0$. But we have seen that $u \equiv 0$ is the unique solution for $\lambda = \lambda_1$. It follows that the u_k cannot be bounded away from zero on any compact subset of D, and the uniqueness follows.

The remainder of this section will be devoted to a proof of Theorem 1 of the introduction.

Step 1 Let (λ,u) be a solution of (1) in R. Then u is uniformly bounded (independent of λ!) on compact subsets of $\overline{R} \setminus \{0,0\}$.

We use the estimate $\int_R \lambda \sinh u \, dxdy < C$ where C is independent of λ. To see this, we write equation (1) in the weak form

$$\int_R \nabla u \cdot \nabla \zeta dxdy = \lambda \int_R \zeta f(u) dxdy \qquad f(u) = \sinh u$$

and set $\zeta = 1 - e^{-\alpha u}$. Then

$$(20) \qquad \int_R \lambda f(u) dxdy = \lambda \int_R e^{-\alpha u} f(u) dxdy + \alpha \int_R e^{-\alpha u} |\nabla u|^2 dxdy \ .$$

Using the symmetry of u, the right hand side of (20) is dominated by

$$C\{\lambda \int_{R_\varepsilon} e^{-\alpha u} f(u) dxdy + \alpha \int_{R_\varepsilon} e^{-\alpha u} |\nabla u|^2 dxdy\} \ .$$

where (as in section (3.2)) R_ε is the subset of R whose points are at least distance ε from R. Hence from estimate (17) we have shown

$$\int_R |\Delta u| dxdy = \lambda \int_R \sinh u \ dxdy < C$$

for a uniform constant C.

To show the uniform boundedness of u at points (x,y) away from the origin consider a point (x_0,y_0) as in the diagram

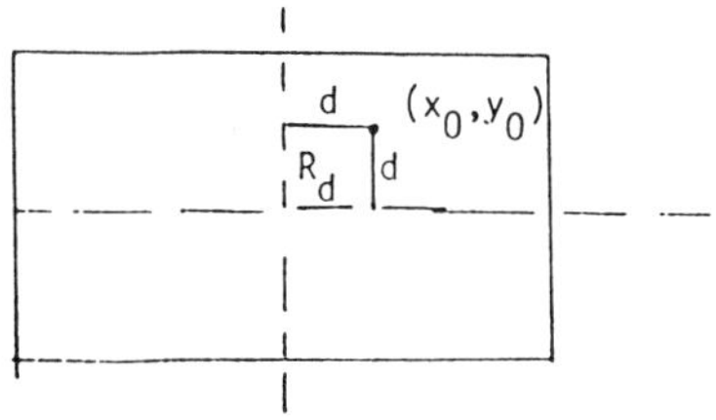

Then by the symmetry properties of u (i.e. $u(x,y) = u(-x,y) = u(x,-y)$, and

$u_x < 0$, $u_y < 0$ in $\{x > 0, y > 0\}$ R) $u(x,y) > u(x_0,y_0)$ for all points (x,y) in R_d. Hence

$$C > \int_R \lambda \sinh u \, dxdy > \int_{R_d} \lambda \sin u \, dxdy > |R_d| \lambda \sinh u\,(x_0,y_0)$$

so that

$$(21) \qquad |\Delta u(x_0,y_0)| < \frac{c}{|R_d|} = \frac{c}{d^2} \quad .$$

Assuming we have a bound for $|\Delta u|$ on compact subsets of $\overline{R} \setminus \{0,0\}$, then since we already know $\Delta u \in L^1(R)$, standard elliptic estimates imply $u \in L^p(R)$ for all p and the uniform bound for $\sup u$ away from $(0,0)$ follows from (21) (see [7].)

Of course, the worst case occurs when (x_0, y_0) is on the x or y axes. Since the argument in both case is the same, assume (x_0,y_0) lies on the positive x axis. We will show using a modification of the method of moving planes that $u(x,y) > u(x_0,0)$ for all points (x,y) in the triangle $T_d : d = x_0 > 0$

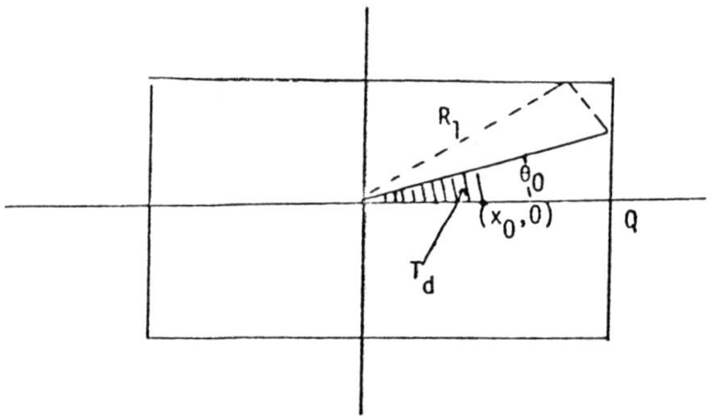

The area of T_d is $d^2/4 \sin 2\theta_0$. The angle θ_0 is determined as follows: Let R_1 be the quarter of R in the first quadrant and consider lines through the origin of slope $\tan\theta$ $\theta > 0$. These lines cut off a "triangular cap" from R_1

(see diagram) and for θ small, the reflection of this triangle about the line is contained in R_1 while for θ too large the reflection leaves $\overline{R}_1$. Then θ_0 is the sup of all θ with the good reflection property. Clearly θ_0 depends only on R.

To prove that $u(x,y) > u(x_0,0)$ for all (x,y) in T_d we consider the family of "planes" $P_\lambda : y = (\tan\theta_0)x - \lambda$, when θ_0 is defined above and use the method of moving planes for this family with respect to the domain R_1. Note that P_0 cuts off the maximal triangular cap Σ_0 by our previous discussion. What makes our argument nonstandard is the fact that we do not have the boundary condition $u = 0$ on the bottom boundary of R_1; instead we use that on this boundary $u_y = 0$, $u_x \leq 0$ with $u_x < 0$ if $x > 0$.

Let $n = (-\cos\theta_0, \sin\theta_0)$ denote the normal directions to the planes P_λ and let λ_0 be the value of λ so that P_λ passes through Q (see diagram) We will show that for $0 < \lambda < \lambda_0$, $\partial_n u = -\cos\theta_0 u_x + \sin\theta_0 u_y > 0$ (note that n points into R_1 at ∂R_1) and $u(X) < u(X^\lambda)$ for $X \in \Sigma_\lambda$ (here $X = (x,y)$ and Σ_λ is the triangular cap cut off by P_λ from R_1).

To get started we observe that for points on R_1 near Q where $u = 0$ then $\partial_n u > 0$ by the Hopf lemma since $\Delta u \leq 0$. On the other hand, for points on ∂R_1 near Q on the x axis,

$$\partial_n u = -\cos\theta_o\, u_x + \sin\theta_o\, u_y = -\cos\theta_o\, u_x > 0 .$$

It follows that for λ close to λ_0, $\lambda < \lambda_0$ we have

$$(22) \qquad \partial_n u > 0 , \quad u(X) < u(X^\lambda) \quad X \in \Sigma_\lambda .$$

We decrease λ until a critical value $\mu \geq 0$ is reached beyond which (22) no longer holds. We show that $\mu = 0$. Suppose $\mu > 0$. Then as in "the old argument" we may apply lemma (4.3) of [5] to conclude that

$$u(X) < u(X^\mu) \text{ in } \Sigma_\mu \text{ and } \partial_n u > 0 \text{ on } R_1 \cap P_\mu .$$

Thus (22) holds for $\lambda = \mu$ and for some $\varepsilon > 0$

(23) $$\partial_n u > 0 \quad \text{in} \quad \Sigma_{\mu-\varepsilon} .$$

From the definition of μ there is a sequence λ^j , $0 < \lambda^j \uparrow \mu$ and points X_j in Σ_{λ^j} such that

$$u(X_j) > u(X_j^{\lambda^j}) .$$

A subsequence of the X_j which we still call X_j converges to a point X_0 in Σ_μ ; then

$$X_j^{\lambda_j} \to X_0^u \quad \text{and} \quad u(X_0) \geq u(X_0^u) .$$

Since (22) holds for $\lambda = \mu$ we must have $X_0 \in \partial\Sigma_\mu$. If $X_0 \in P_\mu$ we reach a contradiction using (23) as in the "old argument". Otherwise the subsequence $X_j \in \Sigma_\mu$ (for j large) and thus

(24) $$u(X_j) > u(X_j) > u(X_j^{\lambda_j}) .$$

It follows from (24) that

(25) $$u(X_0) = u(X_0^\mu) \quad \text{and} \quad \partial_n u(X_0^\mu) = 0$$

and X_0 is a point on $\partial\Sigma_\mu \setminus P_\mu$ where $u(X_0) > 0$. Let $v(X) = u(X^\mu)$ in Σ_μ and set

$$w(X) = v(X) - u(X) \geq 0 \qquad w(X) \not\equiv 0 .$$

Then as before w satisfies an elliptic equation for which the Hopf lemma applies (in fact $\Delta w \leq 0$) and $w(X_0) = u(X_0^\mu) - u(X_0) = 0$ by (25). Hence by the Hopf lemma $\partial_n w(X_0) > 0$. But $\partial_n w(X_0) = \partial_n v(X_0) - \partial_n u(X_0) = \partial_n u(X_0^\mu) - \partial_n u(X_0) =$

$-\partial_n u(X_0) < 0$, a contradiction. The proof of step 1 is complete.

Step 2 Let (λ_k, u_k) be a sequence of solutions of (1) in R with $\lambda_k \downarrow 0$. Then a subsequence of the u_k converges to a multiple of the Green's function of R with pole at the origin. The convergence is uniform away from (0,0) (and in $W^{1,P}(R)$ $1 \leq p < 2$) .

To prove this we first note that the u_k cannot converge uniformly to zero (the trivial solution) for this would imply that $\lambda = 0$ is a Dirichlet eigenvalue of the Laplace operator Δ. From this we conclude that

$$\lambda_k \sinh u_k(0,0) \to +\infty \quad \text{as} \quad \lambda_k \to 0 \quad .$$

For otherwise $|\Delta u_k| \leq C$ in R, $u_k = 0$ on ∂R so that $u_k \to 0$ uniformly in R, a contradiction. Therefore,

$$u_k(0,0) \geq \log \frac{1}{\lambda_k} + 1 \ , \quad k \text{ large} . \tag{26}$$

Later in this section we will show (Proposition 3.3)

$$|\nabla u_k| \leq \frac{C_\varepsilon \log \frac{1}{\lambda}}{\lambda^{1/2 + \varepsilon}} + C_\varepsilon$$

where $\varepsilon > 0$.

Assuming this, we find that for $|X| = \lambda_k$ small

$$u_k(X) \geq u(0,0) - |X| \ \frac{C \log \frac{1}{\lambda_k}}{\lambda_k^{1/2 + \varepsilon}} \tag{27}$$

$$\geq \log \frac{1}{|X|} \quad .$$

Also,

(28) $$u_k \geq \log \frac{1}{|X|} - C \quad \text{on} \quad \partial R$$

Since $\Delta u_k \leq 0$ and $\log \frac{1}{|x|} - C$ is harmonic in $R \setminus B_{\lambda_k}(0,0)$, it follows from the maximum principle (using (27), (28)) that

(29) $$u_k \geq \log \frac{1}{|X|} - C \quad \text{on} \quad R \setminus B_{\lambda_k}(0,0).$$

Since the u_k are uniformly bounded away from (0,0) we may choose a subsequence such that $u_k \to u$ uniformly in R away from (0,0). Evidently, $u \geq 0$ in R,

$$\Delta u = 0 \ \text{ in } \ R \setminus \{0,0\}, \quad u = 0 \ \text{ on } \ \partial R \ \text{ and } \ u \geq \log \frac{1}{|X|} - C \ \text{ in } \ R \setminus \{0,0\}$$

by(29). It follows that u is a positive harmonic in R with a singularity at (0,0). But classical theorems say that $u = c\,G$ is the only possible positive harmonic with an isolated singularity and the proof of Step 2 is complete.

Step 3 The limit u constructed in Step 2 is given by $u = -2 \log|g(z)|^2$ where $g(z)$ is the symmetric conformal map of R onto the unit disk.

We proceed as follows. Define

$$\phi(z,\bar{z}) = u_{zz} - \frac{u_z^2}{2}$$

where $z = x + iy$ and $\frac{\partial}{\partial z}$, $\frac{\partial}{\partial \bar{z}}$ have their usual meaning. In complex notation, if u satisfies (1), then

$$u_{z\bar{z}} + \frac{\lambda}{4} \sinh u = 0 \quad \text{in} \quad R.$$

Therefore,

$$\phi_{\bar{z}} = u_{zz\bar{z}} - u_z u_{z\bar{z}} = - \frac{\lambda}{4} (\cosh u - \sinh u) u_z$$

$$= \frac{\lambda}{4} e^{-u} u_z \quad .$$

Appealing once more to Proposition (3.3) below,

$$|\phi_{\bar{z}}| < c \, \lambda^{1/2 - \varepsilon}.$$

It follows from standard potential theory that $\phi = \psi(z) + \eta(z,\bar{z})$, where $\psi(z)$ is analytic in R, and $|\eta| < c\lambda^{\frac{1}{2} - \varepsilon}$. If fact ψ is just the Cauchy integral of $\phi = u_{zz} - \frac{u_z^2}{2}$ around a simple closed curve $\Gamma \subset R$ containing (0.0). In particular ψ <u>is uniformly bounded independent of</u> λ.

Applying this construction to the (λ_k, u_k) of Step 2, we obtain a sequence ψ_k of analytic functions which are uniformly bounded and form a normal family. Therefore, a subsequence of the ψ_k converge uniformly on compact subsets of R to <u>a regular analytic function</u> ψ. Evidently

$$u_{zz} - \frac{u_z^2}{2} = \psi \tag{30}$$

where $u = \lim u_k$.

We know that $u = -a|\log g(z)|^2$. Then

$$u_z = -a \frac{g'(z)}{g(z)}, \quad u_{zz} = -a \frac{g''(z)}{g(z)^2} + a \frac{g'(z)^2}{g(z)^2} \quad . \tag{31}$$

Substitution of (31) into (30) gives

$$-a\,\frac{g''(z)}{g(z)} + (a - a^2/2)\left(\frac{g'(z)}{g(z)}\right)^2 = \psi(z) \quad \text{in} \quad R\ .$$

Observe that $\frac{g'(z)}{g(z)}$ has a simple pole at $z = 0$ while $\frac{g''(z)}{g(z)}$ is in fact regular at $z = 0$ since $g(z)$ is odd. Since ψ is regular we must have $a = a^2/2$ or $a = 2$. This completes the proof of Step 3.

We have now completed the proof of Theorem 1 modulo the technical estimate given by

Proposition 3.3 Let $u > 0$ satisfy

$$\Delta u + \lambda \sinh u = 0 \quad \text{in} \quad D \subset R^2$$
$$u = 0 \quad \text{on} \quad \partial D\ .$$

Then

$$|\nabla u| < \frac{c_1}{\lambda^{1/2+\delta}} + c_2 \qquad \text{for any} \quad \delta > 0$$

on compact subsets of D.

Proof Let $v = D_\gamma u$ be any directional derivative of u. Then v satisfies

$$\Delta v + dv = 0 \quad \text{in} \quad D$$

where $d = \lambda \cosh u$.

By standard elliptic theory,

$$(32) \qquad |v|_{L^\infty(D_\varepsilon)} < c|dv|_{L^p(D_\varepsilon)} + \sup_{\partial D_\varepsilon} |v| \qquad p > 1$$

where D_ε is chosen as in Section 3.2. Since u is uniformly bounded in a neighborhood of ∂D, it follows (again by regularity theory) that $\sup_{\partial D_\varepsilon}|v| < C$, a uniform constant.

By Holder's inequality, for $1 < p < 2$

(33) $$|dv|_p < |d|_{\frac{2p}{2-p}} |v|_2$$

and by L_p interpolation

(34) $$|d|_q < |d|_1^\lambda |d|_r^{1-\lambda} \qquad \text{where } \frac{1}{q} = \frac{\lambda}{1} + \frac{(1-\lambda)}{r} .$$

Choosing $q = \frac{2p}{2-p}$, we use the estimates

$$|d|_{L^1(D)} < c, \quad |d|_{L^r(D)} < \frac{c_1}{\lambda^{\frac{\alpha}{1-\alpha}}} + c_2 .$$

(This last estimate is just (19) from Section 3.2) which gives

(35) $$|d|_q < \frac{\bar{c}_1}{\lambda^{\frac{\alpha}{1-\alpha}(1-\lambda)}} + \bar{c}_2 .$$

We note that $q > 2$ can be made arbitrarily close to 2 (by taking p close to 1) and that $\lambda < \frac{1}{q} < 1/2$ can be made arbitrarily close to $1/2$ by choosing r large enough. In particular, we may take $1 - \lambda = 1/2 + \delta$. Also, by taking $\alpha > \frac{1}{2}$ close to $1/2$, $\frac{\alpha}{1-\alpha} > 1$ can be made arbitrarily close to 1. Thus

(36) $$\frac{\alpha}{1-\alpha}(1 - \lambda) = 1/2 + \delta \qquad \delta > 0 .$$

Finally,

$$|v|^2_{L^2(D)} \le \int_D |\nabla u|^2 dxdy = \int_D u\lambda \sinh u \; dxdy$$

$$\le |u|_{L^\infty(D)} \int_D \lambda \sinh u \; dxdy$$

$$\le c_1 \log \frac{1}{\lambda} + c_2$$

which is negligible compared to powers of $\frac{1}{\lambda}$. Thus from (32)-(36) the proposition follows

4. The Symmetry Method Without Decay at Infinity

In this section we prove that a positive solution of

$$(37) \qquad \Delta u + u^{\frac{n+2}{n-2}} = 0 \quad \text{in } R^n \qquad n > 2$$

is radially symmetric about some origin. It then follows easily that $u = u(|x-x_0|)$ is regular at infinity and given explicitly by

$$u(x) = \frac{(n(n-2)\lambda^2)^{\frac{n-2}{4}}}{(\lambda^2+|x-x_0|^2)^{\frac{n-2}{2}}} \qquad \lambda > 0$$

We will assume that the reader has a familiarity with [5].

Choose an arbitrary point, which for convenience we take as $x=0$ and expand u in a power series

$$u(x) = a_0 + a_i x_i + \frac{1}{2} a_{ij} x_i x_j + 0(|x|^2)$$

$$u_i(x) = a_i + a_{ij} x_j + 0(|x|)$$

where, $a_o = u(0)$, $a_i = u_i(0)$, $a_{ij} = u_{ij}(0)$

Define the change of variables

$$v(y) = |x|^{n-2} u(x) \quad , \quad x = \frac{y}{|y|^2}$$

Equation (37) is invariant (this is not essential to the argument) , i.e.

$$\Delta v + v^{\frac{n+2}{n-2}} = 0 \quad \text{in } R^n \setminus (0) \tag{38}$$

and v is regular at infinity with expansions

$$v(y) = \frac{1}{|y|^{n-2}} \left(a_o + a_i \frac{y_i}{|y|^2} + \frac{1}{2} a_{ij} \frac{y_i y_j}{|y|^2} + 0\left(\frac{1}{|y|^2}\right)\right)$$

$$v_i(y) = -\frac{ny_i}{|y|^n}\left(a_o + \frac{a_j y_j}{|y|^2}\right) + \frac{a_i}{|y|^n} + 0\left(\frac{1}{|y|^{n+1)}}\right))$$

Note that v has a possible isolated singularity at $y=0$. Following [5], we shift the origin to obtain the optimal expansion at infinity. That is, let $y_{oj} = \frac{-a_j}{(n-2)a_o}$ and consider

$$v(y-y_o) = \frac{1}{|y-y_o|^{n-2}} \left[a_o + a_i \frac{y_i - y_{oi}}{|y-y_o|^2} + \frac{1}{2} a_{ij} \frac{(y_i - y_{oi}) \, |y_j - y_{oj})}{|y - y_o|^2} + 0\left(\frac{1}{|y-y_o|^2}\right)\right]$$

$$= \frac{1}{|y|^{n-2}} \left(1 + \frac{n-2}{|y|^2} y_i\, y_{oi} + \dots \right) \left(a_o + a_j \frac{y_j - y_{oj}}{|y - y_o|^2} + \frac{1}{2} \frac{a_{jk}(y_j - y_{oj})(y_k - y_{ok})}{|y - y_o|^2} + 0\left(\frac{1}{|y|^2}\right) \right)$$

$$= \frac{1}{|y|^{n-2}} \left(a_o + \left(\frac{(n-2)a_o\, y_{oi}}{|y|^2} + a_i\right) y_i + \frac{b_{ij}\, y_j}{|y|^4} + 0\left(\frac{1}{|y|^2}\right) \right)$$

Therefore,

$$(40) \qquad v(y-y_o) = \frac{1}{|y|^{n-2}} \left(a_o + b_{ij} \frac{y_i y_j}{|y|^4} + 0\left(\frac{1}{|y|^2}\right)\right.$$

for suitable coefficients b_{ij} . A similar computation gives

$$(41) \qquad v_i(y-y_o) = - \frac{(n-2)}{|y|^n} a_o\, y_i + 0\left(|y|^{-(n+1)}\right)$$

Let $w(y) = v(y-y_o)$. Since v is (possibly) singular at the origin, w has a possible singularity at $y = y_o$. We shall show that w is axisymetric about the axis passing through $y = 0$ and $y = y_o$. For convenience, assume that this axis is the $-y_n$ coordinate axis. Then we want to show that

$$w = w\,(s, x_n) \qquad s = |x'|\ ,\ x' = (x'\ , \dots\ x_{n-1})$$

$$(42) \qquad w_s < 0 \quad \text{for} \quad s > 0$$

For any unit vector δ orthogonal to $(0,\dots,0,\ 1)$ we will prove that w is symmetric in the plane $\delta \cdot y = 0$ and also $\delta \cdot \text{grad}\ w < 0$ if $\delta \cdot y > 0$. By performing a rotation we may suppose that $\delta = (1,\ 0,\ \dots,\ 0)$

From (41) it follows that

$$(43) \qquad w_1 < 0 \quad \text{For } y_1 > \frac{Co}{|y|}, \quad |y| > R_1$$

A consequence of (40) and (43) is

Lemma 4.1 [5] For any $\lambda > 0$, $\exists R = R(\lambda)$ depending only on min $(1,\lambda)$ (and w) such that for $y = (y_1, y')$ $z = (z_1, y')$ satisfying

$$y_1 < z_1, \quad y_1 + z_1 > 2\lambda, \quad |y| > R$$

we have $w(y) > w(z)$

For any $\lambda > 0$ and any $y = (y_1, y')$, we denote by $y^\lambda = (2\lambda - y_1, y')$ the reflection of y in the plane $T_\lambda : y_1 = \lambda$.

Lemma 4.2 [5] There exists $\lambda_0 > 1$ such that $\forall \lambda > \lambda_0$,

$$(44) \qquad w(y) > w(y^\lambda) \qquad \text{if } y_1 < \lambda$$

Lemma 4.2 gives the desired reflection property (44) for planes T_λ with λ large enough. In the proof of lemma 4.2 we need only to observe that w is a positive superharmonic with possible singularity at $y = y_0$ so that $w(y) > c_0 > 0$ for $|y| < R_1$ and the proof of lemma 4.2 in [5] goes through without change.

Lemma 4.3 [5] Assume that for some $\lambda > 0$, $w(y) > w(y^\lambda)$, $w(y) \neq w(y^\lambda)$ if $y_1 < \lambda$. Then $w(y) > w(y^\lambda)$ if $y_1 < \lambda$ and

$$(45) \qquad w_1(y) < 0 \qquad \text{on } T_\lambda$$

The proof of Lemma 4.3 in [5] goes through unchanged because either w is bounded near y_0 (and the singularity is removable) or else w tends to $+\infty$

at y_0 because w is a positive superharmonic.

Lemma 4.4 [5] The set of positive λ for which (44) holds is open

Lemmas (4.2) - (4.4) imply that (44) , (45) hold for all λ in some <u>maximal</u> open interval $0 \leq \lambda_1 < \lambda < \infty$. Then

$$(46) \qquad w_1(y) < 0 \ , \ y_1 > \lambda_1$$
$$w(y) > w(y^{\lambda}) \ , \ y_1 < \lambda$$

Suppose $\lambda_1 > 0$. By lemma (4.3) we have either

$$w(y) \equiv w(y^{\lambda 1}) \qquad \text{for } y_1 < \lambda_1$$

or else property (44) holds for $\lambda = \lambda_1$.

By lemma 4.1 the former possibility cannot occur while the latter possibility cannot occur by lemma 4.4 and the definition of λ_1 .

Hence for any vector δ orthogonal to $(0,\ldots,0, 1)$ we have shown that $\lambda_1 = \lambda_1(\delta) = 0$. It follows from (46) that w is symmetric about each plane $y \cdot \delta = 0$. Thus w is axisymmetric and (42) holds.

Note that if $y_0 = 0$, that is $|\nabla u(0)| = 0$, then $v \equiv w$ is radially symmetric about the origin and so

$$(47) \qquad u(x) = |y|^{n-2} v(y) \ , \qquad y = \frac{x}{|x|^2}$$

is also radially symmetric about the origin.

The axisymmetry of w about the axis through 0 and y_0 immediately gives the axisymmetry of v about the same axis since

$$v(y) = w(y + y_0)$$

Finally $u(x)$ is also axisymmetric about this same axis. To see this, let $e = \frac{y_0}{|y_0|}$.

If x is any point such that

$$x \cdot e = z \, , \quad |x|^2 - (x \cdot e)^2 = s^2 \qquad \text{with} \quad z, \ s > 0 \ \text{fixed}$$

then $y = \frac{x}{|x|^2}$ satisfies

$$y \cdot e = \frac{z}{s^2 + z^2} \, , \quad |y|^2 - (y \cdot e)^2 = \frac{s^2}{(s^2 + z)^2} \quad , \quad |y|^2 = \frac{1}{s^2 + z^2}$$

and so $u(x)$ is the same for all such points from formula (47).

To complete the proof we choose any other point x_0 as origin and repeat the previous procedure. Then u is axisymmetric about an axis through the vector $\nabla u(x_0)$.

Case i $\qquad \nabla u(x_0)$ is not parallel to $\nabla u(0)$

Assume for simplicity that $\nabla u(0)$ is parallel to the e_n direction. Then

$$u = u(s, x_n)$$

$$\nabla u(x_0) = \frac{u_s}{s} x_{oi} e_i + u_n e_n \, , \quad u_s \neq 0 \, , \quad s > 0$$

and the line through x_0 in the direction of $\nabla u(x_0)$ intersects the x_n axis in the point

$$P = (0, \ldots, 0, x_{on} - \frac{s}{u_s} u_n) \, , \quad \nabla u(P) = 0$$

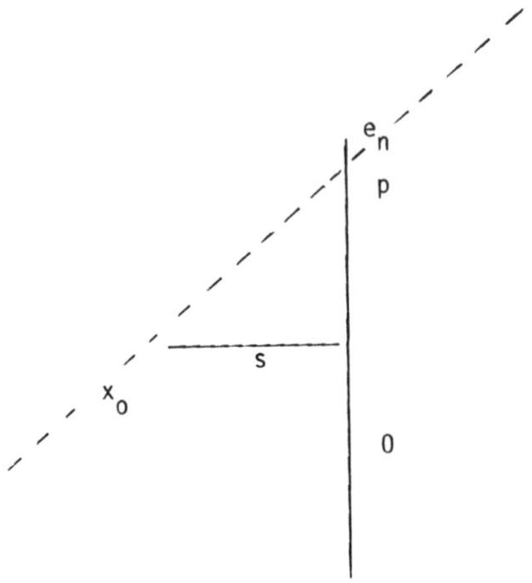

and u is radially symmetric about P.

Case ii ∇u always point in a fixed direction.

For simplicity suppose ∇u is always parallel to the e_n direction. Then $u = u(x_n)$ satisfies

$$u'' + u^{\frac{n+2}{n-2}} = 0 \tag{48}$$

Since u is positive and concave, it is easy to see that there are no global positive solutions of (48) and the proof is complete.

References

[1] A.D. Alexandrov, Uniqueness theorems for surfaces in the large, V. Vestnick, Leningrad Univ. 13, No. 19 (1958), 5-8 . AMS Trans (series 2) 21, 412- 416.

[2] L. Caffarelli, B. Gidas, J. Spruck, in preparation

[3] M. Crandall and P. Rabinowitz, Bifurcation from Simple Eigenvalues, Journal Functional Analysis 8 (1971) 321-340.

[4] L.P. Eisenhart, A. Treatise on the Differential Geometry of Curves and Surfaces, Dover Reprint 1960.

[5] B. Gidas, W. Ni and L. Nirenberg, Symmetry and related properties via the maximum principle, Comm. Math Physics 68, No. 3 (1979), 309-343.

[6] B. Gidas and J. Spruck, Apriori bounds for positive solutions of nonlinear elliptic equations, Comm. in P.D.E (1981), 883-901.

[7] D. Gilbarg and N.S. Trudinzer, Elliptic Partial Differential Equations of Second Order, (Second Edition) Springer-Verlag (1983)

[8] H. Hopf Lectures on Differential Geometry in the large, Standford Lecture notes.

[9] P. Rabinowitz, A global theorem for nonlinear eigenvalue problems and applications. Contributions to Nonlinear Functional Analysis, Academic Press 1971.

[10] J. Serrin, A symmetry method in potential theory, Arch. Rational Mech. Anal. 43 (1971), 304-318

[11] H. Wente, Counterexample to a conjecture of H. Hopf. Pacific J 121 (1986) 193-243)

Tables of contents from other volumes from the program in Continuum Physics and Partial Differential Equations

Homogenization and effective moduli of materials

October 22 - October 26, 1984

J. L. Ericksen
D. Kinderlehrer
R. Kohn
J.-L. Lions

Conference Committee

Amorphous polymers and non-newtonian fluids

March 4 - March 8, 1985

C. Dafermos
J. L. Ericksen
D. Kinderlehrer

Editors

Bird, R.	Mathematical problems in the kinetic theory of polymeric fluids
Caswell, B.	Lagrangian concepts for the numerical analysis of viscoelastic flow
Dafermos, C.	Solutions with shocks for conservation laws with memory
Joseph, D. D.	Hyperbolic dynamics in the flow of elastic liquids
Kearsley, E.	Rubbery liquids in theory and experiment
Marcus, M. and Mizel, V.	Dynamical behavior under random perturbation of materials with selective recall
Nohel, J. and Renardy, M.	Development of singularities in nonlinear viscoelasticity
Rabin	Macromolecules in elongational flow: metastability and hysteresis
Renardy, M. and Hrusa, W.	Propagation of discontinuities in linear viscoelasticity
Wool, R.	Strength and entanglement development at amorphous polymer interfaces

Oscillation theory, computation, and methods of compensated compactness

April 1 - April 4, 1985

C. Dafermos
J. L. Ericksen
D. Kinderlehrer
M. Slemrod

Conference Committee

Chacon, T. and Pironneau, O.	Convection of microstructures by incompressible and slightly compressible flows
DiPerna, R.	Oscillations in solutions to nonlinear differential equations
Forest, M.G. and Lee, J.-L.	Geometry and modulation theory for the periodic nonlinear Schrödinger equation
Harten, A.	On high-order accurate interpolation for non-oscillatory shock capturing schemes
Lax, P.	On the weak convergence of dispersive difference schemes
Majda, A.	Nonlinear geometric optics for hyperbolic systems of conservation laws
McLaughlin, D.	On the construction of a modulating multiphase wavetrain for a perturbed KdV equation
Nunziato, J., Gartling, D. Kipp, M.	Evidence of nonuniqueness and oscillatory solutions in computational fluid mechanics
Osher, S. and Chakravarthy, S.	Very high order accurate T V D schemes
Rascle, M.	Convergence of approximate solutions to some systems of conservation laws: a conjecture on the product of the Riemann invariants
Schonbek, M.	Applications of the theory of compensated compactness
Serre, D.	A general study of the commutation relation given by L. Tartar
Slemrod, M.	Interrelationships among mechanics, numerical analysis, compensated compactness, and oscillation theory
Venakides, S.	The solution of completely integrable systems in the continuum limit of the spectral data
Warming, R. and Beam, R.	Stability of finite-difference approximations for hyperbolic initial value boundary value problems
Yee, H.	Construction of a class of symmetric T V D schemes

Metastability and incompletely posed problems

May 6 - May 10, 1985

S. Antman
J. L. Ericksen
D. Kinderlehrer
I. Müller

Conference Committee

Antman, S. and Malek-Madani, R.	Dissipative mechanisms
Ball, J.	Does rank-one convexity imply quasiconvexity?
Brezis, H.	Metastable harmonic maps
Calderer, M.	Bifurcation of constrained problems in thermoelasticity
Chipot, M. and Luskin,M.	The compressible Reynolds lubrication equation
Ericksen, J.	Twinning of crystals I
Evans, L. C.	Quasiconvexity and partial regularity in the calculus of variations
Goldenfeld, N.	Introduction to pattern selection in dendritic solidification
Gurtin, M.	Some results and conjectures in the gradient theory of phase transitions
James, R.	The stability and metastability of quartz
Kenig, C.	Continuation theorems for Schrodinger operators
Kinderlehrer, D.	Twinning of crystals II
Kitsche, W., Müller, I., and Strehlow, P.	Simulation of pseudoelastic behaviour in a system of rubber balloons
Lions, J.-L.	Asymptotic problems in distributed systems
Liu, T.P.	Stability of nonlinear waves
Maderna, C., Pagani, C., and Salsa, S.	The Nash-Moser technique for an inverse problem in potential theory related to geodesy

Mosco, U.	Variational stability and relaxed Dirichlet problems
Pitteri, M.	A contribution to the description of natural states for elastic crystalline solids
Rogers, R.	Nonlocal problems in electromagnetism
Vazquez, J.	Hyperbolic aspects in the theory of the porous medium equation
Vergara-Caffarelli, G.	Green's formulas for linearized problems with live loads
Wright, T.	Some aspects of adiabatic shear bands

Dynamical problems in continuum physics

June 3 - June 7, 1985

J. Bona
C. Dafermos
J. L. Ericksen
D. Kinderlehrer

Conference Committee

Amick, C. and Kirchgässner, K.	Solitary water-waves in the presence of surface tension
Beals, M.	Presence and absence of weak singularities in nonlinear waves
Beatty, M.	Some dynamical problems in continuum physics
Beirao da Veiga, H.	Existence and asymptotic behavior for strong solutions of the Navier Stokes equations in the whole space
Bell, J.	A confluence of experiment and theory for waves of finite strain in the solid continuum
Boczar-Karakiewicz, B., Bona, J., and Cohen, D.L.	Shallow water waves and sediment transport
Chen, P.	Classical piezoelectricity: is the theory complete?
Glassey, R. and Strauss, W.	On the dynamics of a collisionless plasma
Keller, J.	Acoustoelasticity
McCarthy, M.	One dimensional finite amplitude pulse propagation in electroelastic semiconductors
Morawetz, C.	Weak solutions of transonic flow by compensated compactness
Müller, I.	Extended thermodynamics of ideal gases
Pego, R.	Phase transitions in one dimensional nonlinear viscoelasticity: admissibility and stability
Roytburd, V. and Slemrod, M.	Dynamic phase transitions and compensated compactness
Shatah, J.	Recent advances in nonlinear wave equations
Spagnolo, S.	Some existence, uniqueness, and non-uniqueness results for weakly hyperbolic equations in Gevrey classes